Lecture Notes in Networks and Systems

Volume 902

The series "Lecture Notes in Networks and Systems" publishes the latest developments in Networks and Systems—quickly, informally and with high quality. Original research reported in proceedings and post-proceedings represents the core of LNNS.

Volumes published in LNNS embrace all aspects and subfields of, as well as new challenges in, Networks and Systems.

The series contains proceedings and edited volumes in systems and networks, spanning the areas of Cyber-Physical Systems, Autonomous Systems, Sensor Networks, Control Systems, Energy Systems, Automotive Systems, Biological Systems, Vehicular Networking and Connected Vehicles, Aerospace Systems, Automation, Manufacturing, Smart Grids, Nonlinear Systems, Power Systems, Robotics, Social Systems, Economic Systems and other. Of particular value to both the contributors and the readership are the short publication timeframe and the world-wide distribution and exposure which enable both a wide and rapid dissemination of research output.

The series covers the theory, applications, and perspectives on the state of the art and future developments relevant to systems and networks, decision making, control, complex processes and related areas, as embedded in the fields of interdisciplinary and applied sciences, engineering, computer science, physics, economics, social, and life sciences, as well as the paradigms and methodologies behind them.

Indexed by SCOPUS, INSPEC, WTI Frankfurt eG, zbMATH, SCImago.

All books published in the series are submitted for consideration in Web of Science.

For proposals from Asia please contact Aninda Bose (aninda.bose@springer.com).

Gonzalo Fernando Olmedo Cifuentes ·
Diego Gustavo Arcos Avilés ·
Hernán Vinicio Lara Padilla
Editors

Emerging Research in Intelligent Systems

Proceedings of the CIT 2023 Volume 1

Springer

Editors
Gonzalo Fernando Olmedo Cifuentes
Department of Electrical, Electronic
and Telecommunications
Universidad de las Fuerzas Armadas ESPE
Sangolquí, Ecuador

Diego Gustavo Arcos Avilés
Department of Electrical, Electronic
and Telecommunications
Universidad de las Fuerzas Armadas ESPE
Sangolquí, Ecuador

Hernán Vinicio Lara Padilla
Department of Energy and Mechanics
Universidad de las Fuerzas Armadas ESPE
Sangolquí, Ecuador

ISSN 2367-3370 ISSN 2367-3389 (electronic)
Lecture Notes in Networks and Systems
ISBN 978-3-031-52254-3 ISBN 978-3-031-52255-0 (eBook)
https://doi.org/10.1007/978-3-031-52255-0

This Springer imprint is published by the registered company Springer Nature Switzerland AG
The registered company address is: Gewerbestrasse 11, 6330 Cham, Switzerland

Paper in this product is recyclable.

Organization

General Chairs

Gonzalo Fernando Olmedo Cifuentes, Universidad de las Fuerzas Armadas ESPE, Ecuador
Diego Gustavo Arcos Avilés, Universidad de las Fuerzas Armadas ESPE, Ecuador
Hernán Vinicio Lara Padilla, Universidad de las Fuerzas Armadas ESPE, Ecuador

Local Committee

Alicia Marifernanda Zavala Calahorrano, Medical Sciences, Universidad de las Fuerzas Armadas ESPE, Ecuador
Ana Gabriela Haro Báez, Earth and Construction Sciences, Universidad de las Fuerzas Armadas ESPE, Ecuador
Edgar Alfonso Arauz Sanchez, Security and Defense, Universidad de las Fuerzas Armadas ESPE, Ecuador
Francisco Javier Montero Moya, Energy and Mechanics Sciences, Universidad de las Fuerzas Armadas ESPE, Ecuador
Fausto Lenin Granda Gutiérrez, Electrical, Electronics, and Telecommunications, Universidad de las Fuerzas Armadas ESPE, Ecuador
Freddy Mauricio Tapia León, Computer Sciences, Universidad de las Fuerzas Armadas ESPE, Ecuador
Luis Enrique Simbaña Taipe, Economic and Administrative Sciences, Universidad de las Fuerzas Armadas ESPE, Ecuador
Mauro Hernán Ocaña Garzón, Human and Social Sciences, Universidad de las Fuerzas Armadas ESPE, Ecuador

Sergio Alberto Castillo Páez, Basic Sciences, Universidad de las Fuerzas Armadas ESPE, Ecuador
Wilmer Edison Pozo Rivera, Life Sciences and Agriculture, Universidad de las Fuerzas Armadas ESPE, Ecuador

Program Committee

Alexander Eduardo Vicente Ferreira, Inter-American Defense College, USA
Alisva De Los Ángeles Cárdenas Pérez, Universidad de las Fuerzas Armadas ESPE, Ecuador
Ana Lucía Andrade Carrión, Universidad Nacional de Loja, Ecuador
Anabell Del Rocío Urbina Salazar, Universidad de las Fuerzas Armadas ESPE, Ecuador
Armando Reyna Bello, Universidad de las Fuerzas Armadas ESPE, Ecuador
Bladimir Bacca Cortes, Universidad del Valle, Colombia
Carla Nathaly Villacis Nunez, University of Michigan, USA
Carlos Bran, Universidad Don Bosco, El Salvador
Carlos Gabriel Helguero, Escuela Superior Politécnica del Litoral, Ecuador
Carlos Reinoso Jerez, Yachay Tech, Ecuador
Cecilio Angulo Bahón, Universitat Politècnica de Catalunya, Spain
Christian Zúñiga, University of Louisville, USA
Claudio Burgos, Universidad de O'Higgins, Chile
Claudio Canizares, University of Waterloo, Canada
Cristián Henríquez, Pontificia Universidad Católica de Chile, Chile
Cristian Rabanal, Universidad Nacional de Río Cuarto Universidad Nacional de Villa Mercedes, Argentina
Diego Esteban Fernández Olivo, Universidad Nacional de Educación, Ecuador
Edison Gonzalo Espinosa Gallardo, Universidad de las Fuerzas Armadas ESPE, Ecuador
Eduardo Caicedo Bravo, Universidad del Valle, Colombia
Emilia Vázquez Rozas, Universidad de Santiago de Compostela, Spain
Emrah Tasdemir, Bilecik Seyh Edebali University, Turkey
Ender Enrique Carrasquero Carrasquero, Universidad Dr. José Matías Delgado, Panama
Erick Ramírez-Cedillo, Tecnológico de Monterrey, México
Fernando Tello Oquendo, Escuela Superior Politécnica de Chimborazo, Ecuador
Freddy Ordoñez Malla, Universidad Politécnica Nacional, Ecuador
Gabriel Comparato, Universidad Nacional de La Plata, Argentina
Galileo Violini, Centro Internacional de Física, Colombia
German Eduardo Rodríguez Galán, Universidad de las Fuerzas Armadas ESPE, Ecuador
Gustavo Luis Palazzo, Universidad Tecnológica Nacional, Argentina

Ivón Patricia Escobar Anchaguano, Universidad de las Fuerzas Armadas ESPE, Ecuador
Jaime Alberto Buitrago, Universidad del Quindío, Colombia
Jan A. Lammel Lindemann, Instituto Tecnológico y de Estudios Superiores de Monterrey, Mexico
Javier Roca Pardiñas, University of Vigo, Spain
Javier Vazquez-Armendariz, Tecnológico de Monterrey, Mexico
John W. Castro, Universidad de Atacama, Chile
Jorge Humberto Erazo Aux, Institución Universitaria Antonio José Camacho, Colombia
Jorge Lino Alves, Faculdade de Engenharia da Universidade do Porto, Portugal
Jorge Ramió Aguirre, Universidad Politécnica de Madrid, Spain
Jorge Saúl Sánchez Mosquera, Universidad de las Fuerzas Armadas ESPE, Ecuador
José Emilio Sánchez García, Universidad Autónoma Indígena de México, Mexico
José Gustavo Casas Álvarez, Universidad Autónoma de Tlaxcala, Mexico
José Luis García Dorado, Universidad Autónoma de Madrid, Spain
José Luis Pou, Universidad Nacional de Rosario, Argentina
José Gioberty Bucheli Andrade, Universidad de las Fuerzas Armadas ESPE, Ecuador
Juan Francisco Vivero, Universidad de Investigación e Innovación, Mexico
Juan Francisco Correa Jácome, Universidad de las Fuerzas Armadas ESPE, Ecuador
Nahir Yerely Dugarte Jiménez, Universidad de las Fuerzas Armadas ESPE, Ecuador
Julian Carrillo, Universidad Militar Nueva Granada, Colombia
Katalina Barreiro Santana, Instituto de Altos Estudios Nacionales, Ecuador
Lenin Eduardo Guerra, Universidad de Medellín, Colombia
Luis Enrique Sánchez Crespo, University of Castilla-la Mancha, Spain
Luis Tello Oquendo, Universidad Nacional de Chimborazo, Ecuador
Luis Lenin Recalde Herrera, Universidad de las Fuerzas Armadas ESPE, Ecuador
Manuel Sánchez Rubio, Universidad Internacional de la Rioja, Spain
María-Luisa Chas-Amil, University of Santiago de Compostela, Spain
Mario Sarcinelli-Filho, Universidade Federal do Espírito Santo, Brazil
Miguel Angel Sicilia, Universidad de Alcalá, Spain
Oscar Dieste, Universidad Politécnica de Madrid, Spain
Oswaldo Moscoso, Universidad Tecnológica Equinoccial, Ecuador
Pablo Velarde Rueda, Universidad Loyola Andalucía, Spain
Ramón Alberto Martín Fernández, Universidad de La Habana, Cuba
Raquel Tejeda-Alejandre, Tecnológico de Monterrey, Mexico
Rodrigo Sierra, Geoinformática & Sistemas, Independiente
Roger Clotet Martínez, Universidad Internacional de Valencia, Spain
Ruby Mejía de Gutiérrez, Universidad del Valle, Colombia
Salvador Naya, University of A Coruña, Spain
Santiago Daniel Puma Araujo, Politecnico di Torino, Italy
Suelia de Siqueira Rodrigues Fleury Rosa, University of Brasilia, Brazil
Vera Gelashvili, Rey Juan Carlos University, Spain
Victor Hinojosa, Universidad Técnica Federico Santa María, Chile

Virginia Sahores Avalís, Universidad Nacional de La Plata, Argentina
Vladimir Bonilla, Universidad Yachay Tech, Ecuador
Wilfredo Alfonso-Morales, Universidad del Valle, Colombia
Williams Aranguren Alvarez, Universidad de Carabobo, Venezuela
Yosman José Valderrama Blanco, Universidad de Los Andes, Venezuela

Research Management

Rolando Patricio Reyes Chicango, Universidad de las Fuerzas Armadas ESPE, Ecuador
Tatiana Marisol Gualotuña Álvarez, Universidad de las Fuerzas Armadas ESPE, Ecuador
Carina Alexandra Haro Granizo, Universidad de las Fuerzas Armadas ESPE, Ecuador
Laura María Cedamano García, Universidad de las Fuerzas Armadas ESPE, Ecuador
Pedro Xavier Chinga Marmol, Universidad de las Fuerzas Armadas ESPE, Ecuador
Wendy Estefany Castillo Vega, Universidad de las Fuerzas Armadas ESPE, Ecuador

Preface

We are pleased to present the first volume of the proceedings from the XVIII International Multidisciplinary Congress on Science and Technology (CIT 2023). This significant event marked a milestone in the convergence of bright minds, innovative ideas, and substantial advances in various scientific and technological disciplines. This congress unfolded over five days in November 2023 at the Universidad de las Fuerzas Armadas ESPE in Sangolqui, Ecuador. Scholars, researchers, and professionals worldwide gathered to share knowledge, explore new frontiers, and forge collaborations that will undoubtedly propel the future of science and technology. This volume reflects the diversity and excellence presented during CIT 2023, with each article serving as a testament to the collective commitment to innovation and progress in areas such as Artificial Intelligence, Computational Modeling, Data Communication, Software Engineering, and Security and Cryptography. Furthermore, it encompasses general areas like Computer Sciences, Earth and Construction Sciences, Basic Sciences, Electrical, Electronics, Telecommunications, and Energy and Mechanics Sciences.

Sangolquí, Ecuador
November 2023

Gonzalo Fernando Olmedo Cifuentes
Diego Gustavo Arcos Avilés
Hernán Vinicio Lara Padilla

Contents

Innovation, Development and Data Communication

Proposal Design of an Autonomous Photovoltaic Plant on Potentially Liquefiable Soil in the Isolated Rural Community of Agua Fría, Ecuador

Carmen Morales, Ana G. Haro, Leonardo Jiménez, and Diego Arcos-Aviles

Abstract This paper presents the seismic analysis and design considerations of a foundation for an autonomous photovoltaic generation plant on potentially liquefiable soil. As a case study, the isolated rural community of Agua Fría is selected, which belongs to the rural parish of Majua in Esmeraldas, Ecuador. Considering the population and the electrical load capacity, the plant is designed for a 25-year projection. The power generation system comprises 15 galvanized steel modules as a support structure for the solar panels and a metallic structure warehouse for the batteries and energy converters. The soil characteristics are established based on studies of neighboring areas with similar seismicity and soil conditions. A slab foundation on improved soil by inserting compacted aggregate piers is proposed to mitigate the liquefaction problem. Probability curves for 5, 20, 50, 80, and 95%, which consider the relationship between the cyclic stress CRS and the corrected soil penetration resistance N1.60, are used to probe the soil liquefaction potential of the proposed solution. Finally, a brief environmental impact analysis of the proposed photovoltaic generation plant is described, as well as its advantages against energy generation techniques by exploitation of fossil fuels.

Keywords Photovoltaic generation system · Isolated rural area · Seismic considerations · Foundation · Liquefaction · Compacted aggregate piers

C. Morales · A. G. Haro (✉)
Structures and Constructions Research Group (GIEC), Departamento de Ciencias, La Tierra y La Construcción, Universidad de Las Fuerzas Armadas ESPE, 171-5-231B Sangolquí, Ecuador
e-mail: agharo@espe.edu.ec

L. Jiménez
Strategic Business Unit, Pivaltec S.A, 170505 Quito, Ecuador

D. Arcos-Aviles
Grupo de Investigación en Propagación, Control Electrónico y Networking (PROCONET), Departamento de Eléctrica, Electrónica y Telecomunicaciones, Universidad de Las Fuerzas Armadas ESPE, 171-5-231B Sangolquí, Ecuador

G. F. Olmedo Cifuentes et al. (eds.), *Emerging Research in Intelligent Systems*, Lecture Notes in Networks and Systems 902, https://doi.org/10.1007/978-3-031-52255-0_1

1 Introduction

The growing environmental impact, fueled by obtaining energy through the exploitation of fossil fuels, has made it imperative to seek other sources of natural energy to supply population growth and curb the mentioned impact. Solar and wind energy are the renewable energy sources with the most significant projection globally, considering that the installable power is mainly found in parks or solar farms in rural areas, where land costs are lower, allowing larger surface area for solar plants with much higher powers to achieve [1].

The use of clean energy is gaining strength in Ecuador, with the projection of the use of photovoltaic energy as an alternative to face two problems: environmental pollution and oil scarcity in the long term. An example is the photovoltaic plant on San Cristóbal Island in the Galapagos. In addition to the ecological advantages that these electricity generation systems bring, they are the best alternative to getting electricity to rural areas [2]. However, until 2021, the coverage rate was still reported in Ecuador at 97.29%. That is, 2.71% of the population had no access to electricity. Indeed, this problem is concentrated in rural areas with difficult access [2].

Agua Fría is one of the 15 rural communities located dispersedly on the banks of the Esmeraldas River and the Majua Estuary. This community belongs to the rural parish of Majua-Esmeraldas Canton-Esmeraldas Province, with an approximate population of 200 people. They have partial or non-existent power supply [3]. This forces its inhabitants to resort to unconventional resources, such as using car batteries to generate electricity. Energization is considered a strategic element for developing and evolving productive, commercial, academic, health, and domestic activities. Therefore, it has a direct relationship with the level of poverty and the quality of life of a population [4].

Even though Ecuador is located in a high seismicity region and has officially established design and construction regulations for structures and foundations, there is a lack of consciousness regarding obedience to minimal seismic requirements, especially for non-building structures. This study defines the geotechnical and structural considerations for the foundations of autonomous photovoltaic systems installed on the ground. These considerations are applied in designing the foundation for a photovoltaic generation plant in the rural community of Agua Fría.

The remainder of this paper is organized as follows: Sect. 2 describes the methodology used in this study, the description of the case study, the soil condition, and the main components of the solar PV plant. Section 3 presents the results of the Soil liquefaction control and a brief analysis of the environmental impact. Finally, Sect. 4 summarizes the main conclusions of this study.

2 Methodology

Based mainly on the energy coverage state of communities in rural areas of the Ecuadorian coast, the Agua Fría was selected as a case study. The population demand to satisfy was determined through data registered in the Plan of Development and Territorial Ordinance PDOT of the parish of Majua [5], projecting to 25 years, the reasonable life period of photovoltaic panels.

The critical site seismicity and the soil conditions were also considered revealing parameters at the time of the case study selection. The soil characteristics were determined using data from a soil study for a project in a surrounding area, the Quinindé Canton, where using SPT standard penetration tests by the ASTM D1586 [6], semi-disturbed samples were taken at each meter with a total depth of 12 m and subsequently evaluated in laboratory tests.

3 Case Study

3.1 Location

The implantation area would be in Zone 17, North of Ecuador, with UTM coordinates 0°38′18″ N 79°27′48″ W. The Majua estuary is, on average, 76 m above sea level (MASL). The projected implantation area for the renewable generation system is approximately at an altitude of 116 MASL. Figure 1 shows the details.

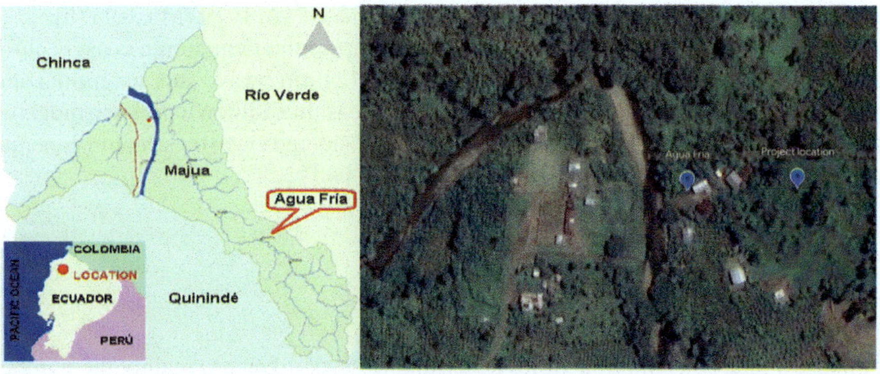

Fig. 1 Implementation area of the renewable generation system in Agua Fría

3.2 Geological Description

The morpho-structural region to which the province of Esmeraldas belongs, which constitutes the western area of the Andes, is the fore-arc sedimentary basin that contains volcanic and sedimentary formations of the Cretaceous-Eocene age. Four sedimentary basins have been formed in this region. One of them is the Esmeraldas Basin, which is divided into three zones: the western zone (Esmeraldas Basin), the elevated central zone, and the subsiding area (Borbón and Tumaco Basin) [7].

The study area is in the Esmeraldas-Borbón basin, whose basement comprises Cretaceous rocks of the Piñón formation and Quaternary terranes with alluvial and colluvial deposits and residual soils [7].

3.3 Seismic Characteristics

The Ecuadorian coast has a complex morpho-structure, relief differences, and fore-arc basins, which are the product of active interplate tectonics. The subduction zone originates from the convergence of two tectonic plates, the Nazca oceanic plate and the South American continental plate, generating the tectonic subsidence zones in this region [8].

The seismic records in the coastal zone of Ecuador report earthquakes magnitudes greater than 7.5, where one of the strongest around the world occurred in the area of Tortuga, Esmeraldas, on January 3rd, 1906, with magnitude 8.8 at 25 km depth. This event is attributed to a displacement of crustal rupture of about 500 km in length, which extends offshore from the limits of Bahía de Caráquez to Tumaco (Colombia); later, it was reactivated in 1942 (Mw 7.9), 1958 (Mw 7.8), 1979 (Mw 8.1) [8].

The province of Esmeraldas in the northern part of the Ecuadorian coast is close to the subduction zone located between 53 and 121 km away. This natural hazard structure can affect the most densely populated areas. In addition to this seismogenic source, crustal geological faults should also be considered since these could generate earthquakes of magnitudes between 6.0 and 7.1 Mw [9].

3.4 Electric Power Consumption

A house's load consumption varies according to several factors, mainly the number of inhabitants, age, and lifestyle. Therefore, it is necessary to determine the number of inhabitants and the daily consumption of a house. Given that this study considers the sizing of a renewable generator system for supplying energy for the next 25 years, a projection of the number of inhabitants is required.

On the one hand, the population projection is calculated by (1) for a period of 25-year, as follows:

Table 1 Load consumption in an isolated residence

Appliances	Quantity	Rated power (W)	Hours per day (h)	Average energy per day (Wh/day)
Light bulbs	4	12	7	336
Refrigerator	1	100	24	2400
Fan	2	60	8	960
Computer	1	90	4	360
Phone charger	2	20	2	80
Radio	1	50	5	250
TV	1	90	7	630
Total (E_L)				5016

$$P_t = P_o \cdot (1 + r)^t \tag{1}$$

where P_t is the final population after t years, P_o is the initial population, r is the annual population growth rate, and t is the projection time. The people and annual growth rate of the Agua Fría community corresponding to 2020 are considered in this study (i.e., $P_o = 223$ and $r = 0.011$), resulting in a population projection of $P_t = 294$ inhabitants. Therefore, assuming four people occupy each house, the total number of residences in this community in 25 years will be $N_H = 74$.

On the other hand, in rural areas, the residences do not have many appliances, so the typical load consumption is usually low [10]. Therefore, this study assumes that each household presents a daily consumption according to the essential appliances in Table 1.

Therefore, considering the inhabitants' projection and the number of houses in 25 years, the daily energy consumption by the community is:

$$E_T = E_L \cdot N_H = 371.184 \text{ [Wh/day]} \tag{2}$$

3.5 *Photovoltaic Generator*

The photovoltaic generator sizing requires determining the available solar resources in the system's implementation area. A simple way to assess the power a photovoltaic generator produces in a given area is through the Peak-Sun Hour (PSH) of this area. Therefore, the Photovoltaic Geographical Information System (PVGIS) provided by the European Commission [11] allows the PSH to be obtained in the Agua Fria community.

The required PV modules are calculated considering a critical PSH of $PSH_c = 3.8$, as follows [10]:

Table 2 Jinko solar JKM300P-72 panel dimensions

Description	Variable	Value	Units
Length	L_{PV}	1.956	m
Width	W_{PV}	0.992	m
Depth	D_{PV}	0.040	m
Weight	$W_{P\text{-}PV}$	26.5	kg

$$P_{PV} = \frac{E_L}{PSH_c} = 97.68\ [\text{kW}] \tag{3}$$

This study considers a 99-kWp photovoltaic array comprising 330 Jinko Solar JKM300P-72 panels [12] connected in 66 strings ($N_p = 66$) of 5 panels each ($N_s = 5$). Table 2 presents the dimensions of the JKM300P-72 panel.

Therefore, the total weight (4) and the surface (5) of the photovoltaic generator are:

$$W_{T-PV} = W_{P-PV} \cdot N_s \cdot N_p = 8745\ [kg] \tag{4}$$

$$A_{PV} = (L_{PV} \cdot N_s) \cdot (W_{PV} \cdot N_p) = 640.32\ [\text{m}^2] \tag{5}$$

3.6 Battery Energy Storage System

The battery capacity can be computed as follows [10]:

$$C_{Ah} = \frac{E_L \cdot N_D}{V_{BB} \cdot DOD} \tag{6}$$

where N_D is the number of autonomous days, V_{BB} is the battery bank voltage, and *DOD* is the permissible depth-of-discharge of the battery system. This study considers a Trojan SOLAR SIND 04 2145 lead-acid battery bank [13], whose characteristics are presented in Table 3.

Considering a maximum $DOD = 50\%$ for lead-acid batteries, an autonomy of one day $N_D = 1$, and a battery bus voltage $V_{BB} = N_{BS} \cdot V_B = 48$ [V], being $N_{BS} = 12$ the number of batteries connected in series, the total capacity of the battery storage system is $C_{\text{Ah}} = 15{,}466$ [Ah]. Therefore, an array of $N_{BP} = 10$ connected in parallel is needed to supply the required current.

Consequently, the total weight (7) and the surface (8) of the battery energy storage system are:

$$W_{T-B} = W_{P-B} \cdot N_{BS} \cdot N_{BP} = 25320\ [\text{kg}] \tag{7}$$

Table 3 Lead-acid battery characteristics

Description	Variable	Value	Units
Voltage	V_B	4	V
Capacity	C_{20}	1647	Ah
Length	L_B	0.691	m
Width	W_B	0.265	m
Height	H_B	0.610	m
Weight	$W_{P\text{-}B}$	211	kg

Fig. 2 Proposed photovoltaic generation plant [14]

$$A_{PV} = (W_B \cdot N_{BP}) \cdot (H_B \cdot N_{BS}) = 19.40\ [\text{m}^2] \tag{8}$$

Figure 2 [14] shows a scheme of the proposed photovoltaic generation plant that comprises 15 galvanized steel modules as a support structure for the solar panels and a metallic structure warehouse for the batteries and energy converters for which a geotechnical solution should be implemented according to the following sections.

3.7 Soil Characteristic

The soil characteristics in the area are defined by the information provided from previous studies in the canton of Quinindé. These studies were carried out in 2020, the most recent in the bibliographic review. The project area is located within the province of Esmeraldas, 35 km away from the community of Agua Fría. Therefore, the same seismicity characteristics are expected. Furthermore, despite having different geological formations, both points are close to the Esmeraldas River. Consequently, they are considered alluvial soils.

Figure 3 shows the typical stratigraphy of the referred study area, where moderately compacted, poorly graded (SP) gray and reddish-brown sands were detected. Sands with few fines are known to be vulnerable to liquefaction during cyclical

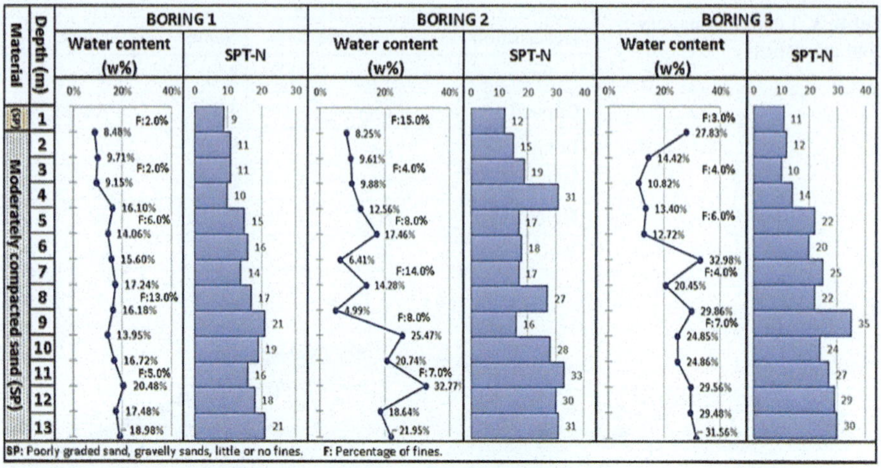

Fig. 3 Typical stratigraphy of the referred study area

loading periods [15]. In addition, these types of soils are susceptible to pore pressure redistribution on a time scale equal to or approximate to that of cyclic loads [16]. Therefore, it is recommended to carry out an analysis to define the liquefiable potential of the soil and subsequent treatment to improve the mechanical properties of the ground and avoid liquefaction problems.

3.8 Liquefaction Potential Analysis

The analysis of the liquefiable potential was performed based on the method described in [17], with several improvements in the correlation of the curves used to define the liquefiable potential. The theoretical foundation is based on relating the coefficient of cyclical shear stress in situ CSR and the corrected soil penetration resistance N1.60.

Table 4 presents the analysis results of the liquefaction potential of the three perforations used as a reference for the proposed project. The liquefaction percentages range from 95 to 20% up to 4 m deep. Based on these results, the need to improve the soil is clear to mitigate the adverse effects of liquefaction and avoid damage that could interrupt the operation of the photovoltaic plant facilities.

3.9 Soil Improvement

To mitigate the liquefaction problems presented by the soil where the project would be implemented, an improvement system based on the construction of compacted gravel piers is proposed. This method was developed in the 1980s and is widely

Table 4 Assessment of the liquefaction potential of different wells

Depth (m)	$N_{1,60CS}$	$CSR*_{eq}$	LP	$N_{1,60CS}$	$CSR*_{eq}$	LP	$N_{1,60CS}$	$CSR*_{eq}$	LP
	Boring 1			Boring 2			Boring 3		
0.5	8.54	0.13	95%	11.42	0.13	80%	10.45	0.13	80%
1.0	10.45	0.13	80%	14.42	0.12	20%	11.42	0.13	80%
2.0	10.45	0.13	80%	18.60	0.12	5%	9.49	0.13	95%
3.0	9.49	0.13	95%	32.06	0.12	<5%	13.41	0.13	50%
4.0	14.42	0.12	20%	16.49	0.12	5%	21.84	0.12	<5%
5.0	15.45	0.12	10%	17.54	0.12	5%	19.67	0.12	<5%
6.0	13.41	0.13	5%	16.49	0.12	5%	25.17	0.12	<5%
7.0	16.49	0.12	5%	27.43	0.12	<5%	21.84	0.12	<5%
8.0	20.75	0.12	<5%	15.45	0.12	5%	36.82	0.12	<5%

used in the United States, Europe, and Asia. The basis of this method is to increase the bearing capacity of the soil with the implementation of compacted aggregate piers (CAP), forming a mesh throughout the foundation area in such a way that the natural stiffness of the matrix soil increases with the excellent rigidity provided by compacted aggregate piers and thus reduce settlements [18].

Inserted compacted aggregate piers considerably improve the resistance of the soil to liquefaction since the piers' rigidity absorbs the effects of cyclical shear caused by seismic actions. Additionally, during the installation of the piers, the compaction process increases the lateral stresses within the adjacent soil mass [19].

For the storage structure of the generation plant, which covers an area of 53.90 m^2, an arrangement of 16 piers with a diameter of 75 cm and a depth of 4 m is used, separated by a distance of 2.40 and 2.00 m along the horizontal axis, as observed in Fig. 4a. In the case of the collecting structure where the solar panels will be placed, the use of 10 compacted gravel piers with 75 cm diameters is projected in an area of 33.99 m^2, with a depth of 4 m and a distance of 2.40 m between piers, as shown in Fig. 4b.

3.9.1 Foundation Structural Design

After the soil improvement, the use of a slab foundation was selected as the best alternative, considering the ease of construction and, consequently, reduction in execution and maintenance time, and the advantage of using a type of shallow foundation settled directly on the improved ground bed with columns of compacted aggregate. However, for the design of the slab, it is essential to consider that when carrying out the improvement, the uniformity of the soil stiffness where the slab sits is substantially altered due to the rigidity provided by the compacted piers concerning the parent soil without modification. For this reason, in [20], design primers for a slab foundation on compacted aggregate piers are developed based on the results of finite

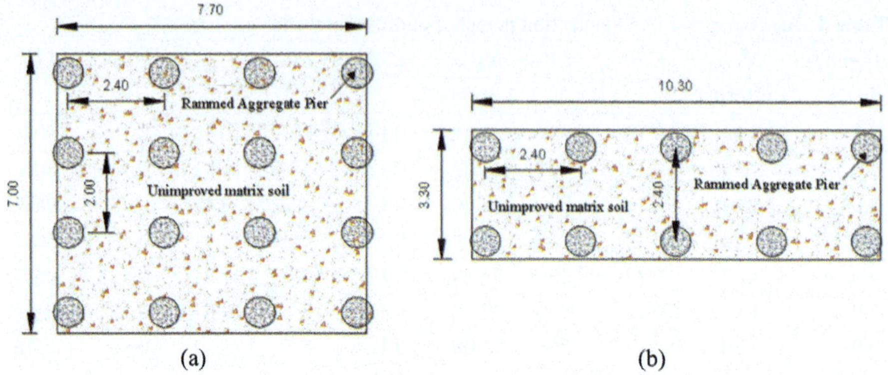

Fig. 4 Arrangement of compacted aggregate piers. **a** storage structure. **b** collecting structure

element models presented in Fig. 5. The blue line represents the value of the relationship between the load on the slab foundation and the concrete resistance measured for the collecting structure. In contrast, the red line is for the storage structure [20]. Through these charts, it is possible to establish the recommended minimum thickness for a slab with a uniformly distributed load based on parameters that depend on the soil improvement design.

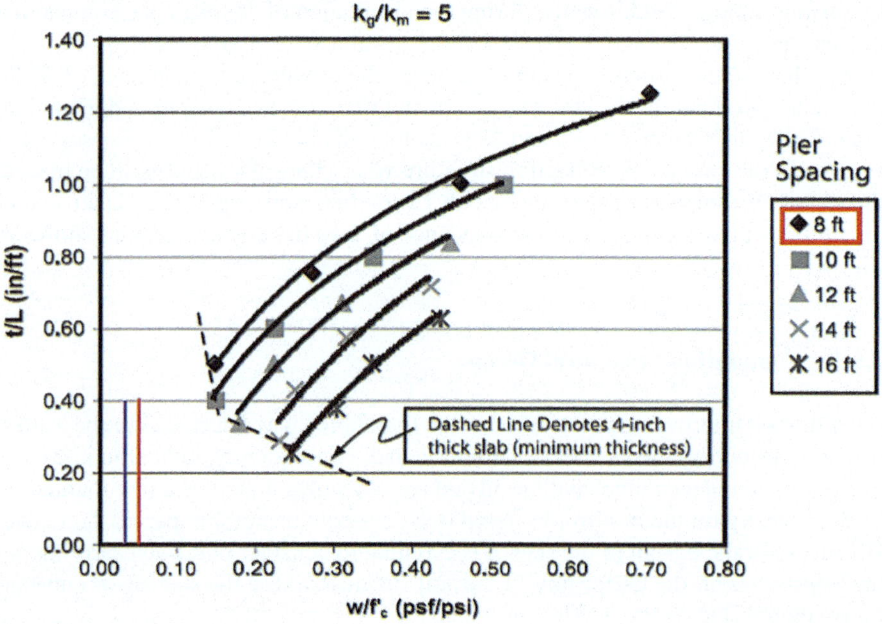

Fig. 5 The standardized thickness of the slab foundation for the stiffness ratio kg/km = 5 [20]

Table 5 Geopier element layout parameters

Parameters	Variable	Storage room	Solar panel zone	Units
Pier center-to-center spacing	L	2.40	2.40	m
Pier stiffness	k_g	7892.61	7892.61	T/m^3
Unimproved matrix soil stiffness	k_m	1578.00	1578.00	T/m^3
Stiffness ratio	$R_s = k_g/k_m$	5.00	5.00	adim
Slab uniform load	w	149.51	90.36	psf
Concrete compressive strength	$f'c$	2986.90	2986.90	psi
Ratio	$w/f'c$	0.05	0.03	psf/psi

The parameters used for the case study are presented in Table 5. Since the vertical loads that rest on the slab foundation are relatively small, the factor $w/f'c$ is below the dotted line that intersects the curve corresponding to the spacing of 8 ft. Consequently, a minimum 4 inches (10.16 cm) thick slab was adopted at this stage.

According to NEC-15 [21] and ACI-318 [22], the minimum cover for surfaces exposed to the elements or in permanent contact with the soil is 75 mm. Finally, a 12 cm thick slab foundation is used for the storage room and the panel area to guarantee the minimum cover for the reinforcing steel.

4 Results

4.1 Soil Liquefaction Control

The liquefaction effects in the soil are associated with the residual shear stress and the magnitude of the shear loads it supports. Therefore, the method used to measure the liquefiable potential of the soil is reduced to the correlation between the cyclic stress ratio CRS and the corrected soil penetration resistance N1.60 measured in a curve with probabilities of 5, 20, 50, 80, and 95%.

It is assumed that shear forces are countered as a function of the element's stiffness. Therefore, a relationship is established between the shear stress resisted by the unimproved soil τ_S and the earthquake-induced shear stress τ. This relationship permits obtaining a cyclic stress reduction factor for the unimproved soil resisting the earthquake demands. According to the sizing of Geopier elements, a replacement area ratio, $Ra = 0.13$, and a stiffness ratio, $Rs = 5$, are established, obtaining a factor of 0.66. Without improvement, the soil will resist 66% of the stresses induced by the earthquake, which reduces the cyclic stress ratio from an average of 0.13–0.08.

The Geopier system provides data from tests on different soil types based on extensive applications. It has shown an increased soil penetration resistance after being improved. The increase in the number of hits, N_{SPT}, implies an increased resistance to penetration into the soil, greater bearing capacity, greater density of the

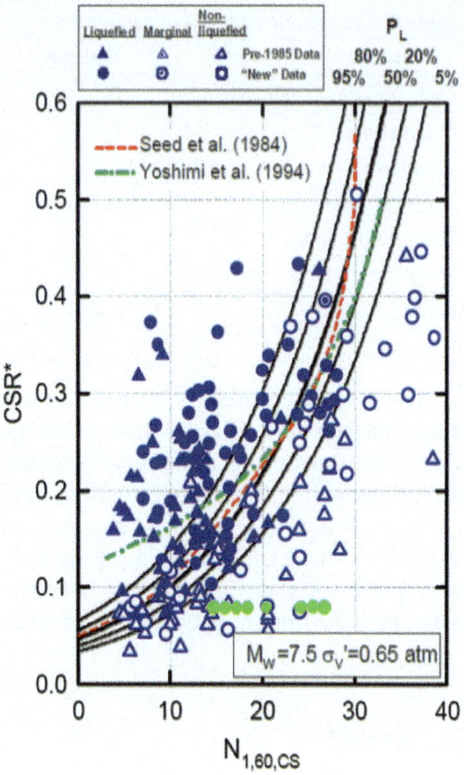

Fig. 6 Measurement of liquefiable potential from borehole one after upgrading [23]

adjacent ground, and increased lateral stresses. By improving these parameters and evaluating them in the probability curves, it is evident that the values at all soil levels are below the 5% curve (see Fig. 6), where the soil does not present liquefaction problems. The blue points represent the values of the series of tests carried out to define the probability curves. The green points represent the values of the parameters obtained based on the tests carried out with the survey's sampling once evaluated after soil improvement for the case study [23].

4.2 Environmental Impact

The use of photovoltaic generation systems for supplying electricity to an isolated community is CO_2 emission-free and noise pollution [24]. Regarding implementation, the construction phases are the ones that have the most significant impact due to the high emission of gases and particles, noise pollution, and waste production [24]. In addition, the power line work entails work with more significant extension than implementing the photovoltaic generation system. However, the United

Table 6 Carbon dioxide emissions from different electricity generation systems

Energy consumption (kWh)		Emissions from energy production		
		Gas (gCO_2/kWh)	Coal (gCO_2/kWh)	Photovoltaic (gCO_2/kWh)
Daily	371.18	148×10^3	74×10^3	11×10^3
Annual	133×10^3	53×10^6	26×10^6	4×10^6
25-years	3340×10^3	133×10^7	66×10^7	10×10^7

Nations considers relative values of carbon dioxide emissions per kWh produced with different approaches. In this regard, photovoltaic energy production emits between 25 and 32 gCO_2/kWh, the energy produced by gas exploitation emits 400 gCO_2/kWh, and the energy produced by coal combustion emits 200 gCO_2/kWh [25]. These values allow the quantification of the environmental impact between renewable and non-renewable energy systems. Table 6 presents the amount of carbon dioxide each generation system would contribute in a projection period of 25 years. Notably, carbon dioxide emission from a photovoltaic generation system is 6.25 times lower than a coal combustion system and 12.5 times lower than the electrical system fed with gas.

5 Conclusions

A geotechnical study provides information to anticipate the possible response of the soil and take the necessary actions to avoid partial or total damage to the foundation structure and the infrastructure, guaranteeing the functionality, in this case, of the photovoltaic generation system. Soils with high liquefaction potential are evident in the selected zone. The Seed et al. [17] method, among those recommended by the Ecuadorian Construction Regulations, is optimal for sand analysis.

Compacted aggregate piers with adequate spacing make it possible to considerably reduce the effects of liquefaction by stiffening the soil layers, increasing lateral rigidity, and densifying the soil.

The analysis of slabs on soils that have been improved using the compacted aggregate piers system changes substantially since the stiffness of the compacted pier is greater than that of the parent soil, which means that, within a numeric model of springs, the slab foundation rests on springs with different stiffnesses, invalidating the traditional analysis based on a uniformly elastic soil. Consequently, the implemented technique in this study employed primers that respond to parameters based on soil improvement.

On the other hand, this type of infrastructure is relatively light regarding the transmitted loads from the structure to the foundation. Therefore, settlement will not be a problem once the soil is improved.

Finally, in the context of the environmental impact, it is once again demonstrated that photovoltaic energy production is less damaging than energy production from fossil fuels. One of the main measures is the carbon footprint. It was found that the production of photovoltaic energy emits 6.25 times less carbon dioxide than the energy production by coal combustion and 12.5 times less carbon dioxide than gas energy production.

Acknowledgements This work is part of the projects 2023-PIS-004 and 2020-EXT-007 "MIRA-ESTE: Specific, innovative microgrids solutions (accounting for environmental, social, technological and economic aspects) for isolated rural areas of Ecuador" from the Structures and Constructions Research Group (GIEC) and Propagation, Electronic Control, and Networking Research Group (PROCONET) at Universidad de las Fuerzas Armadas ESPE and KU Leuven. This work has been developed with the support of VLIR-UOS and the Belgian Development Cooperation (DGD) under the project EC2020SIN322A101.

References

1. Sánchez-Carratalá, C.: Bases para el proyecto de cimentaciones de pilotes hincados para estructuras de apoyo de placas en parques solares fotovoltaicos. XVI Congreso Internacional de Ingeniería de Proyectos 1404–1415 (2012)
2. Agencia de Regulación y Control de Energía y Recursos Naturales No Renovables. Esta-dística Anual y Multianual del Sector Eléctrico Ecuatoriano. 1–178 (2021)
3. Zambrano, R.: Plan de desarrollo y ordenamiento territorial 2019–2023. Esmeraldas (2021)
4. Franco, C., Dyner, I., Hoyos, S.: Contribution of the energy at development of islated communities in not interconnected zones: a case of application of the systems dynamics and sustainable livelihoods in the Colombian Southwest. DYNA **75**(154), 199–214 (2008)
5. GAD Esmeraldas: Plan de desarrollo y ordenamiento territorial PDOT de la parroquia Majua, Cantón Esmeraldas. Esmeraldas (2014)
6. ASTM: Standard Test Method for Standard Penetration Test (SPT) and Split-Barrel Sampling of Soils. ASTM D1586, USA (2022)
7. INECA: Estudio de ingeniería de la carretera Santo Domingo-Esmeraldas y la estructuración jurídica, técnica y económico-financiera (2012)
8. Chunga, K., Pazmiño, N., Martillo, C., Quiñonez, M., Huaman, F.: Estimación de máximos niveles de sismicidad para el Litoral Ecuatoriano a través de la integración de datos geológicos y sismo-tectónicos. Revista Científica y Tecnológica UPSE **1**(2), 1–13 (2013)
9. Huaman, F.: Evaluación de la amenaza sísmica y estructuras sismogénicas para la ciudad de Esmeraldas, costa norte de Ecuador. Universidad Nacional Mayor de San Marcos (2019)
10. Cust, J., Singh, A., Neuhoff, K.: Rural electrification in India: Economic and institutional aspects of renewables. SSRN Electron. J. 1–37 (2007)
11. European Commission: Photovoltaic Geographical Information System. (2019). https://re.jrc.ec.europa.eu/pvg_tools/en/#MR
12. Jinko: Jinko JKM300P-72 (300W) Solar Panel. http://www.jinkosolar.com/
13. Trojan Battery Company: Solar Sind 04 2145. (2019). https://www.trojanbattery.com/pdf/datasheets/SIND_04_2145_DS.pdf
14. Morales, C.: Consideraciones estructurales y geotécnicas para cimentaciones de sistemas de generación fotovoltaica en zonas rurales remotas. Universidad de las Fuerzas Armadas ESPE (2021)
15. Önder Çetin, K., Bilge, H.T.: Recent advances in seismic soil liquefaction engineering. Geotechnical. Geol. Earthq. Eng. Springer, Cham. 585–626 (2014)

16. Castañeda, D., Romero, J.: Evaluación del potencial de licuación por medio del ensayo triaxial cíclico de un suelo en la sabana de Bogotá. Universidad de La Salle (2016)
17. Bolton Seed, H., Tokimatsu, K., Harder, L.F., Chung, R.M.: Influence of SPT procedures in soil liquefaction resistance evaluations. J. Geotech. Eng. **111**(12), 1425–1445 (1985).
18. Fox, N., Weppler, L., Scherbeck, R.: Geopier soil reinforcement system—case histories of high bearing capacity footing support and floor slab support. In: Fifth International Conference on Case Histories in Geotechnical Engineering, pp. 1–5 (2004)
19. Wissmann, K., Lawton, E., Farrel, T.: Behavior of Geopier-Supported Foundation Systems During Seismic Events (1999)
20. Richards, J., Miller, J., Wissmann, K.: Structural Design Considerations for Uniformly-Loaded Floor Slabs Supported by Rammed Aggregate Pier Elements (2016)
21. MIDUVI: Norma Ecuatoriana de la Construcción - Estructuras de Hormigón Armado. Ecuador (2015)
22. ACI: Building Code Requirements for Structural Concrete and Commentary ACI CODE-318-19(22), USA (2019)
23. Seed, R.B., et al.: Recent advances in soil liquefaction engineering: a unified and consistent framework. In: 26th Annual ASCE Los Angeles Geotechnical Spring Seminar (2003).
24. Benito, G., Ruiz, K.: AnáLisis beneficio-costo de la implementación de un sistema de energía solar fotovoltaica en el Campus Aguas Claras de la Universidad Santo Tomás Sede Villavicencio. Universidad Santo Tomás (2018)
25. ONU: La promesa de la energía solar: Estrategia energética para reducir las emisiones de carbono en el siglo XXI. Cronica ONU. https://www.un.org/es/chronicle/article/la-promesa-de-la-energia-solar-estrategia-energetica-para-reducir-las-emisiones-de-carbono-en-el

Seismic Design Considerations for the Installation of Photovoltaic Systems on Flat Reinforced Concrete Roofs: A Case Study in Ecuador

Melanie Cano, Carlos Davila, Ana G. Haro, Emilia Motoasca, and Diego Arcos-Aviles

Abstract Recent earthquakes have reported a significant loss toll due to damage to buildings' structural and non-structural elements. Although non-structural components do not directly affect the structure's integrity, their seismic response is one of the most crucial facts in reducing or extending the function recovery time of a building. Since the current Ecuadorian Construction Standard lacks seismic design provisions for these elements, such as photovoltaic systems, this study seeks to establish minimum requirements to guarantee the proper installation of photovoltaic systems on flat roofs of reinforced concrete buildings. For this purpose, a case study in Sangolquí, Ecuador, is selected to propose a photovoltaic system defined by the size of the roof area, a worst-case load combination scenario, and the maximum possible number of solar panels. As a result, international design standards are adapted, representing a starting point for Ecuador.

Keywords Seismic effects · Photovoltaic system · Flat roof · Non-structural components

---

M. Cano · C. Davila · A. G. Haro (✉)
Structures and Constructions Research Group (GIEC), Departamento de Ciencias, de la Tierra y la Construcción, Universidad de las Fuerzas Armadas ESPE, 171-5-231B Sangolquí, Ecuador
e-mail: agharo@espe.edu.ec

E. Motoasca
Independent Researcher Sustainble Energy Systems and Applications, Ghent, Belgium

D. Arcos-Aviles
Grupo de Investigación en Propagación, Control Electrónico y Networking (PROCONET), Departamento de Eléctrica, Electrónica y Telecomunicaciones, Universidad de las Fuerzas Armadas ESPE, 171-5-231B Sangolquí, Ecuador

G. F. Olmedo Cifuentes et al. (eds.), *Emerging Research in Intelligent Systems*, Lecture Notes in Networks and Systems 902, https://doi.org/10.1007/978-3-031-52255-0_2

1 Introduction

Currently, Ecuador's energy matrix still depends to a large extent on fossil fuels. According to the Ecuadorian electricity sector's annual and multi-year statistics [1], the country has a generation of electrical energy constituted mainly by hydropower 62.62%, biomass 1.68%, photovoltaic 0.33%, wind 0.26%, and 35.02% corresponds to sources that still use fossil fuels. However, attempts have arisen in recent years to replace fossil fuels with clean and renewable energy through photovoltaic systems (PV) in residential buildings and industrial facilities to meet their energy demand.

Furthermore, PV systems are considered a young green technology compared to conventional power generation sources [2]. In addition, this alternative allows the implementation of microgrids, defined as flexible and efficient energy systems comprising distributed generation systems, energy storage systems, and an energy management system aiming at controlling the power flux between all devices within the microgrid [3]. Therefore, their installation contributes to cost savings, reduced carbon footprint, low maintenance costs, and reduced electricity bills. However, installing the different distributed generation systems, specifically photovoltaic generators, presents some challenges that must be faced.

In many countries with a long tradition of solar PV-energy use, photovoltaic installations are placed not only on (flat or inclined) roofs but also integrated with the building envelope or in other infrastructures (bridges, sound barriers, etc.). However, in Ecuador, PV installations have, until now, been mainly placed on roofs [4].

The Agency for the Regulation and Control of Energy and Non-Renewable Natural Resources [1] mentions that to place this type of system, the interested parties will have to carry out an electrical study and provide an open space, be it the patio or the roof, but it does not take into account seismic design and structural construction considerations associated with their installation on roofs since the Ecuadorian Construction Standard NEC-15 [5] lacks analysis procedures for non-structural components. For this reason, considering that Ecuador is located in a high seismic region, it is crucial to conduct a study that allows, as a first effort, the technical implementation of solar panels on roofs under seismic actions. These systems could not only affect the structure but also cause damage to the system itself due to excessive displacements.

This study is supported by international standards such as the American Society of Civil Engineers (ASCE) [6] and the Structural Engineers Association of California (SEAOC) regulations [7] to set a starting point from a case study consisting of a reinforced concrete (RC) building with a flat and underutilized accessible roof where photovoltaic systems could be placed. Previously, a structural assessment of the building was necessary to determine if it complies with current seismic regulations to install the most suitable photovoltaic system, including its support.

The remainder of this paper is organized as follows: Sect. 2 describes the methodology used in this study, the building selected as the case study, and the main components of the solar PV system. Section 3 presents the seismic response of the analyzed building and the seismic design considerations for the two different support conditions PV systems. Finally, Sect. 4 summarizes the main conclusions of this work.

2 Methodology

2.1 Structure Description

The selected building is part of an educational center from around 1983 to 1989 [8], located in Sangolquí, Ecuador (see Fig. 1). Its electricity consumption is high, and using renewable energy could reduce environmental pollution. The structure is a 4-story rectangular classroom block with a total height of 13.6 m. The structural system comprises reinforced concrete frames with 30 cm thick lightened slabs and 15 cm thick infill masonry. The mechanical properties are detailed in Table 1.

It has three spans in the Y direction; the two external spans are 7.10 m long, and the central span is 5.00 m long. Meanwhile, it has eight equal 4.85 m spans in the X direction. The columns are 50 cm by 35 cm. The beams are 30 cm by 40 cm along the short spans and 30 by 60 along the long spans.

Fig. 1 Case study structure [9]

Table 1 Mechanical properties of materials [10]

Detail	Variable	Value	Units
Concrete compressive strength	*f'c*	280	kg/cm^2
Steel yield strength	*fy*	4200	kg/cm^2
Masonry compressive strength	*f'm*	24	kg/cm^2

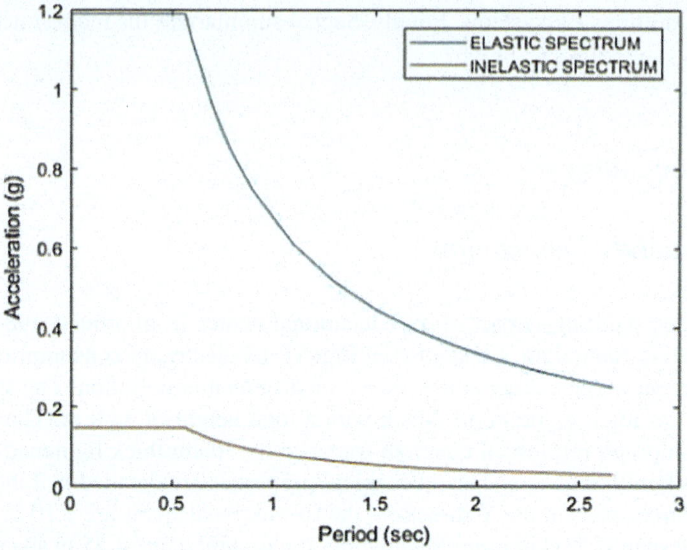

Fig. 2 Design spectrum

2.2 *Design Earthquake Spectra and Spectrum-Compatible Earthquake Records*

Figure 2 shows the elastic and inelastic design spectra calculated according to the recommendations of [5] within the seismic hazard and design chapter NEC-SE-DS. The building is assumed to be on type C soil, with shear velocity Vs30 ranging from 360 to 760 m/s.

According to [6] standard, to carry out a seismic analysis of the structure, at least 11 spectrum-compatible records are necessary. In this case, 12 records were selected using the PEER Ground Motion Database© tool [11]. As a result, the records from Table 2 correspond to a reverse type fault according to the existing failure type suggested by [12], earthquake magnitudes ranging from 4.10 to 6.40 [13], and the corresponding ranges for Vs30.

2.3 *Solar Panels*

In Ecuador, solar PV panels are a relatively new technology, so several companies offer the same brand of panels, mainly from China, e.g. [14]. Even if the provided PV panels are the same type, multiple size choices depend on the project's needs. The sizing of the PV installation (choosing the size of individual panels and the number of needed panels) for a particular project depends on the total power (and subsequent energy) requirements. As multiple combinations of panels may result in the same

Table 2 Seismic records

Seismic event	Location	Year	Magnitude
Lytle Creek	California—USA	1970	5.33
Fruili, Italy-03	Italy	1976	5.50
Fruili, Italy-02	Italy	1976	5.91
Santa Barbara	California—USA	1978	5.92
Coalinga-01	California—USA	1983	6.36
Coalinga-02	California—USA	1983	5.09
Coalinga-04	California—USA	1983	5.18
Coalinga-05	California—USA	1983	5.77
Coalinga-06	California—USA	1983	4.89
Coalinga-07	California—USA	1983	5.21
N. Palm Springs	California—USA	1986	6.06
Whittier Narrows-01	California—USA	1987	5.99

Table 3 520–540 W photovoltaic panel characteristics [14]

Detail	Value	Units	Illustration
Photovoltaic cell type	Monocrystalline	–	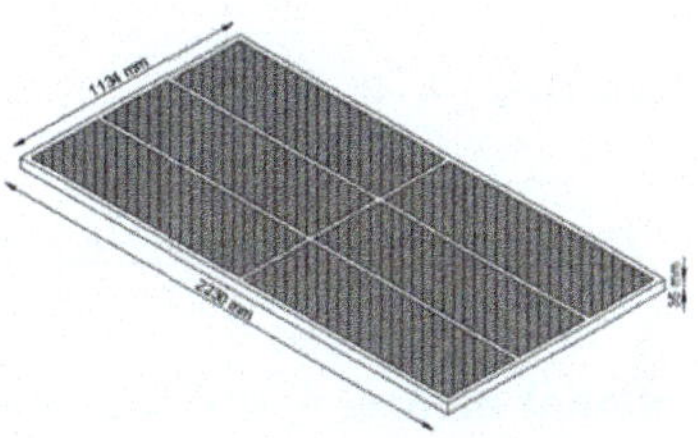
Cells number	144	–	
Length	2230	mm	
Width	1134	mm	
Thickness	35	mm	
Weight	28.90	kg	

energy harvesting capabilities as required, the mechanical loading of the flat roofs should be assessed for all possible combinations of PV panels. At first, however, the panel whose dimensions and weight represent the most significant risk for a flat roof is selected for this study, see Table 3.

2.4 Solar Panels Mounting

Solar panel distribution companies in Ecuador currently do not have technical data sheets for mounting systems, so support systems from foreign companies were taken as a reference. Support structures for solar panels can be installed with anchor bolts directly to the slab or by applying extra weight to the support with concrete blocks called ballast.

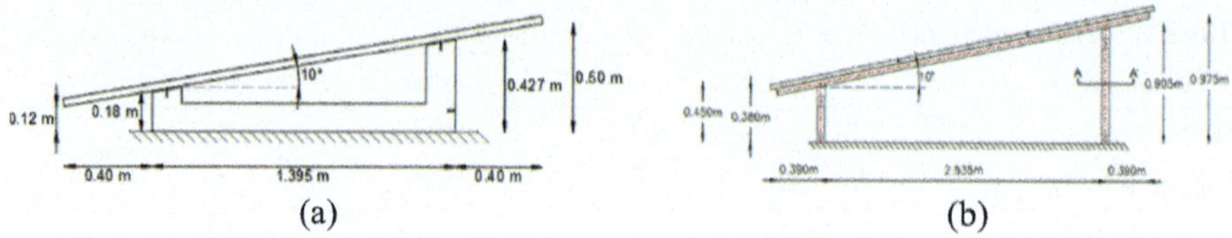

Fig. 3 System geometry: **a** Ballast mounting system, **b** Slab-anchor mounting system

Ballast mounting system. The support is chosen based on the dimensions of the solar panel and the inclination required according to the location. According to [15], the optimal inclination angle to capture sunlight varies from 0 to 30°. For this study, the inclination angle can be 0° since Ecuador, located on the Equator line, receives the solar rays perpendicularly. An inclination of 10° is considered to guarantee self-cleaning so that water, ash, hail, or dust does not accumulate on the surface of the solar panel, see Fig. 3a.

Slab-anchor mounting system. A typical support system and the dimensions of the previously chosen panel are taken as a basis. This mounting system has a 10° inclination to guarantee self-cleaning. See Fig. 3b.

3 Results

3.1 Building Structural Assessment

Vibrational modes. A structure with appropriate modal response presents modal participation of at least 90% of the entire structure's mass in each of the two main horizontal directions, the X and Y axis, whereas, for the third mode, the structure is allowed to rotate around a vertical direction, Z axis. According to Table 4, the building reports an acceptable response due to its regular configuration.

Storey drifts control. The maximum story drifts determined for the X (long direction) and Y (short direction) directions are 0.31 and 0.37% [9], respectively, which do not exceed the 2% maximum story drift for reinforced concrete structures stipulated by [5]. Consequently, the building is rigid enough to prevent lateral displacement-dependent effects due to lateral forces caused by earthquakes.

Beams and columns sections control. The reinforcement ratios and the strong column-weak beam ratios are verified through the software used for the structural

Table 4 Modal participation factors

Mode	X (%)	Y (%)	Z (%)
1	4.29	89.26	5.55
2	95.00	3.17	1.61
3	0.71	7.57	79.24

analysis. The minimum and maximum reinforcement ratios accepted for columns are 1% and 3%, respectively, which were verified. The accepted value for the solid column-weak beam ratio through the specialized software is less than 1.00. In this case, all the columns up to the third story reported ratios under 1.00, except for four columns on the top level that report a warning message, which implies 1.2 times the moment capacity of the beams exceeds the moment capacities of columns at the joint faces. However, this response in connections at the roof level is acceptable since the columns at this level are generally considered the last to yield.

3.2 Maximum Story Responses

A nonlinear time-history analysis is performed to determine the maximum story responses. Since the seismic response of non-structural elements is susceptible to accelerations and displacements, Table 5 shows the top story responses for the 12 compatible events.

3.3 Considerations for Ballast Systems

General structure aspects. According to ASCE/SEI 7-16 [6] chapter "Climate Variability and Its Effect on Flood Risk Through Dynamic Modeling in the Quigushuaycu Stream", solar panels with ballast can only be placed in structures with risk categories I and II (other structures) and III (special occupancy structures). In addition, these

Table 5 Maximum structural responses from the last story

Seismic event	Displacement (m)		Acceleration (m/s^2)	
	X	Y	X	Y
Lytle creek	0.0459	0.0496	20.4447	18.0641
Fruili, Italy-03	0.0422	0.0477	15.0645	18.6645
Fruili, Italy-02	0.0450	0.0508	18.1818	17.7929
Santa Barbara	0.0442	0.0468	16.4354	15.3380
Coalinga-01	0.0521	0.0649	23.0553	22.3637
Coalinga-02	0.0416	0.0556	16.9688	21.6615
Coalinga-04	0.0489	0.0566	19.2868	18.7636
Coalinga-05	0.0458	0.0536	15.6057	20.8546
Coalinga-06	0.0415	0.0467	15.4733	19.1686
Coalinga-07	0.0418	0.0477	14.7092	17.7247
N. Palm Springs	0.0608	0.0469	23.2565	18.3378
Whittier narrows-01	0.0491	0.0553	17.5124	19.1233

structures should be a maximum of 6 stories high, and the roof's slope should be less than or equal to 1:20, which is the case for flat terraces.

Geometric layout. Each panel should be designed to accommodate lateral movement, without impact, instability, or damage, by supporting the relative seismic displacement δ_{mpv} at any edge ledge, curb, or obstruction on the roof surface that prevents its free sliding, where δ_{mpv} is determined according to (1), but must not be less than 0.60 m. In (1), I_e is the importance factor, and S_a is the spectral pseudo acceleration.

$$\delta_{mpv} = 5I_e(S_a - 0.4)^2 \tag{1}$$

Minimum gap between panels [6]. *It must be 0.5·δ_{mpv}, and this gap around the panels must be free of obstructions and delineated with signs or markings. Additionally,* all the edges of the terrace where the panels are placed must be limited by a curb or parapet that is not less than 0.30 m in height and designed to resist concentrated loads of not less than $0.2 \cdot S_a \cdot W_{panel}$, where W_{panel} is the panel weight. Alternatively, the panels can be placed at a minimum distance of $2.0 \cdot \delta_{mpv}$ from any roof edge but not less than 1.22 m (see Fig. 4).

Minimum separation between panels according to the shadow cast between panels. According to [16], it is necessary to leave a minimum distance to avoid shadows. This distance, s, is defined according to (2) and depends on the latitude of the place where the panels will be located and the height of the element that generates shadow (h_{elem}). In the case of the studied building, the latitude is −0.3143 degrees. Figure 5 shows the distances s for the heights of the panel and parapet elements considered in this study (Table 6 and Fig. 5).

$$S = \frac{h_{elem}}{\tan(61 - latitude)} \tag{2}$$

According to [17], a distance of 0.90 m between panels is necessary. Consequently, the adopted separation between panels is 0.90 m. Between the parapet and the panel,

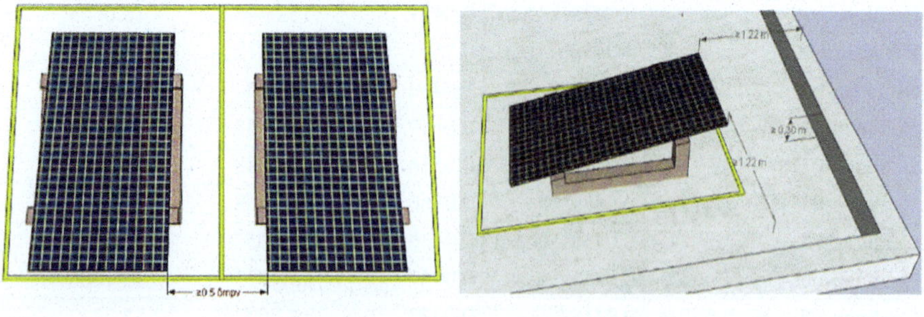

Fig. 4 Distance requirements

Table 6 Minimum distances to avoid shadow: a case study

Element	h_{elem}(m)	Distance (m)
Ledge 1	1.00	0.562
Ledge 2	0.90	0.505
Panel	0.50	0.281

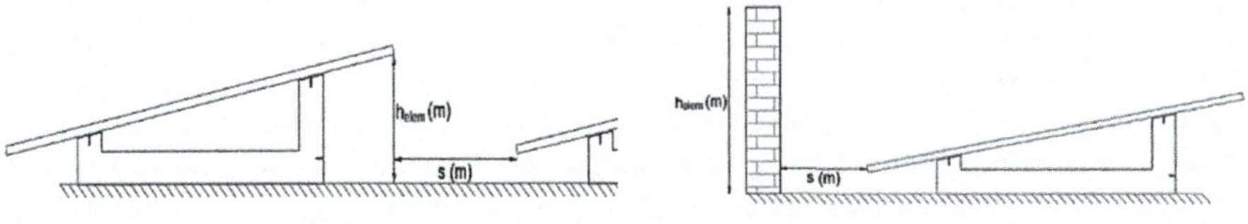

Fig. 5 Minimum distance to avoid shadow between panels and ledges

it will be 1.55 m along the long length of the terrace and 2.00 m along the short distance, as seen in Fig. 6.

Connection force. Each panel shall be interconnected to resist a horizontal force through any section cut in the vertical plane, according to (3), as follows:

$$F_{pb} = 0.20 S_a W_{pi} \tag{3}$$

where F_{pb} is the horizontal seismic force, S_a is the pseudo-spectral acceleration, and W_{pi} is the smaller weight between the two portions of the element. In this study S_a = 1.19 g, W_{pi} = 84.45 kg, and F_{pb} = 20.10 kg.

According to SEAOC [7], component members and connections must resist the F_{pb}. In this case, the design of the bolts was carried out according to the ACI-318 standard [18].

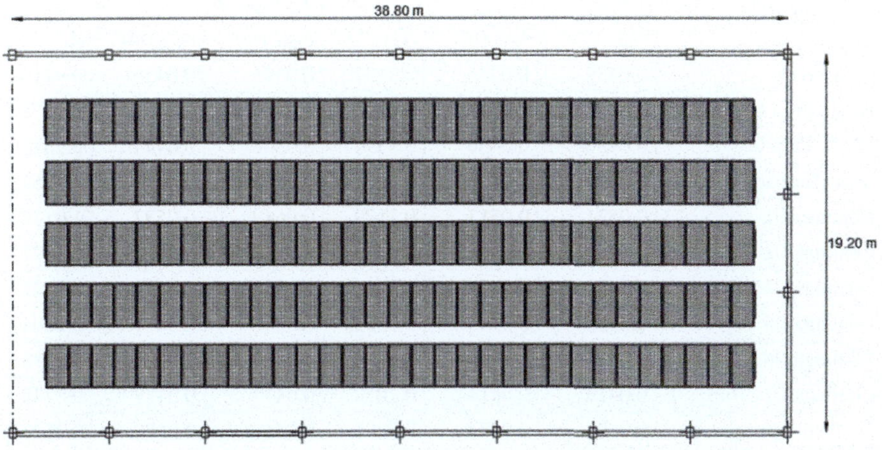

Fig. 6 Distribution of the panels on the roof—assembly for a ballast support system

Frictional force. The panel supports are designed for a seismic force path from the center of mass of each component to the point of frictional resistance equal to the lowest value obtained from:

$$Fp = \frac{a_i a_p W_p}{R_p / I_p} A_x \tag{4}$$

$$F_h = 0.6 W_p \tag{5}$$

where α_i is the maximum ground acceleration, α_p is the component (solar panel) amplification factor that varies from 1.00 to 2.50, W_p is the operating weight of the photovoltaic system component, R_p is the response modification factor of the solar panel, I_p is the solar panel's importance factor, and A_x is the torsional amplification factor.

According to [6], A_x is obtained by (6), where $\delta_{\max}$ is the maximum displacement in the last story, and δ_{avg} is the average between the displacements of the external points of the same story. The value of this factor should not be less than one and should not exceed 3. The quantification of the A_x values for the earthquakes considered in the X and Y directions did not exceed, for the most part, the value of 1, as shown in Table 7.

$$A_x = \left(\frac{\delta_{\max}}{1.2\delta_{avg}}\right)^2 \tag{6}$$

The seismic force differs for the external and internal ballast systems since they support different weights, $W_{\text{pext}} = 84.45$ kg and $W_{\text{pint}} = 98.90$ kg, respectively.

Table 7 Torsional amplification factor—X and Y directions

Seismic event	X-direction			Y-direction		
	$\delta_{\max}$ (m)	δ_{avg} (m)	A_x	$\delta_{\max}$ (m)	δ_{avg} (m)	A_x
Lytle creek	0.0459	0.0446	0.7331	0.0496	0.0426	0.9419
Fruili, Italy-03	0.0422	0.0412	0.7272	0.0477	0.0434	0.8378
Fruili, Italy-02	0.0450	0.0436	0.7394	0.0508	0.0444	0.9085
Santa Barbara	0.0442	0.0438	0.7097	0.0468	0.0429	0.8252
Coalinga-01	0.0521	0.0513	0.7179	0.0649	0.0537	1.0138
Coalinga-02	0.0416	0.0406	0.7290	0.0556	0.0478	0.9397
Coalinga-04	0.0489	0.0474	0.7418	0.0566	0.0479	0.9693
Coalinga-05	0.0458	0.0446	0.7335	0.0536	0.0471	0.9004
Coalinga-06	0.0415	0.0412	0.7059	0.0467	0.0423	0.8488
Coalinga-07	0.0418	0.0414	0.7072	0.0477	0.0438	0.8245
N. Palm Springs	0.0608	0.0594	0.7285	0.0469	0.0450	0.7523
Whittier Narrows-01	0.0491	0.0476	0.7389	0.0553	0.0476	0.9348

Table 8 Ballast system force *Fp*

Spectrum	Inner ballast		Outer ballast	
	Fp_x (kg)	Fp_y (kg)	Fp_x (kg)	Fp_y (kg)
Lytle creek	137.409	121.409	117.333	103.670
Fruili, Italy-03	101.249	125.445	86.456	107.116
Fruili, Italy-02	122.200	119.587	104.346	102.114
Santa Barbara	110.463	103.087	94.323	88.026
Coalinga-01	154.955	152.379	132.315	130.116
Coalinga-02	114.048	145.587	97.385	124.316
Coalinga-04	129.627	126.111	110.688	107.685
Coalinga-05	104.886	140.165	89.562	119.686
Coalinga-06	103.997	128.833	88.802	110.010
Coalinga-07	98.861	119.129	84.417	101.723
N. Palm Springs	156.308	123.249	133.470	105.242
Whittier Narrows-01	117.701	128.529	100.504	109.750
Average	120.975	127.620	103.300	108.974

Considering $I_p = 1$, $a_p = 1$, and $R_p = 1.5$, the seismic force calculated with (4) for the ballasted solar panel is as follows (see Table 8).

According to (5), the F_h seismic force for the internal and external ballasts is $F_{\text{hext}} = 50.67$ kg and $F_{\text{hint}} = 59.34$ kg, respectively. This force must be resisted by friction, which is generated by the contact between the ballast block and the surface of the building terrace. With a friction coefficient of 0.3, which would be the worst case in wet conditions, the friction forces for the inner and outer ballasts are 29.67 and 25.34 kg, respectively. These values indicate that the photovoltaic system would move, so added weight is necessary, increasing the friction force. For this purpose, concrete blocks weighing 98.00 and 107.00 kg are dimensioned for the internal and external ballasts, reporting friction forces of 61.65 and 54.65 kg, thus verifying the system's stability.

3.4 Considerations for Slab Anchored Systems

Geometric distribution. The ASCE/SEI 7-16 [6] does not specify the minimum spaces that must exist between the anchored systems, so for its geometric dimensioning, only the shadow projection computed by (2) will be taken into account (see Table 9) and a distance for circulation and panels maintenance.

According to the dimensions of the roof terrace in the case study, a 0.60 m separation between panels is adopted for the circulation of personnel and maintenance tools. Furthermore, due to the configuration of the selected support structure, distances of

Table 9 Minimum distances to avoid shadow—system anchored to the slab

Element	h_{elem} (m)	Distance s (m)
Ledge 1	1.00	0.562
Ledge 2	0.90	0.505
Panel	0.975	0.547

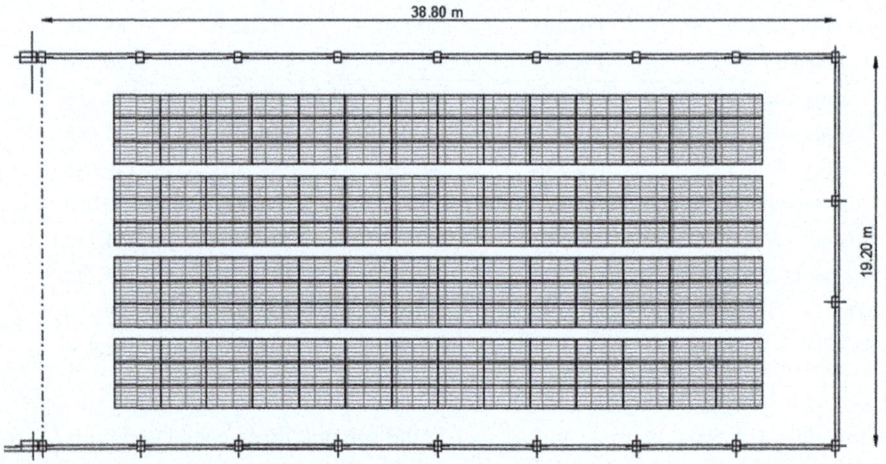

Fig. 7 Distribution of panels anchored to the slab

1.82 and 3.55 m are considered in the short and long directions between the ledge and the panel, respectively (see Fig. 7).

Horizontal seismic force. To determine the design horizontal seismic forces that the element would resist, (4) must be applied with the same I_p, a_p, R_p, and A_x used in the ballasted mounting system and with a total weight of the photovoltaic system $W_p = 1625.06$ kg. The results are presented in Table 10.

These values must be less than those obtained by (7) and greater than (8). Consequently, the condition is satisfied.

$$F_{ps} = 1.6 S_a I_p W_p = 3095.15\,\text{kg} \tag{7}$$

$$F_{pi} = 0.3 S_a I_p W_p = 580.34\,\text{kg} \tag{8}$$

Design of anchor bolts. The anchor bolts of the panel support must resist the actions generated by the horizontal seismic force. Since the support is subjected to a moment, the distance between the tension and compression forces j_d should be determined (see Fig. 8). For design purposes, this calculation uses the elastic theory of beams, considering that this method reports a conservative value.

As a result, an expansion bolt, whose dimensions and properties are provided in Table 11, would be considered.

Table 10 Horizontal seismic forces—X and Y directions

Seismic	a_{iX} (m/s^2)	F_{px} (kg)	a_{iY} (m/s^2)	F_{py} (kg)
Lytle creek	20.445	2257.82	18.064	1994.92
Fruili, Italy-03	15.065	1663.66	18.664	2061.22
Fruili, Italy-02	18.182	2007.92	17.793	1964.97
Santa Barbara	16.435	1815.05	15.338	1693.86
Coalinga-01	23.055	2546.12	22.364	2503.80
Coalinga-02	16.969	1873.97	21.661	2392.20
Coalinga-04	19.287	2129.95	18.764	2072.17
Coalinga-05	15.606	1723.42	20.855	2303.10
Coalinga-06	15.473	1708.80	19.169	2116.90
Coalinga-07	14.709	1624.42	17.725	1957.44
N. Palm Springs	23.256	2568.35	18.338	2025.15
Whittier Narrows-01	17.512	1933.99	19.123	2111.90
Average	18.000	1987.79	18.988	2096.97

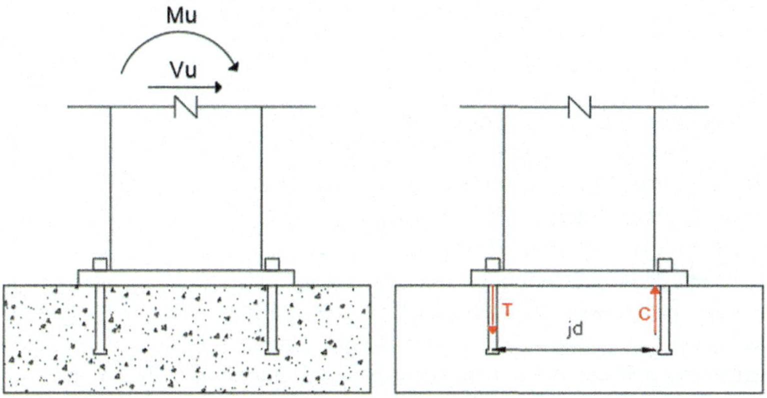

Fig. 8 Bolt behavior under applied forces

Table 11 Anchor bolt characteristics [19]

Detail	Variable	Value	Units
Nominal diameter of the anchor bolt	d_a	1.27	cm
Effective cross-sectional area	A_{se}	0.677	cm^2
Anchor Yield Strength	f_{ya}	6470	kg/cm^2
Anchor Tensile Strength	f_{uta}	8090	kg/cm^2
Effective embedment depth	h_{ef}	8.573	cm

For the design of anchors for non-structural elements, ASCE/SEI 7–16 allows the use of pre-installed and post-installed bolts whose design must comply with the requirements by [17], Chap. 17.

4 Discussion and Conclusions

The seismic response of solar PV panels placed on roofs in Ecuador has been investigated by exemplifying a particular building with a flat terrace. In Ecuador, mounting systems on roofs and/or building envelopes or other infrastructures are not yet technically implemented. That is, a support structure of any accessible material is placed without carrying out studies to verify that it resists the combined forces of the panel's weight and earthquakes. As a result, most customers request only the PV panel and install it with their means, so this study took mounting systems for flat roofs from foreign companies that best fit the characteristics of the chosen panel as a reference.

Although the analyzed building was built with an old regulation that did not contemplate an appropriate seismic resistant design, maximum drifts below 2%, the limit stipulated in the Ecuadorian Construction Standard NEC-15, were registered. Likewise, it was verified that the sections of beams and columns are adequate to support different demands. Therefore, it is concluded that the building is suitable for installing solar panels on its terrace.

American procedures by the ASCE/SEI 7-16 and SEAOC are considered in this study for the correct interaction between a photovoltaic system and a flat roof of a reinforced concrete building. The Colombian NCR-10, Chilean NTM-001, and European CEN standards associated with the seismic design of non-structural elements are also based on the abovementioned procedures.

The photovoltaic system anchored to the slab allows more solar panels in the same space on the terrace surface, with 168 PV panels in 4 rows of 42 panels, each representing 62.15% of the terrace surface. In comparison, with the ballast system, there are 155 panels located in 5 rows of 31 panels each, representing 51.77% of the terrace surface. Consequently, in terms of optimization of the available space, the best option is the system anchored to the slab.

According to the seismic response of the building in terms of accelerations and displacements, it is determined that each of the proposed mounting systems for the considered solar panels reported an appropriate performance. However, since the ballast systems are not anchored to the slab, they need extra weight to prevent excessive displacements or possible overturning effects. In this context, the added weight of the ballasted photovoltaic system and the added concrete blocks would not represent any danger to the slab since it is 57.67% lower than the live load considered for the building since its design.

Finally, regarding possible side effects such as moisture infiltration within the considered roof, the most appropriate option for the supporting system of solar panels would be the ballast system since the bolted supports require a minimum concrete thickness that lightened slabs do not provide.

Accounting for further developments in PV technologies that may also be applied in Ecuador, like, e.g., building integrated PV (BIPV), more research into the seismic performance of PV systems (and of other building added or building integrated energy technologies) and non-structural elements is needed to ensure that their practical deployment will be a success. Consequently, national standards based on international experiences, but mainly on Ecuadorian experimental and analytical program results, are crucial.

Acknowledgements This work is part of the projects 2023-PIS-004 and 2019-EXT-014 from the Structures and Constructions Research Group (GIEC) and Propagation, Electronic Control, and Networking Research Group (PROCONET) at Universidad de las Fuerzas Armadas ESPE. In addition, this work has been developed with the support of the Global Minds Open Faculty Project from KU Leuven.

References

1. Agencia de Regulación y Control de Energía y Recursos Naturales No Renovables. Estadística Anual y Multianual del Sector Eléctrico Ecuatoriano. 1–178 (2021)
2. Ong, N.A.F.M.N., et al.: Development of fire safety best practices for rooftops grid-connected photovoltaic (PV) systems installation using systematic review methodology. Sustain. Cities Soc. **78**, 103637 (2022)
3. Rodriguez, M., Arcos-Aviles, D., Martinez, W.: Fuzzy logic-based energy management for isolated microgrid using meta-heuristic optimization algorithms. Appl. Energy **335**, 120771 (2023)
4. Hamzah, A.H., Go, Y.I.: Design and assessment of building integrated PV (BIPV) system towards net zero energy building for tropical climate. e-Prime Adv. Electr. Eng. Electron. Energy **3**, 100105 (2023)
5. Norma Ecuatoriana de la Construcción. Peligro Sísmico - Diseño Sismo Resistente. Ecuador (2015)
6. ASCE/SEI. ASCE/SEI 7-16, Minimum Design Loads and Associated Criteria for Buildings and Other Structures. Estados Unidos: American Society of Civil Engineers (2017)
7. SEAOC. Structural Seismic Requirements and Commentary for Rooftop Solar Photovoltaic Arrays. USA (2012)
8. Bravo, K. A. (Diciembre de 2014). Historia de la Universidad de las Fuerzas Armadas ESPE. Ecuador: Universidad de las Fuerzas Armadas ESPE, 1–119 (2014).
9. Cano M., Dávila C. (2021). Consideraciones sísmicas para la instalación de sistemas fotovoltaicos en terrazas planas de hormigón armado: Estudio de caso. Sangolquí, Ecuador: Universidad de las Fuerzas Armadas ESPE (2021).
10. Arcentales Naranjo, N.I., Yépez Proaño, V.H.: Evaluación de la respuesta dinámica del Hospital Naval General Hosnag sometido a vibración ambiental. Universidad de las Fuerzas Armadas ESPE, Sangolquí, Ecuador (2020)
11. University of California. PEER Ground Motion Database © 2013. Berkeley, California, Estados Unidos de Norte América. https://ngawest2.berkeley.edu/. Last accessed 25 Nov 2022
12. Ballesteros Salazar, K.S., Caizaguano Montero, D.G.: Guía para la evaluación del grado de vulnerabilidad sísmica de unidades educativas localizadas en la parroquia de Sangolquí, basada en la guía FEMA P-1000. Estudio de caso. Sangolquí, Ecuador: Universidad de las Fuerzas Armadas ESPE (2020)
13. Quinde Martinez, P., Reinoso Angulo, E.: (4 de Abril de 2016). Estudio de peligro sísmico de Ecuador y propuesta de espectros de diseño para la ciudad de Cuenca. Revista de Ingeniería Sísmica **94**, 1–26 (2016)

14. Jinko Solar.: Tiger Pro 72HC 530–550 Watt Mono-facial module (2020). https://www.jinkosolar.com/uploads/5ff587a0/JKM530-550M-72HL4-(V)-F1-EN.pdf, Last accessed 27 Nov 2022
15. Alvarez Lozano, D.A.: Evaluación de la orientación y ángulo de inclinación óptimo de una superficie plana para maximizar la captación de irradiación solar en Cuenca - Ecuador. Universidad Politecnica Salesiana, Sede Cuenca, Cuenca, Ecuador (2017)
16. TecnoSol. Consejos para instalaciones fotovoltáicas: Distancia entre filas de paneles solares para evitar el sombreado. España. https://tecnosolab.com/noticias/distancia-entre-filas-de-paneles-solares/. Last accessed 15 March 2023
17. Sun Ballast. Estructuras para cubiertas planas. España. https://www.sunballast.it/es/estructuras-para-cubiertas-planas/lastre-10l/. Last accessed 05 Dec 2022
18. ACI 318. Requisitos de Reglamento para Concreto Estructural (ACI 318-19). (2019)
19. Simpson Strong-Tie Company, Inc. Mechanical Anchoring Products | Simpson Strong-Tie. Canadá. https://www.strongtie.com/mechanicalanchoringproducts_anchoringsystems/category. Last accessed 10 May 2023

A Brief Guide to the Implementation of Lean Tools in the Ecuadorian Context

Paúl Ayala Taco, Joan Pasaca, and Oswaldo Alexander Ibarra

Abstract This paper aims to develop a brief guide for implementing "Lean Tools" in the context of the Ecuadorian market. For this objective, a review of the state of the art is accomplished to identify the process, most popular tools, complementary approximations to Lean Thinking, and principal issues found during the implementation process of Lean Tools in Ecuador. It is found that Ecuadorian enterprises have implemented these tools during the past seven years. Establishing guidelines derived from successful cases of Lean Tools implementations is possible.

Keywords Lean manufacturing · 5S · Continuous improvement · Lean thinking

1 Introduction

Since the end of the nineteenth century, the principles of what is now known as Lean Manufacturing have been successfully applied in industries of all types, allowing significant competitive advantages and establishing mechanisms to extend this process over the years. This evolutionary nature of the "Lean" philosophy has allowed the emergence of new approaches, methodologies, and tools that are better adapted to the particular needs of organizations and extend the field in which they can be successfully applied. Ecuadorian industry has not been oblivious to the fundamentals proposed by the Lean philosophy. In the country, several institutions have opted for its implementation to reduce costs and, in general, to become more competitive in an increasingly demanding market. This paper aims to provide an overview of how the Ecuadorian industry, in various sectors, has successfully implemented Lean tools, emphasizing the methodology used, the tools with greater acceptance, and the difficulties that this process presented.

P. A. Taco (✉) · J. Pasaca · O. A. Ibarra
Departamento de Eléctrica, Electrónica y Telecomunicaciones, Universidad de las Fuerzas Armadas ESPE, Sangolquí 171103, Ecuador
e-mail: jpayala@espe.edu.ec

G. F. Olmedo Cifuentes et al. (eds.), *Emerging Research in Intelligent Systems*, Lecture Notes in Networks and Systems 902, https://doi.org/10.1007/978-3-031-52255-0_3

2 Methodology

The information on which this study is based is purely documentary. Several secondary sources have been analyzed, detailing the process and results of implementing lean manufacturing tools in Ecuadorian institutions in the manufacturing and services sector. The mentioned works were developed or published in the period 2014–2020.

3 Background

Lean manufacturing is currently presented as an integrating philosophy encompassing techniques that increase an organization's competitiveness and profitability by reducing costs related to its activities. As pointed out in [1], the ultimate goal of lean manufacturing goes beyond the removal of waste (often referred to as "Muda" or waste). This set of techniques gives more value to the product by distinguishing and reducing the activities (within the production chain) that do not add value. Currently, the focus is oriented not to lean manufacturing but to Lean Thinking, which seeks the development of an organizational culture, making each institution member introduce all its actions towards minimizing or eliminating waste [2].

Due to the very nature of the Lean philosophy, which talks about infinite improvement and the constant search for opportunities to optimize processes, today, research and development of strategies continue. Initiatives such as analyzing the impact of lean manufacturing on the sustainability of companies [3] and even the creation of new ways to learn about Lean tools, as in [4] where the process of creating a video game that allows learning the "5S" (sort, straighten, shine, standardize and sustain) tool is detailed.

Since its inception, the Lean philosophy has been presented as an opportunity for improvement and growth at the enterprise level. Several benefits can be achieved thanks to it, highlighting improvement in delivery and production times, reduction of downtime, general decrease in production costs, increase in staff and customer satisfaction rates, improvement of the organizational climate, etc. Interestingly, Lean thinking (or philosophy) is capable of having an impact on aspects that, a priori, are entirely unrelated to the premise of waste reduction, as in the case of [5], where the effect that the implementation of Lean techniques is capable of having on occupational health and safety is explored.

These tools have been widely accepted in Ecuador, especially in recent years. Although their application is not always formal, i.e., it does not follow a proper procedure [6], the fact is that their implementation can be seen in various productive sectors of the country and companies of all "sizes," whether small and midsize enterprises (SMEs) or large companies as detailed in [7].

4 Procedure for the Implementation of Lean Tools

There is a great variety of approaches when discussing a procedure or roadmap for correctly applying lean manufacturing tools. The study in [1] presents a very detailed vision of how this process is carried out, and specific tools and indicators that can be useful in each of the proposed stages are described. As a result of the on-site work carried out in [7], a structure of the process of applying Lean tools is also proposed, and in [6], a roadmap for turning SMEs into Lean companies is discussed. Next, a simplified model of the lean manufacturing tools implementation process is proposed, composed of four stages: Diagnosis, Planning, Execution, and Control, Fig. 1.

4.1 Diagnostic

A point of convergence of all the material covered in this work is the importance of analyzing the initial situation of the company or process where the lean manufacturing tools are to be implemented. The relevance of this stage lies in the fact that it is necessary to know which shortcomings are to be attacked and what repercussions these shortcomings have on the organization. All the possible "mudas" that exist in the process are identified at this point. The aim is to get to the root causes of the waste, so it is expected to reveal that specific problems are, in fact, a consequence of other more profound issues that have not been considered.

The diagnosis analyzes aspects such as process flow, work area design (layout), production times, number of units produced, etc. It should include all those aspects that allow for establishing a baseline on which to work. The outputs of this stage are sources of information that would enable characterizing the "before" of the process and the objectives to be achieved so that its analysis allows correct planning.

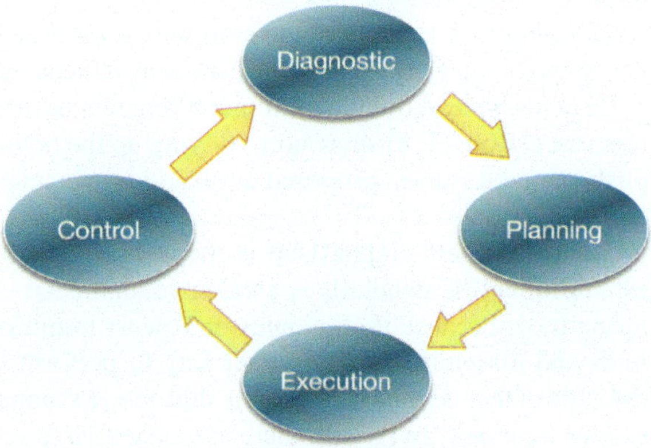

Fig. 1 Procedure for lean tools implementation

4.2 Planning

Once the goal (objectives) has been determined and the starting point (diagnosis) has been identified, the career path must be established, i.e., finding the Lean tools that fit the identified needs and knowing the optimal way to implement them. The necessary tools or tools must be chosen at the planning stage and not before. Many authors start from the tool and adapt it to their needs when the correct thing to do is the other way around. Observing the impact of a given strategy on a process may be of interest to academia. However, for practical purposes, the choice of a Lean tool is subordinated to previously specified objectives. In [5], the importance of adapting the methodology to the company's reality in different instances, such as at the management level and even at the workspace level, is alluded to. Planning not only covers how the objectives will be achieved but should also include establishing mechanisms to evaluate the fulfillment of these objectives. Thus, from the planning process, the tools to be applied, the application process, and the indicators or mechanisms for evaluating results should be obtained as outputs.

4.3 Execution

The execution phase consists of implementing all the strategies defined in previous stages. This is where the Lean methods or tools described as suitable for the company's needs are implemented, and their implications will be highly customized. However, specific recommendations can be considered. Although many lean manufacturing tools apply to the organization's operational and administrative parts, the implementation should be done first in the production processes. This is mainly due to two factors. Rather, modifying managerial procedures can be much more complex, and identifying waste at this level is not so obvious. On the other hand, waste at the operational level usually represents much higher costs and is obvious, so it is imperative to address it first.

In addition, it is advisable to start with all those changes that involve changes in the layout of the work area because these modifications (considering that they are due to Lean tools) lead to imminent improvements in time and efficiency of operational processes. Both [5, 8] show how "tidying up the house" is the first task to be accomplished, and it is even proposed as an additional stage before the implementation of other Lean tools.

Another aspect emphasized is the importance of process standardization. The establishment of standards is another fundamental characteristic of lean manufacturing tools because it eliminates or at least minimizes variability in the tasks and finds and disseminates an optimal way to perform them. Commonly, this is done through manuals, and it is sought that the personnel internalizes the correct way to execute a task. In [9, 10], there are clear examples of how the implementation of process manuals allows this premise to be carried out. They also show that this type of strategy can be implemented not only in manufacturing companies but also in service companies. The outputs of this phase are all the improvements and modifications derived from implementing the Lean shift tool and all the indicators' measurements that will demonstrate the impact of the same.

4.4 *Control*

Most Lean tools are oriented to continuous improvement, so constantly evaluating the organization's points of interest is necessary. The control stage becomes fundamental because it allows comparing the baseline, established in the diagnosis, with thc company's situation after implementing the tools and measuring the degree of impact and benefits obtained. In this phase, the indicators established in the diagnostic and planning stages come into play. Analyzing these indicators makes it possible to demonstrate what advantages are obtained and (perhaps more importantly) to find all those points that can still be improved. At the end of the control stage, qualitative and quantitative data sets are delivered as outputs according to the company's needs. These data are vital since their correct analysis guides the decisions made by top management. It should be emphasized that for implementing Lean tools to have the most significant possible impact, its execution should be extended over time, becoming a cycle instead of ending after the control stage. In the case of [5], it is highlighted that one of the factors why the benefits of these tools cannot be maintained is that they are applied with little formality, and their control stage is omitted or carried out empirically.

5 Lean Tool Selection and Prioritization

As mentioned, the choice of the tool or tools to be implemented must be subordinated to fulfilling particular objectives. These methodologies intrinsically bring advantages or benefits due to their implementation. However, a large part of their success is knowing which tool to use. As stated by several authors, knowing when to apply them is more important than knowing how to do it. For this reason, it is beneficial to know what type of problems can be efficiently addressed with these tools and which are the best depending on the issue.

5.1 *Diagnostic Stage Tools*

Since the purpose of the diagnostic stage is to characterize the current state of the company or process and identify the problems or waste that arise in it, tools such as the Value Stream Map (VSM) [8, 11, 12], Cause and Effect Diagram [2, 11], Checklists

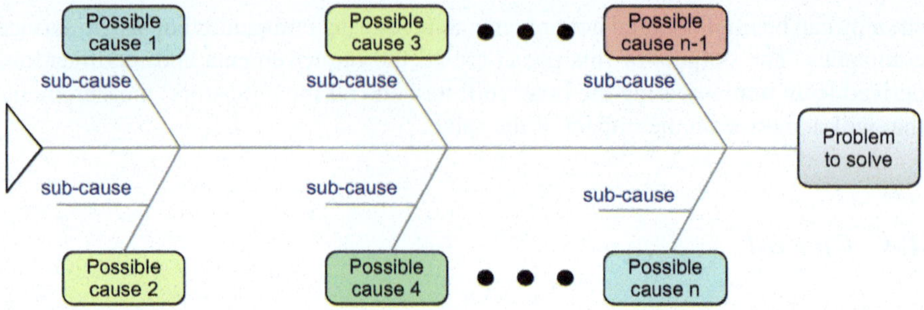

Fig. 2 Procedure for the implementation of Lean tools

[2] and even the Theory of Constraints (TOC) [13] have been used for these purposes. The use of VSM is particularly efficient when it is intended to characterize processes since it provides a panoramic view of how materials flow and is, therefore, extremely useful when optimizing production times, minimizing downtime, or optimizing a value chain. Its use is recommended for purely productive processes.

Cause-effect diagrams, Fig. 2, and checklists are handy for identifying problems that can be characterized qualitatively. They can be applied to both services and production processes. The cause-effect (Ishikawa) diagram has the advantage of identifying the most fundamental causes of problems, allowing the treatment of "the disease and not the symptoms." The theory of constraints is a philosophy in itself (later, we will explore how it can work alongside Lean tools), allowing us to identify "constraints" or "bottlenecks" in a process. It is highly effective in finding physical constraints (machinery, inventory, operations) and can even identify operational or administrative rules. For this reason, this philosophy is ideal for identifying where waste is occurring in the process.

5.2 *Execution Stage Tools*

The information on the different Lean tools is extensive and can be deepened depending on the productive sector where it is applied. For this reason, knowing which tools are most effective in attacking specific problems is extremely useful. According to the analyses proposed in [12, 14, 15], issues related to product quality, high rate of customer dissatisfaction or returns, and the absence of improvement mechanisms can be addressed from the Lean approaches (in general), Kaizen, World Class Manufacturing (WCM) or SHEN model. Emphasizing that, frequently, these approaches are encompassed within Lean thinking. The most prominent tool in this case is "5S", Fig. 3, allowing the introduction of improvements and perpetuating this process to maintain a constant evolution.

SMED (quick tool change) and TPM (Total Productive Maintenance) effectively address waste, especially seen as downtime. Additionally, TPM allows for solving

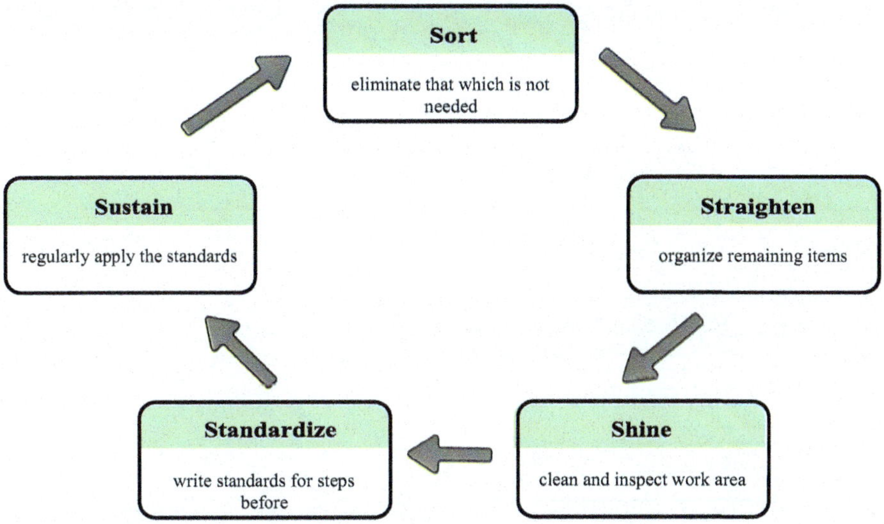

Fig. 3. 5 S sequence

problems related to frequent machinery breakdowns and prioritizing the idea of "zero breakdowns, zero defects, and zero accidents" [2]. TPM is a tool that companies can use to determine the overall equipment effectiveness (OEE) not only in their manufacturing process but also in each machine involved in the productive chain. The TPM´s objective is to enhance the relationship between different manufacturing processes by improving each machine´s process.

OEE contains three components: availability, performance, and quality (see Fig. 3). Each gives a numerical value to identify a possible problem in the production chain. In the world, the best factories have an OEE above 90%, but in Ecuador, most companies do not reach 80%. This percentage shows that Ecuadorian factories have severe problems with availability, performance, quality, and competitiveness.

When issues related to human failures are constantly occurring, it is helpful to consider the implementation of methodologies such as "Poka-Yoke" (avoidance or notification of errors) and "Andon" (use of visual or audible indicators). Finally, when dealing with process-related problems, especially in service companies, the implementation of "5S" continues to lead, followed by "Kanban" (use of cards to identify the status of processes) and "Kaizen" (continuous improvement).

In addition, [15] points out that, with a view to the technification of SMEs in Ecuador, the most appropriate strategies are "5S" and "The SHEN model" since the former represents a strategy scalable to any productive activity that establishes a process of improvement and optimization while minimizing waste. The second option is a model that demonstrates the need to produce according to demand and not to have "stock" as the primary objective. As can be seen, several alternatives are available, so it is necessary to choose one that best suits the company's needs.

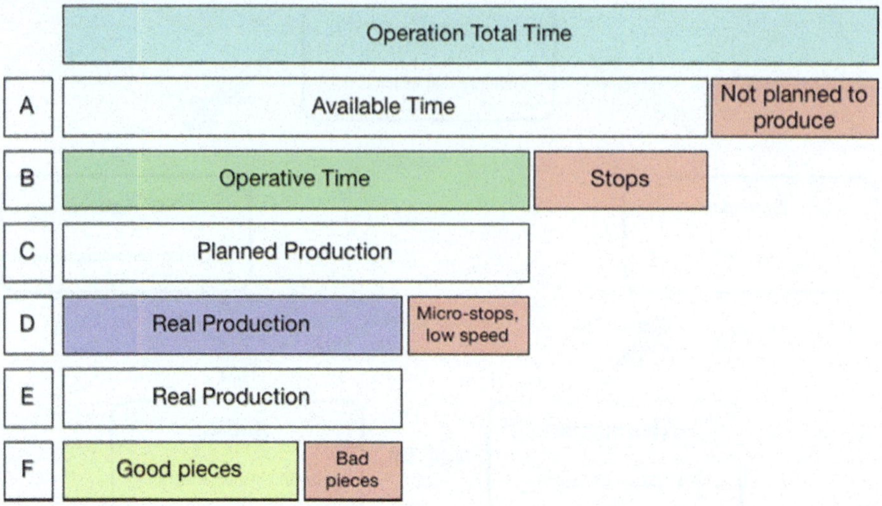

Fig. 4 Overall Equipment Effectiveness

However, even if it is established which drawbacks are to be addressed, several options may fit them (Fig. 4).

According to [12, 14], it is necessary to establish which problems are a priority to choose a set of tools to use. Once this process has been carried out, factors that allow a choice among the tools should be determined if implementing all of them proves impossible. These factors include cost of implementation, ease of implementation, knowledge about the strategy, expected impact, time in which results are expected to be seen, number of issues to be addressed, etc.

A prioritization strategy used, which proves to be intuitive and visual, is the "Hierarchical Analysis Process" (HAP) methodology, which allows for establishing a series of criteria (which may be those mentioned above or others better adjusted to the company's reality), prioritizing these criteria in such a way as to determine which of them is most relevant and finally evaluating the alternatives based on these considerations and priorities.

Finally, it should be noted that "5S" is positioned as the most widely used tool, not only because it represents a low cost and allows identifying problems in a relatively simple way, but also because it has an intuitive process, which has even been applied empirically and, when executed correctly, can be maintained over time, allowing the characteristic of continuous improvement desired in Lean thinking.

6 Complementary Approaches to Lean Philosophy

Lean manufacturing is a collection of tools oriented to competitiveness. However, in the environment, many other equally exciting approaches, as can be seen in [6, 13, 7], can bring considerable benefits when taken hand in hand with Lean strategies. Previously, TOC was mentioned as an excellent mechanism for waste identification. Thus, this approach is complementary to Lean thinking regarding its diagnostic stage. TOC is especially susceptible to being approached together with the characterization of a process or company through a VSM, allowing us to find critical points regarding inventory and process capability [13].

In [7], it is shown how Distribution Requirements Planning (DRP) is a strategy that can give notable results when applied together with Lean tools. Its use is often related to predicting product demand, allowing the necessary stock to be distributed smoothly, and reducing costs at the right time without incurring downtime or inventory overload. This strategy works with Lean tools as a prelude. The demand forecast by the DRP allows for establishing the manufacturing requirements of the company in such a way that it can be determined thanks to it the quantity and timing of the products needed, ideally complement Lean tools such as Just in Time (JIT).

A philosophy currently adapted to Lean Manufacturing is Six Sigma to reduce variability in production processes. Six Sigma aims to reduce or eliminate defects or failures during different product or service production stages. An important input to apply Six Sigma is to have the information related to the variable to be improved. For example, if a chocolate manufacturing process is analyzed, and statistical quantities of raw material waste are available at a particular stage of production, different Lean strategies adapted to Six Sigma can be applied through the application of the DMAIC cycle (define, measure, analyze, improve and control), which allows the product to be within specification limits defined by the company's policies or a national or international standard.

Six Sigma aims to reach a maximum of 6.4 defects per million events or opportunities (DPMO). By applying Six Sigma, production failures are reduced or eliminated, and quality control and production cost savings occur, making the factory more competitive in its environment.

Finally, and remembering that one of the "mudas" referred to in the Lean approach is overproduction [1], it is worth highlighting another way of conceiving the production chain of a company.

The "pull-system" approach proposes that the starting point of production is the demand, pretending that only the necessary material is required or "pulled" to produce a quantity of product determined by the customer's requirements, implying that this good or service will be delivered in the correct amount, at the right time and under precise specifications. This conception, however, requires that the process has been optimized to the maximum since it intends that the production starts if, and only if, a request has been received from the customer, so it requires a system with an extremely short production time, considering that the demands must be satisfied in the shortest possible time. Again, this approach fits very well with Lean techniques such as JIT, and it can be stated that a production line where a "pull" approach is intended will also need to implement JIT.

7 Lean Implementation Problems

When lean manufacturing tools are intended to be implemented, several drawbacks inevitably hinder this process. Two of them occur with alarming frequency in the Ecuadorian context: lack of training and the informality of the application. Other adverse factors that have been documented are the personnel's resistance, the management's lack of commitment, the relegation of the leadership in introducing these tools to the production managers, little control of the operators, and lack of standardization in the processes. All these possible problems ensure the company fails to implement the Lean philosophy.

Many of the Lean tools are easily understood (the difficulty in their implementation depends on the context), but this does not mean that their efficiency will not be diminished if they are implemented without a deep knowledge of them. Something familiar in the Ecuadorian industry, especially at the level of SMEs, is that the starting point of these companies is family enterprises, so often, senior management people do not have training according to their role.

As discussed in [6], the lack of knowledge of this type of production strategy is due to the lack of training of company leaders. Many try to implement it empirically (which is another problem in itself), and because of this, they do not obtain results or cannot maintain them. Hence, in many cases, as in [9], constant training and evaluation is one of the main proposals for the future.

When speaking of informality in implementing these tools, reference is made not only to the cases of empirical performance exposed but also to the lack of commitment, especially with the control stage. In addition to implementing these tools, having a solid theoretical basis, but without a follow-up supported by standardization of processes or mechanisms that allow constant evaluation and even without documenting any of the stages or procedures adopted. At first, even organizations that adopt this practice "half-heartedly" may see favorable results. However, as also discussed in [6], these results can't last if the tool in question is not given due importance.

Finally, the system proposed by ISO 9001, which suggests standardization for quality management processes, has been widely adopted in Ecuador. However, as mentioned in [11], this initiative, which is an excellent roadmap for maintaining process quality and generating added value that translates into competitiveness, often remains only on paper, falling back on the problem of the lack of formality and commitment to production strategies.

8 Discussion

As can be observed, the technification process of the Ecuadorian industry does not align with the implementation of lean manufacturing tools. Although the documentation is not extensive, there is evidence of a trend, especially in the last seven years, to bet on tools, methodologies, or guidelines that, ultimately, allow a company to be more competent in a highly demanding market, such as the current one. However, implementing this type of solution remains informal, in many cases, even at medium and large companies. The origin of these organizations is a possible explanation for this phenomenon. Still, with a view to an increasingly globalized market, the formalization of processes must be more significant. Although this study is limited to the review of documentary material, it is possible to find guidelines that can serve as a future reference or guide (very concrete) of the process and tools to implement Lean thinking. All this is validated with successful cases in the Ecuadorian market.

The scaling of this analysis to a Latin American context, the role of automation in implementing Lean tools, and the proposal of solution strategies to the problems evidenced in this work are possible future works that may derive from this paper.

9 Conclusions

In this work, it is possible to identify a process, divided into four general stages, for implementing Lean techniques, as well as different guidelines to choose and prioritize among the other lean manufacturing tools, considering the problem to be faced. It also identified the most common issues that various Ecuadorian organizations have encountered when implementing these techniques, serving as a guideline for implementing these strategies in Ecuador.

Generally, we can say that the Lean philosophy and its benefits are not known among the different factories of the industrial sector. The proper implementation of the Lean techniques does not ensure the success of the company's processes because the most critical point for real success is the full support from all company employees.

Many of the Lean techniques do not require significant investments in machinery or technology but rather a real commitment from all employees in the company, from the CEO to the plant worker.

References

1. Sundar, R., Balaji, A.N., Kumar, R.M.S.: A review on lean manufacturing implementation techniques. Procedia Eng. **97**, 1875–1885 (2014). https://doi.org/10.1016/j.proeng.2014.12.341
2. Sarmiento Vasquez, C.J.: Incremento de la productividad en el area de produccion de la empresa Mundiplast mediante un sistema de produccion esbelto Lean Manufacturing. Ph.D.

thesis, Escuela Politecnica Nacional, Ecuador (2018). http://bibdigital.epn.edu.ec/handle/15000/19424
3. Bai, C., Satir, A., Sarkis, J.: Investing in lean manufacturing practices: an environmental and operational perspective. Int. J. Prod. Res. **57**(4), 1037–1051 (2019). https://doi.org/10.1080/00207543.2018.1498986, https://doi.org/10.1080/00207543.2018.1498986. Taylor & Francis
4. Quito Sinchi, A.: Diseno y desarrollo en Flash de un juego didactico para el aprendizaje de herramientas Lean Manufacturing. Tesis Maestria, Universidad Politecnica de Valencia, Spain (2017). http://hdl.handle.net/10251/91212
5. Reyes-B, J.V., Aguilar-Sanchez, L.A., Hernandez-Valencia, J.L., Mejias-Acosta,A., Pinero, A.: La MetodologÃa 5S como estrategia para la mejora continua en industrias del Ecuador y su impacto en la Seguridad y Salud Laboral. Polo del Conocimiento **2**(7), 1040 (2017). https://doi.org/10.23857/pc.v2i7.329. Polo de Capacitacion, Investigacion y Publicacion (POCAIP)
6. Margalina, V., Barroso Quinga, A.: La implementacion de Lean Manufacturing y el desempeno de las empresas manufactureras de calzado del canton Ambato. Tesis Maestria, Universidad Tecnica de Ambato, Ecuador (2018). https://repositorio.uta.edu.ec/jspui/handle/123456789/28912
7. Reyes, J., Morales, L., Aldas, D., Quilligana, A., Toasa, R., Reyes, R.: Applying lean manufacturing techniques to distribution requirements planning in ecuadorian flour industry. In: 2018 Congreso Internacional de Innovacion y Tendencias en Ingenieria (CONIITI), pp. 1–5. Colombia (2018). https://doi.org/10.1109/CONIITI.2018.8587069
8. Campoverde, P., Idrovo, A.: Propuesta de produccion del tablero de fibra de carbono utilizando herramientas Lean Manufacturing para el vehiculo monoplaza Formula SAE. Ph.D. thesis, Universidad Politecnica Salesiana, Ecuador (2018). http://dspace.ups.edu.ec/handle/123456789/15754
9. Carrasco Guaita, J.K.: Plan de mejora a los procesos de las areas de servicio del departamento de alimentos y bebidas del hotel Wyndham Quito Airport. Disertacion Doctoral, UNIB. E, Ecuador (2019)
10. Pachacama Gualotuna, D.A.: Mejora de la productividad, en el area de mecanizadotransfer para la fabricacion de griferia en la empresa Franz Viegener, mediante la implementacion de la metodologia Lean Manufacturing. Ph.D. thesis, Escuela Politecnica Nacional, Ecuador (2019)
11. Bautista Villegas, S.D.: Propuesta de mejoramiento aplicando la metodologÃaLean Six Sigma al proceso operativo de traslado de vehiculos y entrega a concesionarios de la empresa Pluslogistics SA, dentro de la ciudad de Quito. Universidad de las Americas, Ecuador, Tesis Maestria (2019)
12. Leyva, L.L.L., Perugachi, E.P.C., Piarpuezan, R.V.S., Orges, C.A.M., Montenegro, E.P.O., Burgos, G.: Lean Manufacturing application in textile industry. In: Proceedings of the International Conference on Industrial Engineering and Operations Management, pp. 26–27. Turkey (2018)
13. Alvarez, K., Aldas, D., Reyes, J.: Towards lean manufacturing from theory of constraints: a case study in footwear industry. In: International Conference on Industrial Engineering, Management Science and Application (ICIMSA), pp. 1–8. IEEE, Korea (2017)
14. Jaramillo, V.H.G.: A Methodology to Transform Small and Medium Companies to Lean Manufacturing Enterprises in Ecuador. Ph.D. thesis, Worcester Polytechnic Institute (2014)
15. Rodriguez, D., Buestan, M.: Identifying an improvement method for Ecuadoriansmall companies. Mexico (2013)

Analytical Study of the Behavior of Tubular Elements Reinforced with Natural Fibers Under Dynamic Loads

Henry H. Iza-Tobar, Shirley K. García, and Inés Iváñez

Abstract Laminates of fiber-reinforced composite materials are a good choice for structural applications due to their high strength and lightweight nature, which enables improved fuel efficiency. The present study examines the behavior of tubular elements reinforced with fibers intended to function as protective components in automobiles. An alternative to structures made of synthetic fiber-reinforced laminates (carbon, glass, etc.) are those reinforced with natural fibers. The high levels of abaca and jute fiber production in Ecuador allow for the use of these fibers in producing passive safety elements in vehicles. A mass-spring model evaluates the dynamic structural behavior of circular section tubular elements reinforced with natural fibers. The analytical model is validated with experimental results obtained from three-point on tubular element reinforced with carbon fiber.

Keywords Natural fibers laminates · Absorption energy · Tubular structure · Mass-spring model

1 Introduction

The regulations on fuel efficiency and gas emissions for cars are currently two fundamental aspects. The best way to increase fuel efficiency without sacrificing safety is to use fiber-reinforced composite materials in structural applications in the transportation industry, as fiber-reinforced composite materials have higher specific strengths than steel or aluminum [1].

H. H. Iza-Tobar (✉)
Department of Energy Sciences and Mechanics, Universidad de las Fuerzas Armadas ESPE, Parroquia Belisario Quevedo, Barrio El Forastero, Latacunga, Ecuador
e-mail: hhiza@espe.edu.ec

S. K. García · I. Iváñez
Department of Continuum Mechanics and Structural Analysis, University Carlos III of Madrid, Avda. de la Universidad 30, 28911 Leganés, Madrid, Spain

G. F. Olmedo Cifuentes et al. (eds.), *Emerging Research in Intelligent Systems*, Lecture Notes in Networks and Systems 902, https://doi.org/10.1007/978-3-031-52255-0_4

The use of laminated composite materials in structural applications continues to grow due to their excellent specific mechanical properties. To design structures using these materials, it is necessary to understand the complex behavior of laminated composite structures [2, 3]. The large number of published works is evidence of the interest generated by this topic and the high level of research activity. Most articles focus on low-velocity impact and their aspects of strength and damage tolerance [4–6].

The thin-walled beam is a representative structural element of great interest in the land and air transport industries and energy production. This beam is used in primary structures such as rotor blades and wing spars of aircraft, where they are primarily subjected to bending. They are also used as energy-absorbing elements in the passive safety structures of automobiles, which are typically subjected to compressive and/or bending loads. Additionally, these structures may experience low-velocity impacts during service life or assembly and maintenance operations [1].

Carbon fiber laminates are widely used to manufacture primary structural components in racing cars. In recent years, their use has extended to the production of passenger vehicles [7], resulting in significant reductions in structural weight [8, 9]. However, using these laminates presents some challenges due to their brittle fracture behavior, which is particularly critical in designing energy-absorbing elements subjected to dynamic loads [10]. Such loads can cause non-visible damage that may significantly alter the structural behavior [11].

The production processes, high manufacturing costs, and carbon fiber availability pose obstacles when manufacturing tubular elements and components for land transport vehicles. Numerous studies on thin-walled tubular parts made with laminates reinforced with syntactic fibers are primarily focused on axial compressive loads or static loads [12–16]. Therefore, an option is to utilize elements made with natural fibers such as cabuya, abaca, and coconut, which are more economical alternatives [17]. Natural fiber composites are mainly used for interior parts such as dashboards, door panels, parcel shelves, seat cushions, backrests, and cabin linings. In contrast, using natural fiber composite parts for exterior applications is minimal [18].

The use of natural fibers for manufacturing passive safety structural elements is limited, especially in Ecuador, where it has mostly been limited to extraction and artisanal use [19]. Previous research has determined that Ecuador is home to nearly 25,000 species of vascular plants, making it the nation with the most plants in Latin America [20]. Among these plants, around 25 species are used for fiber extraction, including cabuya, totora, abaca, coconut, palms, toquilla straw, cotton, loofah, silk, bamboo, vines, pitigua, balsa, mata-palo, rubber, rampira, anona, batea, achiote, majagua, higueron, nettle, wicker, royal vine, and others [17].

The most suitable experimental test for studying the response of these structural elements is the three- or four-point bending test, as it is representative of their loading condition. During a collision, the structure is subjected to dynamic loads. Thus, a proper investigation of this phenomenon requires impact testing. The impact phenomenon is complex, requiring specific experimental devices (gas gun, drop tower, etc.) and sophisticated measurement devices such as high-speed cameras. The numerical models require considerable computing time. Analytical models can solve impact problems quickly and understand the impact's physical behavior [21].

Some authors have studied the energy absorption capacity of tubular elements subjected to bending loads, but they only consider static loads. Palmer et al. [22] investigated the progressive failure of thin-walled beams made of pultruded composite materials. Lim and Lee [1] analyzed square-section tubular elements in a vehicle door using static tests and numerical simulations utilizing different glass/epoxy laminates. Meanwhile, Charoenphan et al. studied the progressive failure occurring in a square-section tube when subjected to static loads in a three-point bending test, observing that failure initiates at the corners of the section [23].

The energy absorbed by laminated composite structures subjected to impact is generated through a combination of failure mechanisms (matrix cracking, fiber/matrix debonding, delamination, fiber breakage, etc.) that can occur separately or in variety. Therefore, predicting the energy absorption capacity of a structure made of laminated composite materials is of great interest, as its response is highly sensitive to numerous parameters [24].

The design of structural elements reinforced with natural fibers could be a viable alternative for manufacturing structural components in the automotive industry. However, studies are required to assess the energy absorption capacity of such structures. Currently, there are no standardized regulations or methodologies available. Therefore, this work employs an analytical mass-spring model to evaluate the dynamic behavior of circular-section elements subjected to a dynamic three-point bending test.

2 Definition of the Cross-Sectional Area

An experimental study is conducted on square and circular tubular elements subjected to dynamic three-point bending tests to determine the optimal cross-section for evaluating the behavior of tubular elements under dynamic loads. The specimens are manufactured at the Spanish National Institute of Aerospace Technology (INTA) by Hexcel Composites' carbon fiber/epoxy (AS4/8552) orthogonal balanced woven prepregs. The specimens are tubes with a square section of 480 mm in length and 2.5 mm in thickness. The works by García-Castillo [25] and Iza-Tobar [26] provide details of the experimental tests. Several low-velocity impact tests are conducted on tubular specimens in a drop-weight tower, CEAST Fractovis. A specific practical device is designed and manufactured for the three-point bending impact test. The supports and the base of the experimental apparatus are of sufficient stiffness to prevent its movement during the impact tests. The hemispheric impactor, with a 30.22 mm radius and 6.128 kg weight, is designed to ensure non-perforation of the tubes.

For example, Figs. 1 and 2 display the force–time records obtained from the dynamic three-point bending tests conducted on the studied tubular elements, square (see Fig. 1) and circular section (see Fig. 2), for different impact energies.

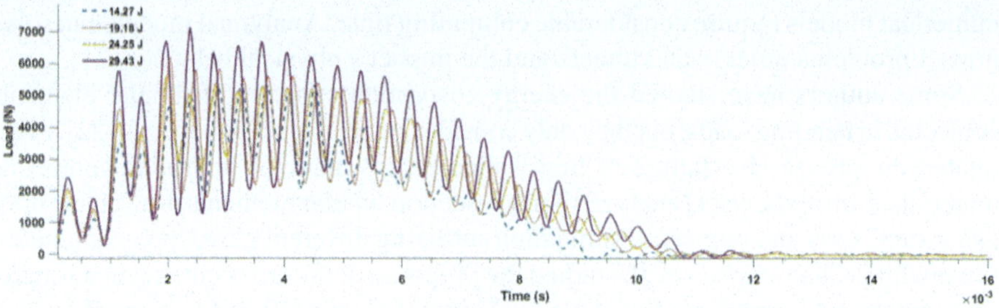

Fig. 1 Load curve versus time for tubular elements of square section for different impact energies

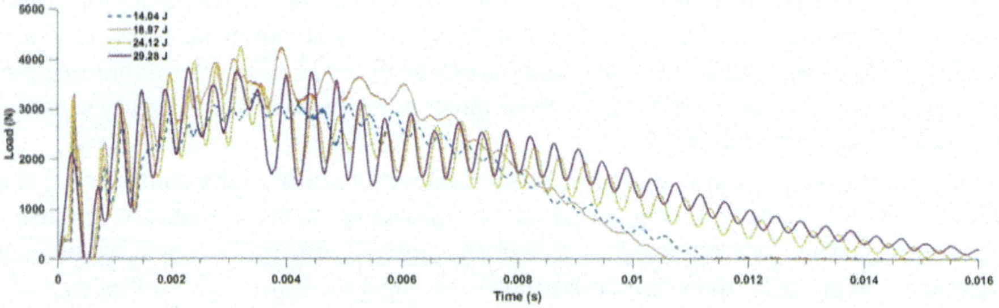

Fig. 2 Load curve versus time for tubular elements of circular section for different impact energies

Figure 3 shows the maximum force versus impact energy record for both circular and square-section tubular elements studied by García-Castillo [25] and Iza-Tobar [26]. The force transmitted in circular tubular elements is lower than in square parts for the different evaluated impact energies. Therefore, in the case of impact, the lower transmitted forces are observed in the circular-section structural elements.

The absorbed energy is calculated from the area under the experimental test's recorded force–displacement curve. Figure 4 shows the energy absorption capacity of the studied tubular elements observed for different impact energies. Both elements

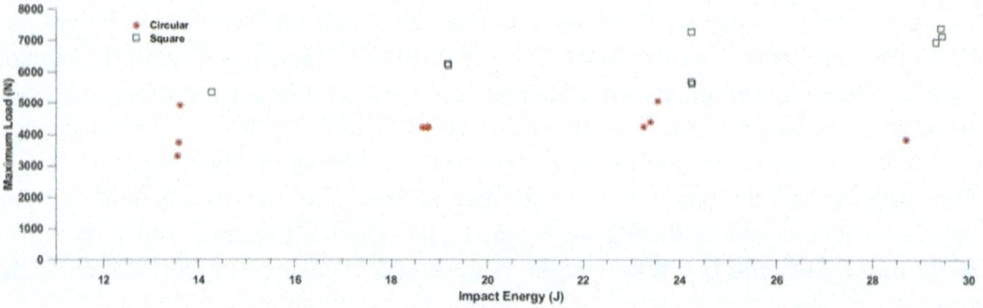

Fig. 3 Maximum load curve versus impact energy for tubular elements with circular and square sections

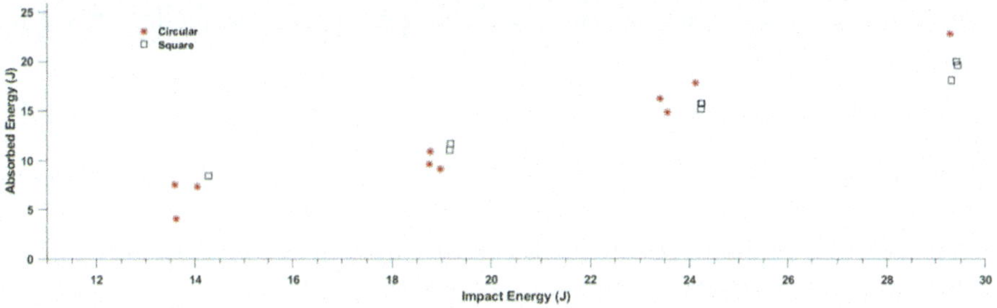

Fig. 4 Energy absorption versus Impact energy for circular and square-section tubular elements

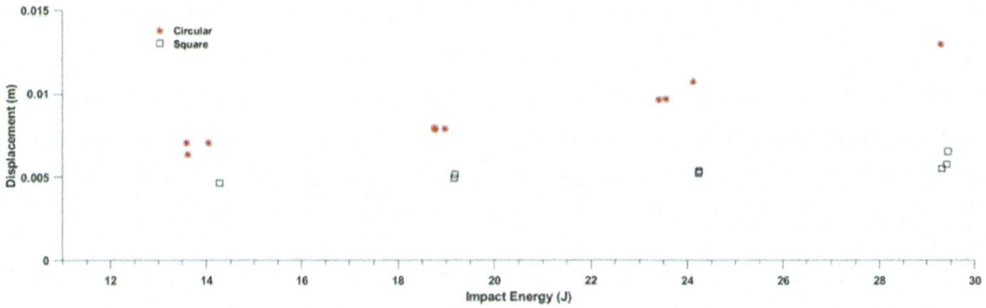

Fig. 5 Displacement versus Impact energy for circular and square-section tubular elements

exhibit similar energy absorption for an impact energy close to 14 J. However, the energy absorption increases with the impact energy in the circular-section elements.

Figure 5 shows the maximum displacement of the evaluated tubular elements, which is higher in the circular-section elements, contributing to their energy absorption capacity.

According to the dynamic three-point bending tests conducted on the tubular elements, it can be concluded that circular-section elements are the most suitable for use as passive safety components in automobiles under dynamic loads.

3 Analytical Model

For the development of the analytical model, circular-section tubular elements are selected due to their lower force transmission (see Fig. 3) and higher energy absorption capacity (see Figs. 4 and 5).

Firstly, the analytical model is developed and validated with experimental results obtained from the practical test for tubular elements of circular section reinforced with carbon fiber/epoxy (AS4/8552) orthogonal balanced woven prepregs [25, 26]. For developing the mass-spring model presented in this work, references are used from [27–29] for plates, beams, and circular elements. The assumptions made for the development of the model are specified below:

- The dynamic three-point bending test can be considered the most representative of a transverse impact at the center of a tubular element.
- The indenter is considered a rigid body because it will maintain shape after the impact occurs.
- The model's development does not consider the fracture or perforation of the tubular element. In other words, damage is not considered.

3.1 Mass-Spring Model

The most suitable mass-spring model for an impact test using a drop tower on a bi-supported circular tube is the two degrees of freedom (2-DOF) model, which consists of two masses connected by two springs. This model is a simplified representation of the actual behavior of the tube. Still, it is sufficient to simulate the impact phenomenon and can provide valuable information about the structural response of the tubular element.

In the 2-DOF model, the first mass represents the mass of the indenter, while the second mass represents the weight of the circular tubular elements. When the drop tower impacts the tube, both masses move, and the springs deform, resulting in vibration in the tube. The amplitude and duration of the vibration depend on the masses, stiffnesses, and magnitude of the impact force.

This model is suitable for this type of test because the tube is supported. The 2-DOF model better captures the actual behavior of the tube by considering longitudinal and transverse vibrations, which can be necessary to interpret test results correctly (see Fig. 6).

The following differential equations can represent the two-degree-of-freedom mass-spring model for the drop tower impact test.

$$m_1 \frac{d^2 y_1}{dt^2} + c_1 \frac{dy_1}{dt} + k_1 y_1 + k_2(y_1 - y_2) = F1(t) \tag{1}$$

$$m_2 \frac{d^2 y_2}{dt^2} + c_2 \frac{dy_2}{dt} + k_2(y_2 - y_1) = F2(t) \tag{2}$$

where m_1 and m_2 are the masses corresponding to the indenter and the tubular elements, respectively, y_1 and y_2 are the positions of the masses along the vertical direction, k_1 and k_2 are the spring constants connecting the masses to the levels of the tower, c_1, and c_2 are the damping constants that represent the resistance to the movement of the masses, $F1(t)$ and $F2(t)$ are the forces applied in the direction of m_1 and m_2 respectively, as functions of time.

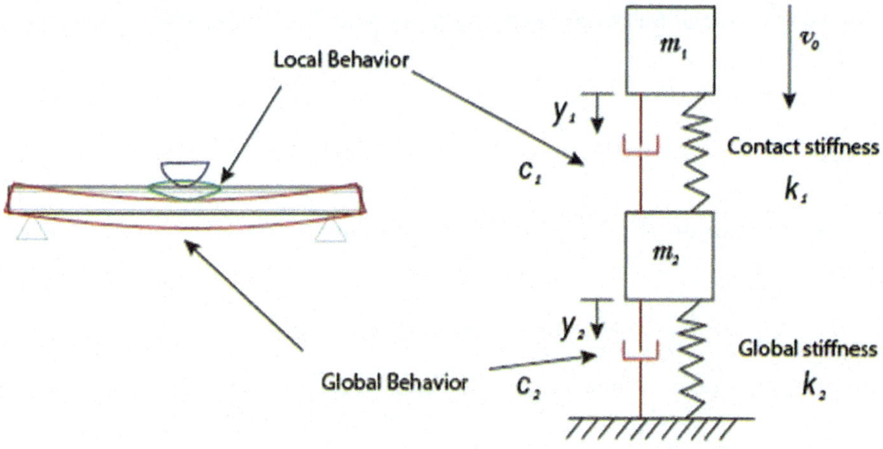

Fig. 6 Diagram of the two-degree-of-freedom mass-spring model

3.2 Validation of the Analytical Model

The validation of the analytical model is a crucial process to ensure the accuracy and reliability of the results. It is performed by comparing experimental data with the model predictions and assessing its ability to reproduce the results accurately. Suppose the model agrees with the experimental data. In that case, the model is validated and can be confidently used to predict the system's behavior. Model validation is essential in research, allowing for informed decision-making and continuous design improvement.

The differential equations are numerically solved using the Runge–Kutta numerical integration method to calculate the maximum force generated during the test. It is assumed that the initial conditions for velocities are zero, while for positions, it is considered that the indenter is located at a height of 0 m above the tube [30].

The natural frequency of the system is calculated as follows:

$$f_{n1} = \frac{1}{2\pi}\sqrt{\frac{k_1}{m_1}} \tag{3}$$

$$f_{n2} = \frac{1}{2\pi}\sqrt{\frac{k_2}{m_2}} \tag{4}$$

Calculating the natural frequency of the two-degree-of-freedom mass-spring system is necessary to predict the system's response to dynamic excitation in a drop tower impact. The natural frequency is when the system will vibrate freely without applying an external force. In a drop tower impact, the system's natural frequency can affect its ability to absorb energy and prevent severe damage. Therefore, the natural frequency is crucial for designing the system optimally and ensuring its capacity to protect the main element.

The equivalent spring constant is calculated as follows:

$$k_{eq} = \frac{m_1 m_2}{m_1 + m_2} f_{n2}^2 + k_1 \tag{5}$$

Also, the tube displacement can be calculated as follows:

$$y_{max} = \frac{F_{max} L_c^3}{3 E_c I} \tag{6}$$

where F_{max} is the maximum force, L_c is the length of the tube measured from the supports, E_c is the elastic modulus of the carbon fiber tube, and I is the second moment of area of the tube, which can be calculated as:

$$I = \frac{\pi}{4}\left(r_o^4 - r_i^4\right) \tag{7}$$

where r_o is the outer radius of the tube, and r_i is the inner radius of the tube.

In addition, the indenter position as a function of time t can be expressed as:

$$y(t) = y_{max}\left(1 - \cos\left(\frac{\pi t}{t_{max}}\right)\right) \tag{8}$$

where t_{max} is the maximum impact time, which can be calculated as:

$$t_{max} = \sqrt{\frac{2 m_1 y_{max}}{k_{eq}}} \tag{9}$$

Once the position of the indenter as a function of time is obtained, the Runge–Kutta numerical integration method can get the position and velocity of mass one and mass two as a function of time.

The equation to calculate the maximum force during the impact test on the drop tower using the two-degree-of-freedom mass-spring model is:

$$F_{max} = k_{eq} \cdot y_{max} \tag{10}$$

The natural frequency of the tubular element is calculated as:

$$f_{n2} = \frac{1}{2\pi}\sqrt{\frac{E_c A_c}{m_2 L_c^3}(1 + \nu_c)} \tag{11}$$

where A_c is the cross-sectional area of the tube, and ν_c is the Poisson's ratio of the tube. For an impact test on the drop tower, the impact energy can be calculated as:

$$E = \frac{1}{2}m_1 v_1^2 = \frac{1}{2}k_{eq} y_{max}^2 \tag{12}$$

where v_1 is the initial velocity of the indenter, and m_1 is the mass of the indenter. Upon substituting the previous expressions into the equation for the maximum force, we have:

$$F_{max} = \sqrt{2Ek_{eq}} \tag{13}$$

The spring constant of the indenter can be calculated from the radius of curvature of the indenter's head using the following equation:

$$k_1 = \frac{4E_s}{3(1 - v_s^2)}\left(\frac{1}{r}\right)^2 \tag{14}$$

where E_s is Young's modulus of steel, v_s is Poisson's ratio of steel, and r is the radius of curvature of the indenter's head. The equivalent spring constant of the carbon fiber tube can be calculated using the following equation:

$$k_2 = \frac{E_c A_c}{L_c}\frac{1}{1 - v_c^2} \tag{15}$$

It is necessary to find the maximum deformation of the tube, denoted as y_{max}. For this purpose, the following equation is used:

$$y_{\max} = \frac{1}{2} \cdot \frac{F_{\max}}{k_{eq}} \cdot \left(\frac{r_o^4 - r_i^4}{r_o^3 r_i}\right) \tag{16}$$

At 14.04 J of impact energy, according to the experimental procedure, the maximum force corresponds to a value of 3310.38 N. In contrast, the two-degree-of-freedom mass-spring model's total load is 3381.6 N (see Fig. 7). The curve behavior is similar for energy levels of 18.97, 24.12, and 29.28 J. The analytical model has a 10% percentage variation compared to the experimental test, validating the model.

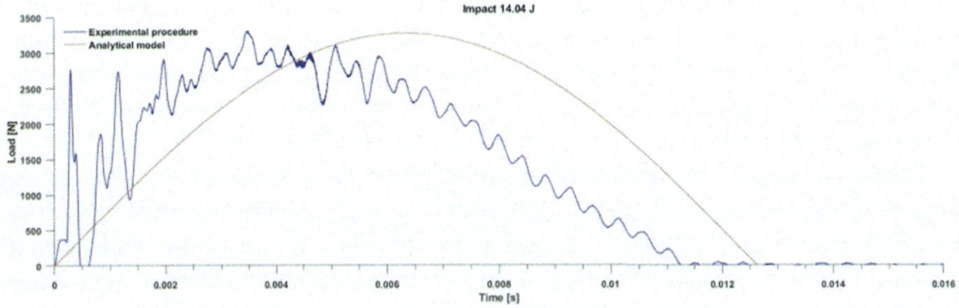

Fig. 7 Load versus time, experimental procedure/TDOF for 14.04 J

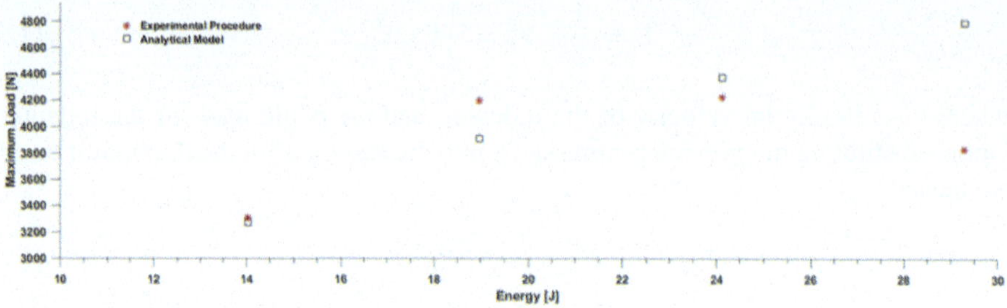

Fig. 8 Comparison of maximum load versus impact energy

The developed model does not consider the rupture of the tubular element. Figure 8 shows that at higher energies, there is a significant variation in the load between the experimental procedure and the analytical model.

4 Dynamic Contact Force on Tubular Elements Reinforced with Natural Fibers

Natural jute and abaca fibers, extracted from plants, exhibit distinctive morphologies that make them valuable in various applications. Jute, derived from the jute plant, is characterized by its long, soft fibers with a fine texture and a golden color. These fibers are intertwined into a braid-like structure, giving them remarkable strength and flexibility. On the other hand, abaca, coming from the Manila banana plant, has long and robust fibers with a more rustic appearance and beige tones. These fibers, interestingly, are hollow, making them especially light and durable. Both natural fibers are appreciated for their versatility in manufacturing clothes, bags, fabrics, and various crafts, standing out for their sustainability and authenticity [17].

Bi-supported tubular elements reinforced with natural fibers are commonly used in constructing structures due to their low cost, high strength, and lightweight nature. However, it is vital to understand their dynamic behavior under extreme conditions, such as impacts, to ensure their safety and effectiveness [17].

When a bi-supported tubular element reinforced with natural jute fibers is subjected to an impact, forces, and vibrations are generated that affect its dynamic behavior. Various factors, including the mass of the tube, the type of jute fiber reinforcement, the height of the drop tower, the velocity, and the impact angle, influence the dynamic response of the bi-supported tubular element.

During the impact, the load is transferred through the material of the tube, causing deformations and stresses within it. Jute fibers reinforce the tube, increasing its load-bearing capacity and reducing the likelihood of failure. However, the tube strength can also be compromised if the threads are damaged during the impact. The analytical model described in Sect. 3.2 is applied to natural jute and abaca fibers, whose properties are presented in Table 1 [17].

Table 1 Mechanical properties of natural fiber lamina

Properties	Abacá	Yute
Longitudinal elastic modulus E_{11}	9 GPa	18 GPa
Transversal elastic moduls E_{22}	9 GPa	18 Gpa
Shear module G_{12}	0.4 GPa	2 GPa
Longitudinal Poisson's ratio v_{12}	0.3	0.11
Transversal Poisson's ratio v_{21}	0.3	0.11
Tensile strength Y_c	400–1289 MPa	393–773 MPa

Figure 9 shows a typical sinusoidal response of mass-spring models, wherein the jute fiber exhibits a higher load within a shorter duration. This behavior can be attributed to the disparate material properties inherent to each material.

All natural fibers have inherent properties that determine their usefulness in the application. The differences in mechanical properties (ultimate strength and longitudinal module) could explain that the tubular structures of jute laminates support the more significant maximum load (see Fig. 10). According to the model, abaca fiber structures transmit a lower force, suggesting that it could be considered as protective elements against impacts, especially in automobiles.

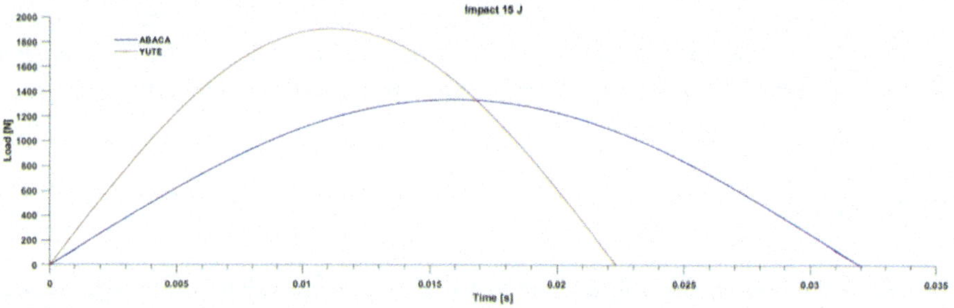

Fig. 9 Load versus time curves for laminates reinforced with abaca and yute fibers

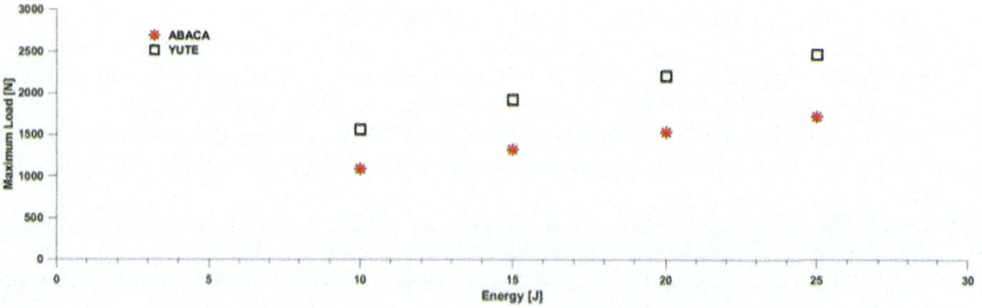

Fig. 10 Maximum load versus time

4.1 Conclusions

The appropriate cross-sectional shape was determined for the application of the analytical model. Circular tubular elements exhibit better energy absorption than square tubular elements, as the corners of the latter form gaps between layers, limiting their absorption capability. The three-point bending test is the most suitable for this type of investigation in the analytical model, and no fiber rupture was considered. The two-degree-of-freedom mass-spring model was selected because it demonstrates the closest actual behavior of the tube. The analytical model was validated with tubular elements of a circular section made of carbon fiber for impact energies between 14.04 and 29.43 J. The maximum difference between the experimental and analytical results did not exceed 10%. The load versus time curves obtained from the analytical model exhibited a sinusoidal shape, characteristic of a mass-spring system, confirming the validity of the chosen model. The analytical model was applied to tubular elements reinforced with natural fibers such as abaca and jute. The analytical model allows for determining the contact force transmitted to the tubular component due to an impact. Jute, a natural fiber, has a higher elasticity modulus than abaca. Therefore, according to the model, the force–time curve exhibited better behavior when using this type of fiber.

The analytical model successfully predicted the behavior of tubular elements reinforced with natural fibers, specifically abaca and jute.

The choice of natural fiber reinforcement played a significant role in the performance of the tubular elements. With its higher modulus of elasticity, jute demonstrated better behavior regarding force–time response than abaca.

References

1. Lim, T., Lee, D.: Mechanically fastened composite side-door impact beams for passenger cars designed for shear-out failure modes. Compos. Struct. COMPOS STRUCT **56**, 211–221 (2002)
2. Akca, E., Gursel, A.: A review on the matrix toughness of thermoplastic materials. Period. Eng. Nat. Sci. **3**, 1–8 (2015)
3. Adhikary, S.D., Li, B.: Simplified analytical models to predict low-velocity impact response of rc beams. Pract. Period. Struct. Des. Constr. **23**, 04018002 (2018)
4. Abrate, S.: Modeling of impacts on composite structures. Compos. Struct. **51**, 129–138 (2001)
5. Alharthi, H., Abdellah, M.Y., Abdo, H.S., Hassan, M.K.: Size effect and fracture toughness of glass fiber composite structures. Int. J. Adv. Sci. Technol. **29**, 8567–8577 (2020)
6. Ribeiro, M.L., Martins, T., Sartorato, M., Ferreira, G.F., Tita, V., Vandepitte, D.: Analysis of low energy impact on filament-wound composite cylinders. Int. J. Veh. Struct. Syst. **4**, 118–122 (2012)
7. Carrera, M., Cuartero, J., Miravete, A., Jergeus, J., Fredin, K.: Crash behaviour of a carbon fibre floor panel. Crash Behav. Carbon Fibre Floor Panel **44**, 268–281 (2007)
8. Botkin, M.E.: Modelling and optimal design of a carbon fibre reinforced composite automotive roof. Eng. Comput. **16**, 16–23 (2000)
9. Cheon, S., Choi, J.H., Lee, D.G.: Development of the composite bumper beam for passenger cars. Compos. Struct. **32**, 491–499 (1995)

10. Feraboli, P., Masini, A., Taraborrelli, L., Pivetti, A.: Integrated development of cfrp structures for a topless high-performance vehicle. Compos. Struct. **78**, 495–506 (2007)
11. Remacha, M., Sánchez-Sáez, S., López-Romano, B., Barbero, E.: A new device for determining the compression after impact strength in thin laminates. Compos. Struct. **127**, 99–107 (2015)
12. Dong, Q.B., Zha, Y., Sun, J., Gan, X., Cai, M., Zhou, T.: Multi-objective optimization for energy absorption of carbon fiber-reinforced plastic/aluminum hybrid circular tube under both transverse and axial loading. J. Mater. Eng. Perform. **29**, 5609–5624 (2020)
13. Ataabadi, P.B., Karagiozova, D., Alves, M.: Crushing and energy absorption mechanisms of carbon fiber-epoxy tubes under axial impact. Int. J. Impact Eng. **131**, 213–225 (2019)
14. Friedrich, K., Almajida, A.: Manufacturing aspects of advanced polymer composites for automotive applications. Appl. Compos. Mater. **20**, 107–128 (2013)
15. Mamalis, A.G., Manolakos, D.E., Ioannidis, M.B., Papapostolou, D.P.: The static and dynamic axial collapse of cfrp square tubes Finite element modelling. Compos. Struct. **74**, 213–225, (2006)
16. Zeng, T., Fang, D.N., Lu, T.J.: Dynamic crashing and impact energy absorption of 3d braided composite tubes. Mater. Lett. **59**, 1491–1496 (2006)
17. Dávila, J.L., Galeas, S., Guerrero, V.H., Pontón, P., Rosas, N.M., Sotomayor, V., Valdivieso, C.: Nuevos Materiales: Aplicaciones Estructurales e Industriales, 1st edn. Escuela Politécnica Nacional, Quito-Ecuador (2011)
18. Peças, P., Carvalho, V., Salman, P., Leite, M.: Natural fibre composites and their applications: a review. J. Compos. Sci. **2**, 66 (2018)
19. Neill, D.A.: ¿Cuántas especies nativas vasculares hay en Ecuador? Revista Amazónica Ciencia y Tecnología **1**, 70–83 (2012)
20. INABIO: Estrategia para la conservación de plantas. http://inabio.biodiversidad.gob.ec. Accessed 23 May 2023
21. Alonso, L., Martínez-Hergueta, F., Garcia-Gonzalez, D., Navarro, C., García-Castillo, S.K., Teixeira-Dias, F.: A finite element approach to model high-velocity impact on thin woven gfrp plates. Int. J. Impact Eng. **142**, 103593 (2020)
22. Yuan, Y., Yao, X., Liu, B., Yang, H., Imtiaz, H.: Failure modes and strength prediction of thin ply cfrp angle-ply laminates. Compos. Struct. **176**, 729–735 (2017)
23. Charoenphan, S., Bank, L.C., Plesha, M.E.: Progressive tearing failure in pultruded composite material tubes. Compos. Struct. **63**, 45–52 (2004)
24. Moure, M., García-Castillo, S.K., Sánchez-Sáez, S., Barbero, E., Barbero, E.J.: Matrix cracking evolution in open-hole laminates subjected to thermo-mechanical loads. Compos. Struct. **183**, 510–520 (2018)
25. García-Castillo, S., Santuiste, C., Sánchez-Sáez, S., Barbero, E.: Estudio de la absorción de energía en elementos tubulares sometidos a impacto de baja energía. AEMAC Actas del IX Congreso Nacional de Materiales Compuestos **10**, 657–662 (2011)
26. Iza-Tobar, H., García-Castillo, S., Iváñez, I.: Análisis experimental de la absorción de energía de elementos tubulares de fibra de carbono de sección cuadrada sometidos a impactos transversales. Energía y Mecánica, Innovación y Futuro **10**, 48–53 (2021)
27. Li, H., Chen, W., Pham, T.M., Hao, H.: Analytical and numerical studies on impact force profile of rc beam under drop weight impact. Int. J. Impact Eng. **147**, 103743 (2021)
28. Shivakumar, K.N., Elber, W., Illg, W.: Prediction of impact force and duration due to low-velocity impact on circular composite laminates. ASME Trans. J. Appl. Mech. **52** (1985)
29. Sun, J., Lam, N., Zhang, L., Ruan, D., Gad, E.: Contact forces generated by hailstone impact. Int. J. Impact Eng. **84**, 145–158 (2015)
30. Rao, S.: Mechanical Vibrations, 5th ed. Prentice Hall (2011)

Landstorfer Antenna Structure Shaping Based on Parameterized Parallel Curves

Aldrin Reyes Narváez and Hernan Barba Molina

Abstract This paper shows a novel method to model a Landstorfer antenna based on the shaping of the constitutive antenna elements obtained from profile curves generated by controlled smoothed mathematical step functions. The parallelism concept of profile curves supports forming a reflector and director. An antenna prototype operating at 904 MHz is optimized and built on a dielectric-characterized FR-4 substrate and measured as a proof of concept of the proposed structure. Measurements show good agreement with simulations. The antenna prototype performs similarly to the original Landstorfer idea but with a more straightforward antenna elements profiling procedure design. The simulated radiation characteristics of the antenna are a directivity of 11.3 dBi, a side-lobe level of −10.6 dBi, and a front-back ratio of 18.7 dB. Furthermore, the structure has an overall length of $0.75\lambda_0$. The results verify the excellent performance of the parallelism approach. Limitations and range of utilization for the method are given as well.

Keywords Landstorfer antenna · Parallel curves · Antenna structure characterization · Radiation pattern

1 Introduction

Since the Yagi-Uda antenna was reported in [1, 2], many structural modifications based on the original operating principles have been proposed, showing novel design procedures that optimize the idea regarding radiation characteristics and structure conformation. Regarding the latter, the Landstorfer antenna describes a 3-element structure having, as a driven element, a reshaped dipole longer than the free-space half-wavelength ($\approx 3\lambda_0/2$, where λ_0 is the free-space wavelength). This approach exhibits radiation characteristics similar to a conventional Yagi-Uda antenna with at

A. R. Narváez (✉) · H. B. Molina
Departamento de Electrónica, Telecomunicaciones y Redes de Información (DETRI), Escuela Politécnica Nacional, 170525 Quito, Ecuador
e-mail: detri@epn.edu.ec

G. F. Olmedo Cifuentes et al. (eds.), *Emerging Research in Intelligent Systems*, Lecture Notes in Networks and Systems 902, https://doi.org/10.1007/978-3-031-52255-0_5

least ten elements [3], demonstrating, beyond any doubt, structural size reduction [4].

Printed Landstorfer antenna implementations have become attractive among applications due to ease of manufacture [5–8]. A well-known technique based on inserting lumped reactances is utilized to achieve shapes with reduced dimensions to obtain more than 50% smaller sizes [5]. Furthermore, a planar antenna array that consists of Landstorfer and conventional Yagi-Uda radiators with directivities of 9.5 dBi and 7.3 dBi, respectively, is implemented for 360-degree azimuthal coverage [6]. In this case, a feed/switching network allows control of the array by activating the radiators one at a time to have a reconfigurable radiation configuration. In another realization, the reflector of the so-called Quasi-Landstorfer antenna is replaced by a modified ground plane that permits a coplanar stripline (CPS) feeding network [7]. Although the antenna's gain decreases compared with the original design (6.6 dBi), this prototype shows an overall size reduction of 44% and an advantage by simplifying the implementation of the feeding network with the CPS structure. So, the Landstorfer-shaped method is applied in a printed log-periodic dipole array antenna with a gain of 10.5 ± 0.5 dBi for Ka-band applications [8].

The structure design proposed by Landstorfer starts by shaping a monopole radiator with linear segments that maximize the effective height ($0.55\ \lambda_0$) and directivity (7.8 dBi) by controlling the number of segments and their tilt angles as part of the structure [4]. Thus, the dipole length reduces to about one-wavelength dipole with a one-quarter-wavelength stub. The reflector and director are optimized experimentally to produce a unidirectional pattern with a reported gain of about 11.5 dBi [4, 9].

One of the disadvantages that a designer must face during the construction of such a structure is the shaping procedure, especially in a wired configuration, as experimented with in a co-supported work by the authors [10]. New technologies offering printed planar structures supported by CAD software have alleviated this issue, providing ease by shaping almost any required forms on substrates. In contrast, substrates are merely utilized as mechanical structure support, even for high-frequency applications where models demand complex realizations.

Different methods have been reported to model antennas' driven, reflector, or director elements based on Landstorfer's original idea. Modeling procedures include geometrical combinations of lines and arcs with different lengths and radii to shape optimized forms [6–8]. Parameterized curves can also shape these forms [11–13].

This paper presents a simple modeling procedure for shaping the driven, reflector, and director elements of a Landstorfer antenna supported by the generation of a controlled reconfigurable smoothed mathematical step function. The proposed structure and theoretical considerations for shaping the driven, reflector, and director profiles are described in the first part. Likewise, the theory related to the generation of parallel curves is detailed. Section 3 shows the design and construction of a proof-of-concept structure. Measurement results are presented. Section 4 shows the discussion of the proposed concept. Conclusions are given in Sect. 5.

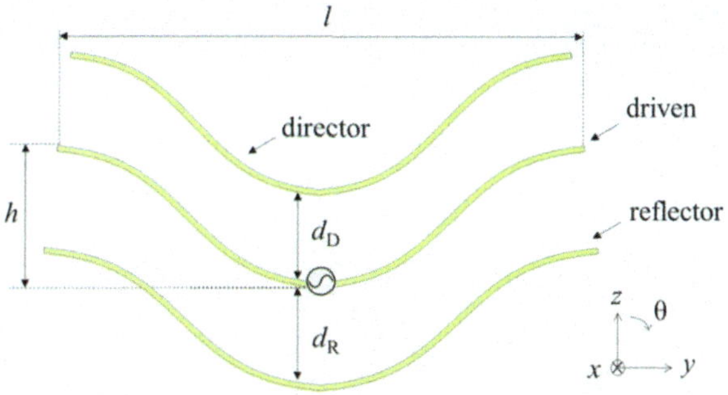

Fig. 1 Proposed structure of the Landstorfer antenna with the driven element modeled by a smoothed step function and director and reflector modeled by parallel-to-driven curves. Projected dimensions l and h are referred to as the driven element

2 Antenna Structure

The theoretical design of the proposed structure is presented in this section. A general description of the Landstorfer antenna is depicted along with a detailed mathematical concept based on a parallelism approach to generate profiles for the constitutive elements of the antenna. This approach is the key focus of this work.

2.1 *Proposed Structure*

Figure 1 illustrates the proposed structure for the Landstorfer antenna. The design has each constitutive element (driven, reflector, and director) obtained from a parameterized function that modifies the form based on its mathematical characteristics. The latter allows to fit the electrical length of the driven to approximately $3\lambda_0/2$, as given in [4], each time the form of the curve changes. Likewise, reflector and director elements, longer and shorter than the driven element, are obtained from profile curves parallel to the one generated for the driven.

2.2 *Driven Generating Curve*

The driven profile is modeled by mirroring the curve obtained by the normalized smoothed step function [14].

$$f(t, \zeta) = \frac{\frac{1}{1+e^{-2t(2\zeta-1)}} - \frac{1}{2}}{\frac{2}{1+e^{-2t}} - 1} + \frac{1}{2} \tag{1}$$

where $0 \leq \zeta \leq 1$ and t controls the smoothness of the curve from a linear transition ($t \to 0$) to a symmetrical step function ($t \to \infty$).

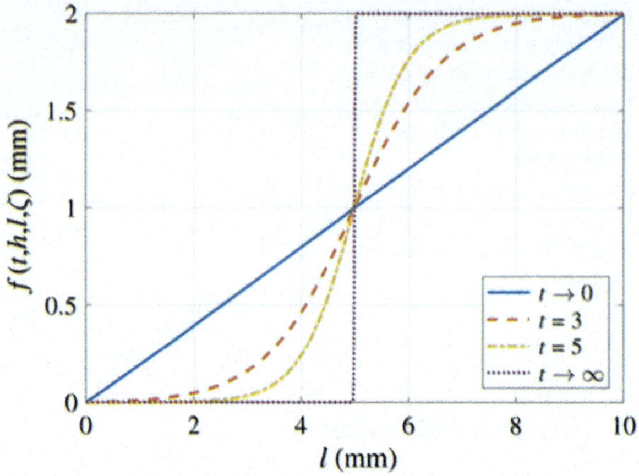

Fig. 2 Modeling example of one side of the driven element for $h = 2$ mm, $l = 10$ mm, and different values of t

As the domain and the range of (1) are restricted to unitary maximum length and height (i.e., $\zeta = 1$ and $f(t, \zeta = 1) = 1$), to match these parameters to a required projected length and projected height of the driven element (i.e., $0 \le \zeta \le l$ and $0 \le f(t, \zeta) \le h$), the function now becomes:

$$f(t, h, l, \zeta) = h\left(\frac{\frac{1}{1+e^{-2t\left(\frac{2}{l}\zeta-1\right)}} - \frac{1}{2}}{\frac{2}{1+e^{-2t}} - 1} + \frac{1}{2}\right) \tag{2}$$

where l is the projected length and h is the projected height of the driven element.

Figure 2 shows an example application of (2) to be utilized for modeling the variation of the structure of one side of the driven element with arbitrary parameters.

2.3 *Reflector and Director Modeling*

As mentioned above, reflector and director profile curves are modeled parallel to the driven curve obtained by (2).

Parallel curves. According to the math described in [15], a curve function $\beta(t)$ can be parallel obtained from a given generating function $\alpha(t)$. The generated function has the general expression provided by:

$$\beta(t) = \alpha(t) + rn(t) \tag{3}$$

where $\alpha(t)$ is the parametric function of any curve, r is the distance at which the parallel curve is located (it can be a positive or negative value), and $n(t)$ is the parametric unit normal vector at any point on $\alpha(t)$ curve.

The expression for the parametric unit normal vector is given by [15]:

$$n(t) = \frac{1}{\|\alpha'(t)\|}\left(-y'(t), x'(t)\right) \tag{4}$$

Considering the parameterization of the curve $\alpha(t)$ and the unit normal vector $n(t)$, the parametric function of the parallel curve is given by:

$$\beta(t) = \left(x(t) - \frac{ry'(t)}{\|\alpha'(t)\|}, y(t) + \frac{rx'(t)}{\|\alpha'(t)\|}\right) \tag{5}$$

Now, by replacing (5) with the function proposed to model the driven structure in (2), the parametric function to obtain the parallel curve results in:

$$\beta(t,h,l,\zeta) = \begin{pmatrix} \zeta - r\left(\frac{\frac{df(t,h,l,\zeta)}{d\zeta}}{\sqrt{\left(\frac{df(t,h,l,\zeta)}{d\zeta}\right)^2+1}}\right) \\ f(t,h,l,\zeta) + r\left(\frac{1}{\sqrt{\left(\frac{df(t,h,l,\zeta)}{d\zeta}\right)^2+1}}\right) \end{pmatrix} \tag{6}$$

where:

$$\frac{df(t,h,l,\zeta)}{d\zeta} = \left(\frac{\frac{4ht}{l}}{\frac{2}{1+e^{-2t}} - 1}\right)\left(\frac{e^{-2t\left(\frac{2\zeta}{l}+1\right)}}{\left(e^{\frac{-4t\zeta}{l}} + e^{-2t}\right)^2}\right) \tag{7}$$

Modeling. The reflector and director structure modeling starts from (6) and (7) to obtain the parallel curve to the generating function of the driven structure at a distance r.

Being the Landstorfer antenna based on the Yagi-Uda antenna, the driven-reflector and driven-director separation is set to about $\lambda_0/4$ [3], which makes the parameter r frequency dependent for this approach. Furthermore, the projected length of the reflector l_r and the projected length of the director l_d must be more prolonged and shorter than the projected length of the driven l, respectively. These considerations are considered for each case to obtain (2) before calculating the parallel curves.

3 Antenna Prototype

As a proof of concept, the first approach of a Landstorfer antenna, based on the parallel-concept procedure presented above, is designed and optimized for an operation frequency of 904 MHz. Analyses and optimizations are done using an electromagnetic field simulation tool (Antenna and RF Toolboxes of MATLAB R2022b).

3.1 Design

The antenna prototype is shown in Fig. 3. The overall structure is printed on a single-layer substrate FR-4 (dielectric characterized with $\varepsilon_r = 4.8$) with dimensions 38.9 cm 28.4 cm. The width for all the strips is set to 4 mm due to ease of manufacturing.

First, the shape of the driven element must be determined by the smoothness parameter t and its projected length l, as well as the projected height h. Simulations reveal that a structure with $t = 1.8$, $1 = \lambda_0$, and $h = 0.6\lambda_0$ (optimized to adjust the electrical length of the driven element to $3\lambda_0/2$) exhibits a maximum directivity of 5.4 dBi. This agrees well with the optimum shape of the monopole in [16]) which has a computed directivity of 7.8 dBi (i.e., 4.8 dBi corresponding to the dipole one).

Next, reflector and director elements are obtained from parallel curves positioned at $d_R = 0.23\lambda_0$ and $d_D = 0.21\lambda_0$ apart from the driven and whose lengths are resized by a 5% longer and shorter factor, respectively.

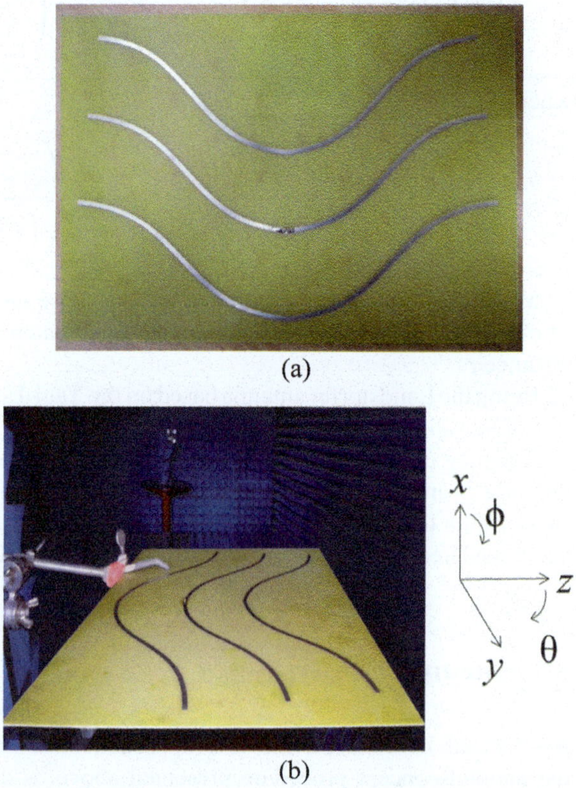

Fig. 3 **a** Photograph of the constructed Landstorfer antenna prototype. **b** Prototype mounted in the anechoic chamber together with the definition of coordinates

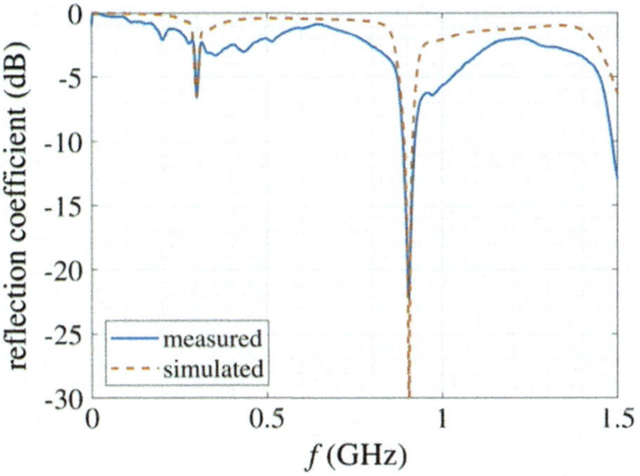

Fig. 4 Measured (solid blue) and simulated (dashed red) reflection coefficient magnitude of the proposed Landstorfer antenna

3.2 *Antenna Measurements*

Figure 4 shows the simulated and measured reflection coefficient magnitude of the proposed Landstorfer antenna when radiating into free space. It can be seen that the antenna operates at the second resonance, which verifies the effective electric length of the structure of about $3\lambda_0/2$. There is a good agreement between simulations and measurements.

The radiation pattern measurements are done in the anechoic chamber of the University's facilities, allowing power measurements in the far field using an RF generator, a spectrum analyzer, and other complementary equipment. Figure 5 shows the antenna's simulated and measured radiation pattern at 904 MHz radiating into free space. The agreement between simulation and measurement is good. Thus, the directivity of the structure is obtained by simulation of 11.3 dBi. Note that this value is received by a conventional Yagi-Uda antenna with an overall length of $0.8\lambda_0$ (i.e., five elements) [3]. Here, the overall length of the three-element structure is $0.75\lambda_0$. The simulated side-lobe level and front-back ratio are -10.6 and 18.7 dB, respectively.

4 Discussion

The curve shape for the driven element with $t = 1.8$ generates a dipole with a directivity of about 5.4 dBi. Compared with the directivity of a linear dipole of $3\lambda_0/2$ [3], this value is greater than 2 dB, improving the antenna's performance. Besides, the structural concept of the whole antenna is based on a single one-parameter mathematical function for profiling, which takes advantage of experimental optimizations or different parameterization procedures.

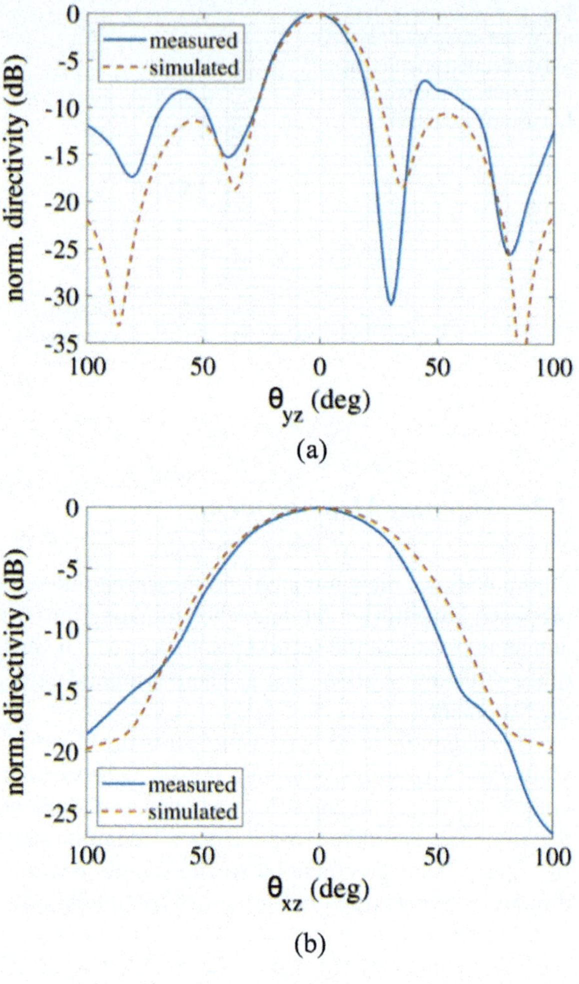

Fig. 5 Measured (solid blue) and simulated (dashed red) radiation pattern of the Landstorfer antenna prototype for $f = 904$ MHz. Results are invalid in a sector between 100 and 180° due to the measurement setup. **a** E-Plane pattern ($\phi = 90°$), **b** H-Plane pattern ($\phi = 0°$)

During the optimization process of the proposed approach, a natural limitation is found by generating the parallel curves. This lies in the parameters that cause a small curvature radius in the generating function of the parallel curves, namely, the curve factor t and the ratio l/h, which allows the electrical length optimization of the driven element. Note that the structure proposed in this work does not depend on the frequency of operation.

Figure 6a illustrates an example of the generated curves for driven, reflector, and director elements operating at 5 GHz with $t = 3$ and $l/h = 2$. At the same time, Fig. 6b shows the same curves for $t = 9$ and $l/h = 3$. Error generation can be appreciated encircled. A slight curvature radius in the parallel curve-generating function causes the reflector and director structure to intersect. Thus, a determined range can only model the structure before its generation procedure collapses.

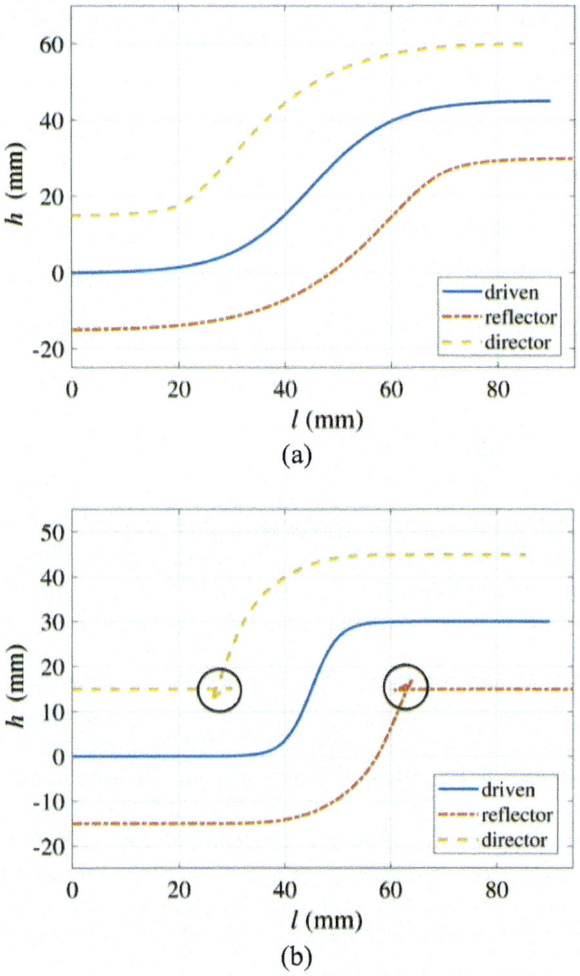

Fig. 6 Obtained curves for a 5 GHz Landstorfer antenna structure. **a** $t = 3$, $l/h = 2$. **b** $t = 9$, $l/h = 3$. Error generation in director and reflector curves are encircled

Table 1 summarizes the feasible combinations of the parameters that limit the structure proposed in this work without causing an intersection or loss of continuity of the curvature radius in the structure of the reflector and the director. Thus, this is a failure in the modeling process. The maximum value proposed for the ratio l/h is 20 since, with higher values, the structure of the proposed driven resembles that of a flat driven, losing the optimization of the antenna directivity described in [4].

Table 1 Summary of parameters that limit the structure of parallel curves

Curve factor t	l/h (range)
1	1 … 20
2	1 … 20
3	1 … 20
4	1 … 20
5	3 … 20
6	8 … 20
7	12 … 20
8	15 … 20
9	19,20

5 Conclusions

A method to shape an antenna based on the Landstorfer idea using the concept of parallel curves is proposed. The driven element (dipole) is shaped by a parameterized smoothed step curve function, which works at the second resonance, i.e., a length of $\approx 3\lambda_0/2$. At the same time, the reflector and director are generated parallel by the generating curve of the driven, longer and shorter than this, respectively, according to the classic Yagi-Uda concept. The method does not depend on the operation frequency. Thus, it can be applied to any wired structure. Furthermore, a proof-of-concept prototype shows an 11.3 dBi simulated directivity at an operation frequency of 904 MHz. Measurements and simulations agree well and reveal the excellent performance of the parallelism approach. In addition, a natural mathematical limitation caused by the parallel procedure is depicted. For that, feasible utilization ranges are given for modeling the structure. Since this approach is based on a well-known smoothed step function, modeling a wired structure becomes easy when supported by CAD software.

Although realizations of the Landstorfer antenna idea have been conceived utilizing different methodologies, the procedure described here for shaping the antenna elements contributes consistently to the antenna design because of the parameterization of profiles using a single tunable one-parameter generating mathematical function. The overall performance of the antenna prototype can be used as terminal antennas in sensor networks and eventually extended to more complex realizations such as printed structures for dual-band operation, taking advantage of its first resonance and wide bandwidth antennas containing this procedure.

References

1. Uda, S.: Wireless beam of short electric waves. J. IEE 273–282 (1926)
2. Yagi, H.: Beam transmission of ultra short waves. Proc. IRE **16**, 715–740 (1928)

3. Balanis, C.A.: Antenna Theory: Analysis and Design, 3rd edn. Wiley Inc (2005)
4. Landstorfer, F.: Zur optimalen form von linearantennen. Frequenz **30**, 1 (1976)
5. Reljíc, B.: A novel printed full-wave yagi-uda antenna. Microw. Rev. 10–12 (2008)
6. Mak, A., Rowell, C., Murch, R.: Low cost reconfigurable landstorfer planar antenna array. IEEE Trans. Antennas Propag. **57**, 3051–3061 (2009)
7. Aziz, M.A., Roy, S., Braaten, B.D.: A new printed quasi-landstorfer antenna. IEEE Trans. Antennas Propag. **60**, 2531–2536 (2012)
8. Chen, Y., Wang, S., Shi, S., Jiang, M., Ding, J., Gao, J., Zhai, G.: Landstorfer printed log-periodic dipole array antenna with enhanced stable high gain for 5g communication. IEEE Trans. Antennas Propag. **69**, 8407–8414 (2021)
9. Kraus, J.D., Marhefka, R.J.: Antennas for all Applications, 3rd edn. McGraw- Hill (2003)
10. Espín, G., Romero, B., Carrera-Suarez, L.-F.: Diseño, simulación e imple-mentacion de las antenas de transmisión, recepción y radioenlace de una estación de radioaficionado (2020)
11. Páez-Rueda, C.-I., Bustamante-Miller, R.: Diseño óptimo de antenas yagui-uda. Revista de Ingeniería 10–17 (2002)
12. Kataja, J., Nikoskinen, K.: The parametric optimization of wire dipole antennas. IEEE Trans. Antennas Propag. **59**, 350–356 (2011)
13. Kataja, J.: On shape optimization of wire dipole antennas. IEEE **8**, 269–271 (2010)
14. Jacobson, A.: Smoothly parameterized ease curve (2013). [Online]. https://www.alecjacobson.com/weblog/?p=3255
15. Ruiz, C.P.: Curvas paralelas. Miscelánea Matemática **21**, 29–52 (1994)
16. Landstorfer, F.: A new type of directional antenna in 1976 Antennas and Propagation Society International Symposium, vol. 14, pp. 169–172. IEEE (1976)

Solving the Generation and Transmission Capacity Expansion Planning Framework Modeling the AC Power Flow Problem and Incorporating Reactive Compensation Devices

Nicolás Solís-Llanos, Victor H. Hinojosa, and Diego Ortiz-Villalba

Abstract The capacity expansion planning (CEP) problem requires an adequate representation of the physical and electrical constraints faced by Independent System Operators (ISO) in practice. Nowadays, researchers employ simplified linear DC-based models to solve the transmission power flow problem. The main challenges for current planning methodologies involve the application of appropriate models to accurately represent the electrical network and incorporate all technical operational variables (voltage and reactive power flow magnitudes). Therefore, this study introduces a detailed framework to model generation and transmission capacity expansion planning considering the alternating current (AC) power flow problem, so that power system engineers can ensure secure and safe operational conditions for long-term scenarios. The study reports the results and effects of nonlinear modeling, contrasting the accomplished model with the traditional DC-based planning approach. Finally, extensive computational experiments are conducted to assess the performance of the capacity planning problem. The solver simulation time and the number of variables and constraints are also included to demonstrate the mathematical complexity.

Keywords Generation and transmission expansion planning · Reactive power planning · AC power flow problem · Mixed-integer nonlinear models

N. Solís-Llanos · V. H. Hinojosa (✉)
Universidad Técnica Federico Santa María, Valparaíso, Chile
e-mail: victor.hinojosa@usm.cl

N. Solís-Llanos
e-mail: nicolas.solis@alumnos.usm.cl

D. Ortiz-Villalba
Universidad de las Fuerzas Armadas ESPE, Sangolquí 171103, Ecuador

G. F. Olmedo Cifuentes et al. (eds.), *Emerging Research in Intelligent Systems*, Lecture Notes in Networks and Systems 902, https://doi.org/10.1007/978-3-031-52255-0_6

Nomenclature

Indices

g	Index of generation units.
v	Index of virtual generation units.
i,j,k	Index of electrical buses.
ij,jk	Index of transmission lines.
r	Index of reactive compensation units.
d	Index of demand.
s,t	Index of blocks and years steps over the planning horizon.

Sets

Ω_N	Set of electrical buses.
Ω_S	Set of blocks in the planning horizon.
Ω_T	Set of years in the planning horizon.
Ω_j	Set of components connected to bus j.
Ω^G/Ω^{G^+}	Set of existing and candidate generator units.
Ω^V	Set of loss of load unit (virtual generation).
Ω^L/Ω^{L^+}	Set of existing and candidate transmission lines.
Ω^{R^+}	Set of candidate reactive compensation unit.
Γt	Discount economic factor at planning stage t.

Parameters

I_g	Investment cost of candidate generator g.
$O\&M_g$	Operation and maintenance fixed cost of generator g.
F_g	Fuel cost of generator g.
$VoLL$	Value of load loss.
P_g^m/P_g^M	Minimum/maximum active power generation of unit g.
Q_g^m/Q_g^M	Minimum/maximum reactive power generation of unit g.
I_{ij}	Investment cost of candidate line ij.
$O\&M_{ij}$	Operation and maintenance fixed cost of line ij.
M	A large enough number (big-M).
G_{ij}/B_{ij}	Conductance/susceptance of line ij.
B_{ij}^s	Shunt susceptance of line ij.
Fs_{ij}^M	Maximum power capacity (thermal limit) of line ij.
I_r	Investment cost of candidate reactive compensations r.

Q_r^m/Q_r^M	Minimum/maximum reactive power generation of unit r.
P_{st}^d	Active power demand at block s in year t.
Q_{st}^d	Reactive power demand at block s in year t.
$\overline{P}_t^d$	Maximum active power demand in year t.
RM	Capacity reserve margin criterion.
G_j^s/B_j^s	Shunt conductance/susceptance of bus j.
H_s	Hourly demand duration at block s.
V_j^m/V_j^M	Minimum and maximum voltage magnitude of bus j.
Z	Interest money rate.

Variables

$x_{g,t}$	Binary variable for candidate generator selection in year t.
$x_{ij,t}$	Binary variable for candidate transmission line selection in year t.
$x_{r,t}$	Binary variable for candidate reactive compensation selection in year t.
$p_{g,s,t}/q_{g,s,t}$	Active/reactive power generation of unit g at block s in year t.
$fp_{ij,s,t}/fq_{ij,s,t}$	Active/reactive power of transmission line ij at block s in year t.
$q_{r,s,t}$	Reactive power generation of compensation unit at block s in year t.
$v_{j,s,t}$	Voltage magnitude of bus j at block s in year t.
$\phi_{ij,s,t}$	Phase angle difference across line ij at block s in year t.

1 Introduction

The capacity expansion planning (CEP) problem solves the following issues—(1) how much generation and transmission capacity requires the power system, (2) when and what type of new power units and new transmission elements need the power system, and (3) where locate these new elements (generation units, transmission transformer, and lines) to adequately supply the long-term energy of the customers [1, 2].

In this study, the CEP problem is formulated as a cost-minimization problem considering both generation and transmission investment candidates [2]. Power system researchers recognize that the CEP problem achieves more safe and financial benefits [3]. However, Generation Expansion Planning (GEP) [4, 5] and Transmission Expansion Planning (TEP) [6, 7] problems are carried out as two separate optimization problems because of the large-scale mathematical complexity. Notice that the operational problem has been traditionally formulated using a DC-based power flow (DCOPF) problem [5, 7, 8].

Gradually, new challenges such as sustainable growth, penetration of renewable energies, or development of new technologies, complement the expansion problem, making the planning power system capacity an area extensively studied in the technical literature [9]. Review reference [10] to know more information about CEP issues.

On the one hand, researchers have utilized the classical DC-based approach to model the transmission power flow problem [2, 11–13]. Recently, efforts have led to the inclusion of security constraints in the optimization problem [5, 8, 14, 15]. However, those linear approaches cannot fully capture the correct impact of the nonlinear power flow problem. Therefore, the ISO or the Regulator company could require complementary actions, like dispatch or new investment decisions, to achieve convergence and a feasible operational AC condition as well as power quality (voltage profile) [16]. On the other hand, in the technical literature, researchers have considered the reactive power flow problem and the voltage profile using several approaches. In [17, 18], the authors present a linear approach based on the Taylor methodology and disjunctive techniques to solve the planning problem. In [19, 20], authors apply a second-order cone to carry out a planning problem achieving a mixed-integer second-order cone programming problem (MISOCP). In [21], a branch and bound algorithm based on the semidefinite programming relaxation is applied to solve the AC planning problem. It is worth pointing out that these studies do not include the reactive planning problem.

Using the AC nonlinear power flow problem is expected to yield a real power system condition [16]. Notwithstanding, modeling the optimization problem as mixed-integer nonlinear programming (MINLP) poses a complex challenge due to its large-scale, mixed-integer, nonlinear and non-convex characteristics. By reviewing the technical literature, it could be mentioned that few capacity planning studies have been published using nonlinear formulations [16, 22]. Therefore, it is considered that developing new planning methodologies is still an open issue in the power system planning field.

Based on our best technical knowledge, this is the first time authors have integrated the CEP problem modeling nonlinear AC constraints and incorporating reactive power devices as planning options. Additionally, it would be interesting to conduct several analyses to determine the performance of the AC-based CEP methodology solving different-scale power systems. This study compares simulation results using the classical DC-based formulation to demonstrate performance and convergence conditions.

The remainder of the paper is organized as follows. Section 2 describes the proposed CEP AC formulation. Numerical results are provided and discussed in Sect. 3. Finally, Sect. 4 details the main conclusions.

2 Mathematical AC Formulation

The optimization problem is formulated as a multi-stage expansion planning problem. From a central planner point of view [2], the CEP problem can be formulated as the minimization of generation and transmission investment cost as well as operational costs. In addition, shunt devices are modeled as electrical generators limited to generating continuously reactive power [23].

2.1 Objective Function: FO

The CEP problem minimizes the total cost associated with investment options and operational scenarios as shown the objective function given in (1). This study considers generation $x_{g,t}$, transmission $x_{ij,t}$, and reactive investment shunt devices $x_{r,t}$, where binary variables are used instead of integer variables for investment decisions.

$$\sum_{t\in\Omega_T}\sum_{s\in\Omega_S}\Gamma_t\left[\sum_{g\in\Omega^{G+}} P_g^M\left(I_g x_{g,t} + O\&M_g\sum_{t^*=t}^{\Omega_T} x_{g,t^*}\right) + \sum_{g\in\Omega^{G\cup G+}} F_g H_s p_{g,s,t} + \cdots\right.$$
$$\left.\sum_{ij\in\Omega^{L+}} Fs_{ij}^M\left(I_{ij}x_{ij,t} + O\&M_{ij}\sum_{t^*=t}^{\Omega_T} x_{ij,t^*}\right) + \sum_{r\in\Omega^{R+}} Q_r^M I_r x_{r,t} + VoLL\sum_{v\in\Omega^V} H_s p_{v,s,t}\right] \tag{1}$$

where $\Gamma_t = \frac{1}{(1+Z)^t}$ is the monetary updated factor.

The optimization problem is subject to the following constraints:

2.2 Power Balance and Reserve Margin

Nodal active and reactive balance equations are defined in (2) and (3). Also, the planning reserve margin given in (4) is designed to supply the expected demand in the event of contingency operations.

$$\sum_{g\in\Omega_j^{G\cup G+\cup V}} p_{g,s,t} - G_j^s v_{j,s,t}^2 - P_{j,s,t}^d = \sum_{(i,j)\in\Omega_j^{L\cup L+}} fp_{ij,s,t} - \sum_{(j,k)\in\Omega_j^{L\cup L+}} fp_{jk,s,t}$$
$$\forall j\in\Omega_N, \forall s\in\Omega_S, \forall t\in\Omega_T \tag{2}$$

$$\sum_{g\in\Omega_j^{G\cup G+}} q_{g,s,t} + \sum_{g\in\Omega_j^{R+}} q_{r,s,t} + B_j^s v_{j,s,t}^2 - Q_{j,s,t}^d = \sum_{(i,j)\in\Omega_j^{L\cup L+}} fq_{ij,s,t} - \sum_{(j,k)\in\Omega_j^{L\cup L+}} fq_{jk,s,t}$$
$$\forall j\in\Omega_N, \forall s\in\Omega_S, \forall t\in\Omega_T \tag{3}$$

$$\sum_{g\in\Omega^{G+}} P_g^M x_{g,t} + \sum_{g\in\Omega^G} P_g^M \geq (1+RM)\overline{P}_t^d \qquad \forall t\in\Omega_T \tag{4}$$

2.3 Transmission Network Modeling

Using the AC power flow formulation [24], the study includes nonlinear active and reactive equations for existing transmission elements using (5) and (6). To improve notation, this study uses the following definitions $R_{ij,s,t} = v_{i,s,t} v_{j,s,t} \cos(\phi_{i,s,t} - \phi_{j,s,t})$ and $L_{ij,s,t} = v_{i,s,t} v_{j,s,t} \sin(\phi_{i,s,t} - \phi_{j,s,t})$.

Constraint (7) models the transmission thermal limit, and (8) and (9) limits the active and reactive power flow variables. Constraints (10) and (11) show the power flow problem for candidate transmission (lines and transformers) components. These constraints are based on a disjunctive formulation–Big M [25]. Constraint (12) models the thermal limit, and (13) and (14) limit the power flow variables for candidate elements.

$$fp_{ij,s,t} = G_{ij} v_{i,s,t}^2 + G_{ij} R_{ij,s,t} - B_{ij} L_{ij,s,t} \quad \forall ij \in \Omega^L, \forall s \in \Omega_S, \forall t \in \Omega_T \tag{5}$$

$$fq_{ij,s,t} = B_{ij}^s v_{i,s,t}^2 - B_{ij} v_{i,s,t}^2 + B_{ij} R_{ij,s,t} - G_{ij} L_{ij,s,t} \quad \forall ij \in \Omega^L, \forall s \in \Omega_S, \forall t \in \Omega_T \tag{6}$$

$$fp_{ij,s,t}^2 + fq_{ij,s,t}^2 \le \left(Fs_{ij}^M\right)^2 \quad \forall ij \in \Omega^L, \forall s \in \Omega_S, \forall t \in \Omega_T \tag{7}$$

$$\left|fp_{ij,s,t}\right| \le Fs_{ij}^M \quad \forall ij \in \Omega^L, \forall s \in \Omega_S, \forall t \in \Omega_T \tag{8}$$

$$\left|fq_{ij,s,t}\right| \le Fs_{ij}^M \quad \forall ij \in \Omega^L, \forall s \in \Omega_S, \forall t \in \Omega_T \tag{9}$$

$$\left|fp_{ij,s,t} - G_{ij} v_{i,s,t}^2 - G_{ij} R_{ij,s,t} + B_{ij} L_{ij,s,t}\right| \le \left(1 - x_{ij,t}\right) M \quad \forall ij \in \Omega^{L^+}, \forall s \in \Omega_S, \forall t \in \Omega_T \tag{10}$$

$$\left|fq_{ij,s,t} - B_{ij}^s v_{i,s,t}^2 + B_{ij} v_{i,s,t}^2 - B_{ij} R_{ij,s,t} + G_{ij} L_{ij,s,t}\right| \le \left(1 - x_{ij,t}\right) M \quad \forall ij \in \Omega^{L^+}, \forall s \in \Omega_S, \forall t \in \Omega_T \tag{11}$$

$$fp_{ij,s,t}^2 + fq_{ij,s,t}^2 \le \left(Fs_{ij}^M\right)^2 \quad \forall ij \in \Omega^{L^+}, \forall s \in \Omega_S, \forall t \in \Omega_T \tag{12}$$

$$\left|fp_{ij,s,t}\right| \le x_{ij,t} Fs_{ij}^M \quad \forall ij \in \Omega^{L^+}, \forall s \in \Omega_S, \forall t \in \Omega_T \tag{13}$$

$$\left|fq_{ij,s,t}\right| \le x_{ij,t} Fs_{ij}^M \quad \forall ij \in \Omega^{L^+}, \forall s \in \Omega_S, \forall t \in \Omega_T \tag{14}$$

2.4 Power Generation and Unserved Energy Limits

The expression for the lower and upper bounds of existing and candidate generation power units are presented in (15)–(16) and (17)–(18), respectively. The maximum and minimum values for the reactive compensation devices are limited in (19). Unserved energy limits are given in (20).

$$P_g^m \le P_{g,s,t} \le P_g^M \qquad \forall g \in \Omega^G, \forall s \in \Omega_S, \forall t \in \Omega_T \tag{15}$$

$$Q_g^m \le Q_{g,s,t} \le Q_g^M \qquad \forall g \in \Omega^G, \forall s \in \Omega_S, \forall t \in \Omega_T \tag{16}$$

$$P_g^m x_{g,t} \le P_{g,s,t} \le x_{g,t} P_g^M \qquad \forall g \in \Omega^{G^+}, \forall s \in \Omega_S, \forall t \in \Omega_T \tag{17}$$

$$Q_g^m x_{g,t} \le Q_{g,s,t} \le x_{g,t} Q_g^M \qquad \forall g \in \Omega^{G^+}, \forall s \in \Omega_S, \forall t \in \Omega_T \tag{18}$$

$$Q_r^m x_{r,t} \le Q_{r,s,t} \le x_{r,t} Q_r^M \qquad \forall r \in \Omega^{R^+}, \forall s \in \Omega_S, \forall t \in \Omega_T \tag{19}$$

$$0 \le P_{v,s,t} \le P_{v,s,t}^d \qquad \forall v \in \Omega^V, \forall s \in \Omega_S, \forall t \in \Omega_T \tag{20}$$

2.5 Electrical Power System State-Variables

Constraints (21) and (22) limit the maximum and minimum value of voltage magnitude and voltage angle for each electrical bus.

$$V_j^m \le v_{j,s,t} \le v_j^M \qquad \forall j \in \Omega_N, \forall s \in \Omega_S, \forall t \in \Omega_T \tag{21}$$

$$\left|\phi_{j,s,t}\right| \le \frac{\pi}{2} \qquad \forall j \in \Omega_N, \forall s \in \Omega_S, \forall t \in \Omega_T \tag{22}$$

2.6 Investment Constraints

Generation, transmission and reactive devices are presented using binary decision variables $x_{g,t}$, $x_{ij,t}$, and $x_{r,t}$, respectively. Constraints (23)–(25) are included in the mathematical formulation to avoid multiple investments over the long-term planning horizon.

$$\sum_{t \in \Omega^T} x_{g,t} \le 1 \qquad \forall g \in \Omega^{G^+}, \forall t \in \Omega_T \tag{23}$$

$$\sum_{t \in \Omega^T} x_{ij,t} \leq 1 \quad \forall ij \in \Omega^{L^+}, \forall t \in \Omega_T \tag{24}$$

$$\sum_{t \in \Omega^T} x_{r,t} \leq 1 \quad \forall r \in \Omega^{R^+}, \forall t \in \Omega_T \tag{25}$$

It should be noted that the nonlinear component is provided by the description of AC power flows. The rest of the problem can be correctly modeled in a linear form. The DC-based power flow methodology simplifies the introduced model, where Equations (5) and (6) could be linearized. More details about the DC-based formulation could be reviewed in [2].

3 Simulation Results

This study compares the achieved AC-based planning methodology (AC^R) with two alternative CEP formulations. In the first formulation (AC), the investment options in reactive power shunt devices are not allowed. This is accomplished to understand the technical and financial benefits of co-optimizing active and reactive power. The second formulation (DC) is the classical DC-based methodology [2]. The first (MINLP problem) and second (MILP problem) optimization problems are included to compare costs, feasibility, and simulation time (performance). To assess the operational feasibility of the investment plan solution, power system investments obtained by each methodology are verified through an AC power flow simulation for each time step. The three mathematical formulations are written in Pyomo [26]. In addition, a solver, Bonmin [27], for the MINLP problem and Gurobi [28] for the MILP problem are used in this study.

Three electrical test power systems are used to asses the scalability and determine the performance of the three CEP methodologies. Technical information for these networks (3-bus, IEEE 30-bus, and IEEE 118-bus power system) can be found in MatPower [29].

3.1 3-Bus Power System

The power system is presented in Fig. 1. Transmission and load data has been included in the diagram. A long-term horizon of 11 years is considered to solve the three CEP methodologies. The load demand grows by 4% each year from the second year. The load data is multiplied by 8760 hours (one year) to obtain the yearly energy.

Table 1 shows technical information for new power units (G) and reactive shunt devices (R). Table 2 presents transmission line information for two candidate projects.

The voltage bus range is set from 0.97% to 1.03%. The cost of load shedding (*VoLL*) is 10,000 [$/MWh]. In this case, the convergence solver criterion (GAP) is 1 $\times$ 10^{-9} [pu].

Table 3 presents investment and operational costs for the three planning formulations, including the investment period (year) information in the third column.

Both AC-based optimization problems have similar objective function costs, and generation investments are also made in the first year. The main difference lies in the solver transmission or shunt investment decision. While the AC^R algorithm decides the shunt device R_2 at year 8, the AC algorithm reinforces the transmission line $L_{1,3}$ in the same year.

Notice that the DC-based formulation obtains the lowest FO cost. This effect is caused by reactive power and $Q - v$ simplifications [24]. Therefore, the solver only decides to invest in the power unit G_2 in the third year because power losses are ignored in the model, so power system operators could have problems adequately supplying the customers total energy.

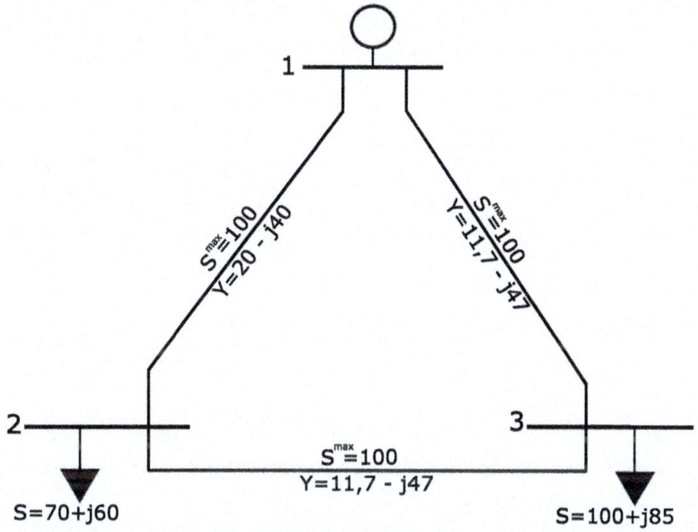

Fig. 1 Topology of the 3-bus power system

Table 1 Candidate generator and reactive shunt data for the 3-bus system

Unit #	Bus #	P_g^m [MW]	P_g^M [MW]	Q_g^m [Mvar]	Q_g^M [Mvar]	F_g [$/MWh]	I_g [$/$P_g^M$]	I_r [$/$Q_r^M$]	$O\&M_g$ [$/$P_g^M$]
G	2	10	120	−100	100	14.36	150,000	–	20,000
R	2	–	–	−100	150	–	–	30,000	–

Table 2 Candidate transmission lines data for the 3-bus system

Line	Fs_{ij}^{M} [MVA]	b_{ij}^{s} [pu]	g_{ij} [pu]	b_{ij} [rad]	I_{ij} [$\$/Fs_{ij}^{M}$]	$O\&M_{ij}$ [$\$/Fs_{ij}^{M}$]
1–2	100	0.04	11.7	−47	100,000	12,000
1–3	100	0.04	11.7	−47	100,000	12,000

Table 3 Optimal cost for the three CEP methodologies applied to the 3-bus system

Case	Unit	Year	Operational	Investment	Total cost
AC^R	G_E	–	107,848,898.01	–	107,848,898.01
	G_2	1	86,983,513.31	35,146,961.05	122,130,474.36
	$L_{1,2}$	–	–	–	–
	$L_{1,3}$	–	–	–	–
	R_2	8	–	2,309,211.53	2,309,211.53
	$\sum$		194,832,411.32	37,456,172.58	**232,288,583.91**
AC	G_E	–	86,887,079.85	–	86,887,079.85
	G_2	1	107,484,898.01	35,146,961.05	142,631,859.06
	$L_{1,2}$	–	–	–	–
	$L_{1,3}$	8	–	7,594,740.15	7,594,740.15
	$\sum$		194,371,977.86	42,741,701.20	**237,477,679.06**
DC	G_E	–	66,555,329.47	–	66,555,329.47
	G_2	4	137,365,219.53	24,105,338.21	161,470,557.74
	$L_{1,2}$	–	–	–	–
	$L_{1,3}$	–	–	–	–
	$\sum$		203,920,549.60	24,105,338.21	**228,025,887.21**

Transmission losses for each operational scenario using power flow simulations are needed to demonstrate the advantages and limitations of each CEP methodology. These results are summarized in Table 4.

The AC^R solution presents an increase in power losses of 23.5% concerning the AC solution. It is worth highlighting that the R_2 device generates reactive power at bus 2, leading to an increase in the current magnitude in the transmission line and, subsequently, transmission losses (Joule effect). The AC solution achieves lower power losses due to the installation of the new transmission line $L_{1,3}$ which reduce the total impedance between bus 1 and 3, consequently decreasing power losses.

It is worth mentioning that the DC planning solution does not converge in the second and third operational periods. Consequently, it is impossible to compare the total power losses. This information validates the incorrect DC-based solution (feasibility criterion) and ratifies the coherence in the investment decisions achieved by both AC-based CEP methodologies.

Table 4 Transmission power losses in the 3-bus system for each CEP methodology

Model year	AC^R [MW]	AC [MW]	DC [MW]
1	0.415	0.415	1.768
2	0.453	0.453	–
3	0.497	0.497	–
4	0.547	0.547	0.547
5	0.602	0.602	0.602
6	0.663	0.663	0.663
7	0.730	0.730	0.730
8	0.745	0.458	0.802
9	0.808	0.458	0.881
10	0.884	0.541	0.965
11	1.062	0.588	1.055
Total	7.406	5.992	–

Table 5 Simulation performance for each CEP methodology applied to the 3-bus system

Case	Constraints	Binary var.	Continuous var.	Solver time [s]
AC^R	1,225	44	374	1,542.15
AC	1,224	33	363	425.12
DC	587	33	198	1.18

Table 5 presents the relationship between variables and constraints, and the solver simulation time achieved by the three CEP formulations (AC^R, AC, and DC).

AC-based methodologies have approximately twice constraints and continuous variables compared to the DC formulation. Besides, the AC^R model has eleven extra binary variables corresponding to the reactive power shunt devices. Between the AC and AC^R nonlinear formulations, the simulation time increases by approximately 362.8% (1,117 s).

The AC^R methodology saves MM$ 5.189, equivalent to 2.19% of the AC objective function. Consequently, it obtains the best technical and economical solution because the solver has better investment options improving the reactive power control and the voltage profile.

3.2 IEEE 30-Bus Power System

This study considers a long-term horizon of 6 years to solve the CEP problem. Starting from the second year, the load demand grows by 20%. To estimate the yearly energy, a load duration curve is proposed, comprising four demand periods:

peak (1825 h), mean (2555 h), rest (1460 h), and valley (2920 h), which are obtained multiplying the original load of the customers by a factor of 1.0, 0.9, 0.8 and 0.5, respectively.

Table 6 shows technical information for new power units (G) and reactive shunt devices (R). Table 7 includes transmission line information.

The voltage bus range is setting from 0.95 to 1.05%, except in buses 2, 13, 22, 23, and 27, where the maximum voltage is 1.1. The cost of the load-shedding is 10,000 [$/MWh], and the GAP is 1×10^{-4} pu.

Table 8 resumes the investment and operational costs for the three formulations.

The total cost of the *DC* model is the lowest, at 50% compared to the AC-based models. This effect is associated with the DC-based simplifications, and the impossibility of capturing the nonlinear *AC* power flow conditions.

Both AC-based methodologies present two transmission investments, where the first is accomplished in the second year. The *AC* model complements the capacity planning problem by adding the generator G_{21}. The AC^R model includes two reactive investments (R_9 and R_{21}), which allows a more economic generation option (G_{11}). On the contrary, the DC-based methodology invests only one transmission line in the third year and one power unit in the fourth year.

Figure 2 shows the voltage profile for the peak period of year six obtained by both AC-based methodologies.

It is worth highlighting that a better voltage profile is obtained when the solver includes two shunt devices, one at bus 9 (R_9) and one at bus 21 (R_{21}).

Therefore, the new power unit G_{21} obtained by the *AC* solution can be substituted efficiently by distributed reactive shunt resources.

The behavior of power losses can be seen in Table 9.

The *DC* methodology presents three scenarios where the investment decisions do not allow convergence in the AC power flow. This fact occurs in the peak and

Table 6 Candidate generation and reactive shunt data for the IEEE 30-bus system

Unit #	Bus #	P_g^m [MW]	P_g^M [MW]	Q_g^m [Mvar]	Q_g^M [Mvar]	F_g [$/MWh]	I_g [$/P_g^M$]	I_r [$/Q_r^M$]	$O\&M_g$ [$/P_g^M$]
G_{11}	11	0	120	−10	20	3.36	100,000	–	20,000
G_{21}	21	0	200	−100	100	7.36	200,000	–	25,000
R_9	9	–	–	−100	200	–	–	30,000	–
R_{21}	21	–	–	−100	200	–	–	30,000	–

Table 7 Candidate transmission data for the IEEE 30-bus system

Unit	$i-j$ # - #	Fs_{ij}^M [MVA]	b_{ij}^s [pu]	g_{ij} [pu]	b_{ij} [rad]	I_{ij} [$/Fs_{ij}^M$]
$L_{6,8}$	6–8	50	0.03	11.7	−47	100,000
$L_{8,22}$	8–22	50	0.00	11.7	−47	100,000

Table 8 Optimal cost for the three CEP methodologies applied to the IEEE 30-bus system

Case	Unit	Year	Operational	Investment	Total cost
AC^R	G_E	–	1,487,137.61	–	1,487,137.61
	G_{11}	4	15,835,434.17	13,948,376.6	29,783,810.77
	G_{21}	–	–	–	–
	$L_{6,8}$	6	–	3,104,606.61	3,104,606.61
	$L_{8,22}$	2	–	4,545,454.54	4.545,454.54
	R_9	6	–	3,725,527.94	3,725,527.94
	R_{21}	6	–	3,725,527.94	3,725,527.94
	$\sum$		17,321,662.67	29,049,493.64	**46,371,156.31**
AC	G_E	–	1,053,615.21	–	1.053.615.21
	G_{11}	–	–	–	–
	G_{21}	4	16,824,716.82	40,328,839.93	57.153.553.75
	$L_{6,8}$	6	–	3,104,606.61	3.104.606.61
	$L_{8,22}$	2	–	4,545,454.54	4.545.454.54
	$\sum$		17,878,234.10	47,978,901.09	**65.857.135.19**
DC	G_E	–	1.172.475.28	–	1.172.475.28
	G_{11}	4	15.426.147.46	13.948.376.6	29.374.524.06
	G_{21}	–	–	–	–
	$L_{6,8}$	3	–	4.132.231.4	4.132.231.4
	$L_{8,22}$	–	–	–	–
	$\sum$		18.080.608.01	16.598.622.73	**34.679.230.74**

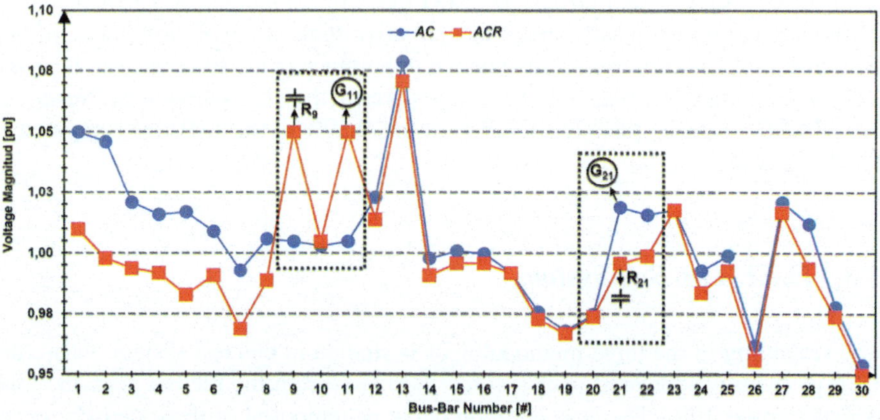

Fig. 2 Voltage profile for the IEEE 30-bus system

Table 9 Transmission power losses in the IEEE 30-bus system for each CEP methodology

Model Year	AC^R [MW]	AC [MW]	DC [MW]
1	11.012	11.016	11.011
2	15.081	15.130	–
3	19.170	19.180	19.316
4	19.676	20.861	20.888
5	22.639	23.577	23.528
6	23.301	22.175	–
Total	110.879	111.939	–

Table 10 Simulation performance for each CEP methodology applied to the IEEE 30-bus system

Case	Constraints	Binary var.	Continuous var.	Solver time [s]	Total cost
AC^R	21,180	36	6,720	211,390	46,371,156.31
AC	21,178	24	6,672	29,570	65,857,135.19
DC	9,889	24	3,696	0.68	34,679,230.74

average period of year two and the peak period of year six. Consequently, the DC-based methodology solution cannot be considered a lower bound for the capacity planning problem, as mentioned in [21].

Table 10 shows the variables and constraints for each planning methodology.

Simulation results verify that the resolution time depends on the problem size (number of variables and constraints) and the programming solver (LP or NLP). In particular, in large-scale mixed-integer algorithms, the solution space of the nonlinear problem acquires transcendental importance in the solution search, so the AC^R methodology increases 7.14 times compared to the AC methodology. Nevertheless, the AC^R capacity planning solution saves 29.59% of the total cost compared to the AC solution.

3.3 IEEE 118-Bus System

The scalability of the three methodologies is studied in this test system, where the CEP problem is addressed over a 4-year horizon. From the second year, the load demand grows 10%. The load duration curve is composed of three periods: peak, average, and valley, and it considers for the peak period a factor of 1.0 and 1825 h, for the mean period a factor of 0.75 and 4015 h, and for the valley period a factor of 0.5 and 2920 h. In addition, seven transmission candidates, eight power unit candidates and five reactive shunt devices are included. The voltage bus range is set from 0.94 to 1.06%. The $VoLL$ is 10,000 [\$/MWh], and the GAP is 3×10^{-2} pu, i.e., 3%.

Table 11 Simulation performance for each CEP methodology applied to the IEEE 118-bus system

Case	Constraints	Binary var.	Continuous var.	Solver time	Total cost
AC^R	35,303	60	11,241	178,085	3,508,099,634.68
AC	35,208	45	11,196	43,309	3,544,830,841.00
DC	18,009	45	6,138	1,153	3,419,304,156.91

Table 11 summarizes the computational performance for the three CEP methodologies.

The simulation time for the AC^R methodology is 4.11 times higher than that of the AC methodology. However, the AC^R capacity planning solution saves 1.05% of the total cost compared to the AC solution, equivalent to MM$36.731.

It is worth mentioning that the DC methodology cannot capture the effects of reactive power distribution and power losses, so the AC power flow does not converge for the peak period at year 3.

4 Conclusions

This study has presented an AC-based CEP methodology to formulate large-scale transmission networks. This study also compared three CEP methodologies (DC, AC, and AC^R). Several analyses have been conducted to determine the effects and insights into solving these planning methodologies with different electrical power systems. The proposed CEP methodology has resulted in more cost-effective investment plans. The outstanding result of this study is the optimal operational condition achieved by the shunt solutions, which can significantly optimize existing and future active and reactive resources. The operating power system feasibility has been demonstrated using AC power flow simulations for each capacity planning solution. From the ISO perspective, the DC-based solution cannot be considered a lower bound of the investment planning solution. It has been demonstrated that certain operational conditions do not achieve convergence due to power losses and $Q - v$ simplifications. The ISO must accomplish a generation re-dispatch in these cases or propose new investment generation, transmission, and shunt devices to ensure safe and secure operational power system conditions. These actions are decoupled from the CEP problem, which generates poor performance, and mainly planning and operating over costs.

Acknowledgements This work is supported by the Universidad Técnica Federico Santa María, Chile, under project USM PI LIR 2022 03.

References

1. Conejo, A.J., Baringo, L., Kazempour, S.J., Siddiqui, A.S.: Investment in electricity generation and transmission, vol 106. Springer International Publishing, Cham Zug, Switzerland (2016)
2. Hinojosa, V.H., Sepúlveda, J.: Solving the stochastic generation and transmission capacity planning problem applied to large-scale power systems using generalized shift-factors. Energies **13**(3), 516 (2020)
3. Tohidi, Y., Olmos, L., Rivier, M., Hesamzadeh, M.R.: Coordination of generation and transmission development through generation transmission charges? a game theoretical approach. IEEE Trans. Power Syst. **32**(2), 1103–1114 (2017)
4. Koltsaklis, N.E., Dagoumas, A.S.: State-of-the-art generation expansion planning: a review. Appl. Energy **230**, 563–589 (2018)
5. Hinojosa, V.H., Velásquez, J.: Stochastic security-constrained generation expansion planning based on linear distribution factors. Electr. Power Syst. Res. **140**, 139–146 (2016)
6. Vahidinasab, V., Tabarzadi, M., Arasteh, H., Alizadeh, M.I., Mohammad Beigi, M., Sheikhzadeh, H.R., Mehran, K., Sepasian, M.S.: Overview of electric energy distribution networks expansion planning. IEEE Access **8**, 34750–34769 (2020)
7. Hinojosa, V.H., Sepúlveda, J.: Solving the stochastic transmission capacity expansion planning problem based on generalized partial transmission distribution factors. In: 2019 IEEE Milan PowerTech, pp. 1–6 (2019)
8. Gutíerrez-Alcaraz, G., Díaz-López, B., Arroyo, J.M., Hinojosa, V.H.: Largescale preventive security-constrained unit commitment considering n−k line outages and transmission losses. IEEE Trans. Power Syst. **37**(3), 2032–2041 (2022)
9. Quintero, J., Zhang, H., Chakhchoukh, Y., Vittal, V., Heydt, G.T.: Next generation transmission expansion planning framework: models, tools, and educational opportunities. IEEE Trans. Power Syst. **29**(4), 1911–1918 (2014)
10. Gacitua, L., Gallegos, P., Henriquez-Auba, R., Lorca, A., Negrete-Pincetic, M., Olivares, D., Valenzuela, A., Wenzel, G.: A comprehensive review on expansion planning: models and tools for energy policy analysis. Renew. Sustain. Energy Rev. **98**, 346–360 (2018)
11. Tohidi, Y., Hesamzadeh, M.R., Regairaz, F.: Sequential coordination of transmission expansion planning with strategic generation investments. IEEE Trans. Power Syst. **32**(4), 2521–2534 (2017)
12. Li, J., Li, Z., Liu, F., Ye, H., Zhang, X., Mei, S., Chang, N.: Robust coordinated transmission and generation expansion planning considering ramping requirements and construction periods. IEEE Trans. Power Syst. **33**(1), 268–280 (2018)
13. Rad, V.Z., Torabi, S.A., Shakouri, H.: Joint electricity generation and transmission expansion planning under integrated gas and power system. Energy **167**, 523–537 (2019)
14. Dom´ınguez, A.H., Escobar, A.H., Gallego, R.A.: An milp model for the static transmission expansion planning problem including hvac/hvdc links, security constraints and power losses with a reduced search space. Electr Power Syst Res **143**, 611–623 (2017)
15. Macedo, L.H., Montes, C.V., Franco, J.F., Rider, M.J., Romero, R.: Milp branch flow model for concurrent ac multi-stage transmission expansion and reactive power planning with security constraints. IET Gener. Transm. Distrib. **10**(12), 3023–3032 (2016)
16. Solís-Llanos, N., Calle, I.A., Hinojosa, V.H.: New approach applied to the generation expansion planning considering an ac modeling. In: 2017 CHILEAN Conference on Electrical, Electronics Engineering, Information and Communication Technologies (CHILECON), pp. 1–7 (2017)
17. Olfatinezhad, N., Vahidinasab, V., Ahmadian, M., Arasteh, H., Aghaei, J., Mehran, K.: Flexible two-stage robust model for moving the transmission and reactive power infrastructures expansion planning towards power system integration of renewables. IET Renew. Power Gener. **14**(11), 1921–1932 (2020)
18. Akbari, T., Bina, M.T.: Approximated milp model for ac transmission expansion planning: global solutions versus local solutions. IET Gener. Transm. Distrib. **10**(7), 1563–1569 (2016)

19. Zhou, Z., He, C., Liu, T., Dong, X., Zhang, K., Dang, D., Chen, B.: Reliabilityconstrained ac power flow-based co-optimization planning of generation and transmission systems with uncertainties. IEEE Access **8**, 194218–194227 (2020)
20. Ghaddar, B., Jabr, R.A.: Power transmission network expansion planning: a semidefinite programming branch-and-bound approach. Eur. J. Oper. Res. **274**(3), 837–844 (2019)
21. Mehrtash, M., Cao, Y.: A new global solver for transmission expansion planning with ac network model. IEEE Trans. Power Syst. 1–1 (2021)
22. Yasasvi, P.N., Mohapatra, A., Srivastava, S.C.: Efficient sensitivity based approach for ac transmission expansion planning. IET Gener. Transm. Distrib. **14**(12), 2378–2388 (2020)
23. Wolgast, T., Ferenz, S., Nieße, A.: Reactive Power Markets: A Review. IEEE Access (2022)
24. Wood, A.J., Wollenberg, B.F.: Power Generation, Operation and Control. John Wiley & Sons (1996)
25. Moreira, A., Pozo, D., Street, A., Sauma, E.: Reliable renewable generation and transmission expansion planning: co-optimizing system's resources for meeting renewable targets. IEEE Trans. on Power Syst. **32**(4), 3246–3257 (2017)
26. Hart, W.E., Laird, C., Watson, J.-P., Woodruff, D.L.: Pyomo–Optimization Modeling in Python, vol. 67. Springer Science & Business Media (2012)
27. Bonami, P., Lee, J.: Bonmin user's manual. Numer. Math. **4**, 1–32 (2007)
28. Gurobi, I.: Optimization, "Gurobi optimizer reference manual" (2015). http://www.gurobi.com (2016)
29. Zimmerman, R.D., Murillo-Sánchez, C.E., Gan, D.: Matpower: a matlab power system simulation package, vol. 1. Manual, Power Systems Engineering Research Center, Ithaca NY, (1997)

Reconstruction of the Boundary Condition of the Convection–Diffusion–Reaction Equation with Automatic Selection of the Step Length

Cristhian Troya, Geovanny Gordillo, and Atal Vivas

Abstract Predicting water quality in shallow bodies of water can become very complex due to the information required by the different existing models. Sometimes, the data coming from the measurement stations is incomplete due to the stations' failures. To overcome this drawback, a method is presented that reconstructs the boundary condition of the one-dimensional convection–diffusion and reaction using the adjoint equation. The approach also incorporates the automatic choice of the necessary step length in the descending gradient. The results demonstrated that the implemented technique allows reconstruction and recovery of the boundary condition and the target, respectively, with little computational effort. The convergence of the results tends to be immediate due to the optimal choice in each iteration of the step length, avoiding the difficult task of doing it by trial and error. The results also showed that the more discontinuous the boundary condition, the more complicated its reconstruction will be.

Keywords Adjoint · Gradient-descent · Step length · Transport equations

1 Introduction

Predicting the surface water quality with conventional computational simulation models sometimes results in extensive and tedious work. To evaluate the water quality, these models require multiple inputs, such as the water temperature, the

C. Troya · G. Gordillo (✉)
Pontificia Universidad Católica del Ecuador, Sede Manabí, Portoviejo, Ecuador
e-mail: bggordillo@espe.edu.ec

C. Troya
e-mail: ctroya3523@pucesm.edu.ec

G. Gordillo · A. Vivas
Departamento de Ciencias de la Vida, Universidad de lsas Fuerzas Armadas ESPE, Sangolquí 171103, Ecuador

G. F. Olmedo Cifuentes et al. (eds.), *Emerging Research in Intelligent Systems*, Lecture Notes in Networks and Systems 902, https://doi.org/10.1007/978-3-031-52255-0_7

water flow, and the biology of the aquatic ecosystem, in addition to other factors like the ambient temperature, wind speed, and roughness coefficient.

The Water Quality Analysis Simulation Program (WASP) is one of them. This is very useful because it allows the evaluation of different state water quality variables [1]. The model has also been used in environmental impact studies to evaluate the impact of infrastructure projects and other developments on water quality. However, the accuracy of these models is countered by the quality of the input information [2]. Different alternatives have been developed to overcome this drawback. For example, Van Griensven et al. [3] presented a methodology for automatically calibrating an integrated water quality model in river basins. Similar works are considered by Afshar et al. [4], where the particle swarm technique of interacting particles found the best possible solution. Machine learning techniques, such as artificial neural networks and genetic programming, are also analyzed by [5, 6] to model and predict the dynamics of algal blooms and chlorophyll.

Autonomous learning techniques have also been used in the prediction of water quality index [7] and variables such as dissolved oxygen (DO), biochemical oxygen demand (BOD), chemical oxygen demand, and other parameters [8–11]. In contrast, the models cited previously have helped in predicting water quality. However, these models do not work well if input data are absent [12].

On the other hand, works like the one presented by Sun [13] simultaneously determined the boundary condition and the physical properties of the participating medium in the heat transfer by radiation conduction through a hybrid optimization technique. Likewise, in [14], they determined the boundary conditions of the state variables of a water quality model at a confluence with a constant step length.

From this point of view, the objective of this work is based on the reconstruction of the boundary condition of a state variable using a target set downstream through a variable step length. This reach length is automatically selected through the golden section method implemented in the gradient descent process in an artificial channel. The following section presents the main features of the method used to solve the given system. The results of the proposed method are presented in Sect. 3. In Sect. 4, the main discussions of the study are highlighted. Finally, the conclusions are specified in Sect. 5.

2 Materials and Methods

This section develops the methodology used in this work. First, the section describes the one-dimensional transport equation. Then, the importance of initial and boundary conditions is presented. After that, the forward problem and the adjoint equation are considered and analyzed as the main characteristics to implement the method. Finally, this section ends by considering the gradient descent method and the technique used to find the step length automatically.

2.1 *Transport Equation*

The transport of contaminants in surface water bodies is governed by three principles: convection, diffusion, and reaction. These terms determine the temporality of the pollutant in the water. The one-dimensional transport equation determines the expression representing this process:

$$\frac{\partial \phi}{\partial t} = \frac{D^2 \phi}{\partial x^2} - u\frac{\partial \phi}{\partial x} - k\phi \tag{1}$$

where ϕ is the solute concentration expressed as mass per unit volume (ML^{-3}), t is the time (T), D is the longitudinal dispersion coefficient ($L^{-2}T^{-1}$), x is the longitudinal coordinate (L), u is the longitudinal flow velocity (LT^{-1}), and k is the first-order decomposition rate (T^{-1}).

2.2 *Initial and Boundary Conditions*

The mathematical solution of (1) requires specific initial and boundary conditions. Consequently, the boundary conditions must be accurately described or established before the numerical solution. Generally, these concentrations are expressed in terms of the amount of solute per unit volume of water. Therefore, selecting the boundary conditions has required considerable attention [15].

On the other hand, the initial condition can be an arbitrary function in which the flow has a constant background concentration. That is, it is the distribution of the amount of solute concentration at an initial moment $t = 0$.

2.3 *Problem Forward*

The first-order upwind method is the numerical method to solve the transport equation forward (1). This technique is used because obtaining an analytical or semi-analytical solution is impossible in some complex cases. The method discretely approximates the partial derivatives in finite volumes through a mesh of points. One of the main features of the upwind plan is its easy implementation. However, its efficiency in solving problems with complex geometries and boundary conditions is prominent. In addition, this method can be easily generalized to higher orders to obtain more accurate approximations. The reader can find more procedure details in [16, 17].

2.4 Backwards Problem

As the present study's objective is to implement a method that reconstructs the boundary condition through the adjoined technique with a variable step length, it becomes necessary to analyze certain vital aspects of the development.

As mentioned, a first-order upwind scheme can be easily modified according to our needs. Consequently, the technique used in this study is the adjoint-backward problem technique. This algorithm has a mathematical approach that allows the calculation of gradients of objective functions that depend on solutions to partial differential equations. The main feature of the method is its use in optimizing nonlinear designs or parameters.

The adjoint method is based on the idea that to calculate the gradient of the objective function, the derivatives of the solution of the partial differential equation concerning the design variables or problem parameters must be calculated. These derivatives are known as direct gradients. Subsequently, the solutions of an adjoint problem are calculated, which in this study is the adjoint transport equation. The solution of the adjoint problem provides the derivatives of the objective function concerning the solution of the partial differential equation. These derivatives are known as adjoint gradients. More details can be found in Gordillo et al. [2] to obtain the adjoint equation.

The adjoint method is considered efficient, accurate, conservative, and stable for calculating gradients in problems involving partial differential equations, as it avoids the need to solve multiple adjoint equations for different design variables or parameters. Another advantage of the adjoint is that it calculates all gradients at once. This approach can benefit high computational cost optimization problems requiring multiple objective function evaluations.

2.5 Objective Function

In this study, the objective function is based on the mean square error defined as:

$$J = \frac{1}{2N_T} \sum_{p=1}^{N_T} \left(\phi_p - \phi_{o,p}\right)^2 \tag{2}$$

being N_T the total number of pairs observed-calculated over time, ϕ_p the calculated concentrations, and ϕ_o the concentrations observed at the measurement point.

2.6 Gradient Descending Method

The equation belonging to the descending gradient is used To minimize (2) as follows:

$$p^{m+1} = p^m + \in^m (\Delta J)^m \tag{3}$$

where m indicates the iteration level, $\in$ the step length, ΔJ is the gradient of J, which is represented by:

$$\Delta J = \frac{\delta J}{\delta p} \tag{4}$$

The primary basis of this method is to iteratively fit the model parameters so that the function that measures the discrepancies between the model predictions and the observed values is minimal. With the gradient found through the adjoint equation, the opposite direction of the gradient is taken to minimize the loss function. Since the gradient is a vector that indicates the maximum slope at each point, and when directed in the opposite direction, it is possible to follow a downward trajectory on the surface of the function, allowing a local minimum to be reached.

This process is repeated iteratively, adjusting the parameters in each iteration based on the gradient of the objective function until reaching a level of convergence. The learning rate that controls the size of the steps taken in the direction of the gradient is calculated in this study based on the golden section method.

2.7 Golden Section Method

This work uses the variable step length in each iteration, achieved by implementing the golden section method. The technique aims to find a function's minimum value in a given interval. The basic idea of the method is constituted in the golden ratio. The algorithm is initialized with an interval $[a, b]$ containing the minimum searched. Subsequently, the interval is divided into two parts so that the ratio between the length of the greater interval and the lesser interval is equal to the golden ratio. Next, the function at the extremes of the two subintervals is evaluated, and the section that contains the minimum is determined. The process is repeated, dividing the new segment into two parts and respecting the golden ratio. The algorithm stops when a desired precision or a maximum number of iterations is reached. The expression that represents the golden section is determined as:

$$x = a + (1 - \Phi)(b - a) \tag{5}$$

where x is the point that divides the interval $[a, b]$ into two parts and respecting the golden ratio, a and b are the endpoints of the initial interval, and Φ is the golden ratio, approximately equal to 1.618.

The point x is computed at each iteration of the algorithm and is used to determine which of the two subintervals should be discarded at the next iteration [18].

3 Results

Two cases are presented in this study to verify the method's efficiency. Each contemplates a pollutant's convection–diffusion and reaction transport equation over a rectangular section channel.

3.1 Case 1, Instantaneous Discharge of a Contaminant

Bodies of water generally carry different types of pollutants. This physical process can be replicated with the example presented in Chapra [19].

Over a prismatic channel 8000 m long and with an area of 10 m^2, 5 kg of solute is spilled along the boundary for 5 min. The solute is transported in a flow of 2 m^3/s for 10 h. This case can be evaluated with the following analytical solution:

$$\phi(x, t) = \frac{m_p}{\sqrt{4\pi E t}} e^{-\frac{(x-ut)^2}{4Et} - kt} \tag{6}$$

where m_p is the total mass of the particles normalized to the area of the cross-section.

On the other hand, the optimization process should consider the following: (a) the case is solved numerically with all the known variables. A position is previously set downstream at $x = 4000$ m to record a target throughout the simulation. (b) The optimization process begins for this treatment. In the first instance, a forward simulation is executed again, but this time with a boundary condition $x_0^n = 0$. (c) Subsequently, the adjoint equation (backward problem) is solved to calculate the gradient. With this value and the step length given between $[0{:}1 \times 10^5]$, the descending gradient method and the golden section technique calculate the new boundary conditions with which the forward problem will be solved (transport equation).

It should be noted that the golden section method must choose the optimal step length for that iteration within the given interval. (d) the forward problem is calculated again with the new conditions obtained from the adjoint problem, and then the function is compared. If the error is within a tolerance, the process stops. Otherwise, the process is repeated from literal c until the functional value is within the set tolerance or until the number of iterations is equal to that proposed.

Figure 1 shows the evolution of the target and the reconstruction of the boundary condition in Fig. 2 in only two iterations. The method can perfectly recover the target

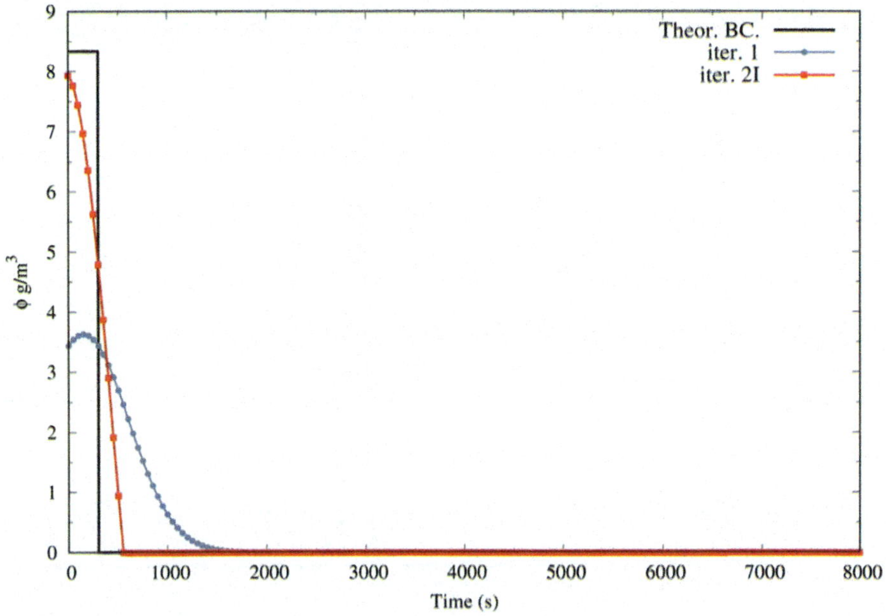

Fig. 1 Case 1. Evolution of the boundary condition reconstruction

at the position $x = 4000$ m (see Fig. 2). The technique is also stable and efficient in each iteration.

Figure 1 represents the reconstruction of the boundary condition. In this case, the boundary condition reconstructed in the second iteration closely approximates the theoretical boundary condition.

On the other hand, Fig. 3 represents the functional and the step length 4. The technique allows the function to reach a value of approximately 0.030387 (see Fig. 3). This value is considered acceptable due to the previously analyzed results.

Regarding the step length ($\in$), it can be seen that the method searches for the best value within the established range (Fig. 4). In the first iteration, the optimal value is 1.097694, while in the second iteration, the ideal weight to reduce the function is 4.959968. This work aims to achieve the objective of automatically choosing the value ($\in$) within an assigned interval.

3.2 *Case 2, Solute Pulse*

It is considered a spillage of a solute over a prismatic channel. The presented case is similar to the previous one, with the difference in the solute pulse at the entrance. The contaminant is spilled over the channel for 1000 s. The flow is steady and uniform, with a cross-sectional area of $A = 10$ m^2. and $Q = 1$ m^3/s. The velocity of the flow

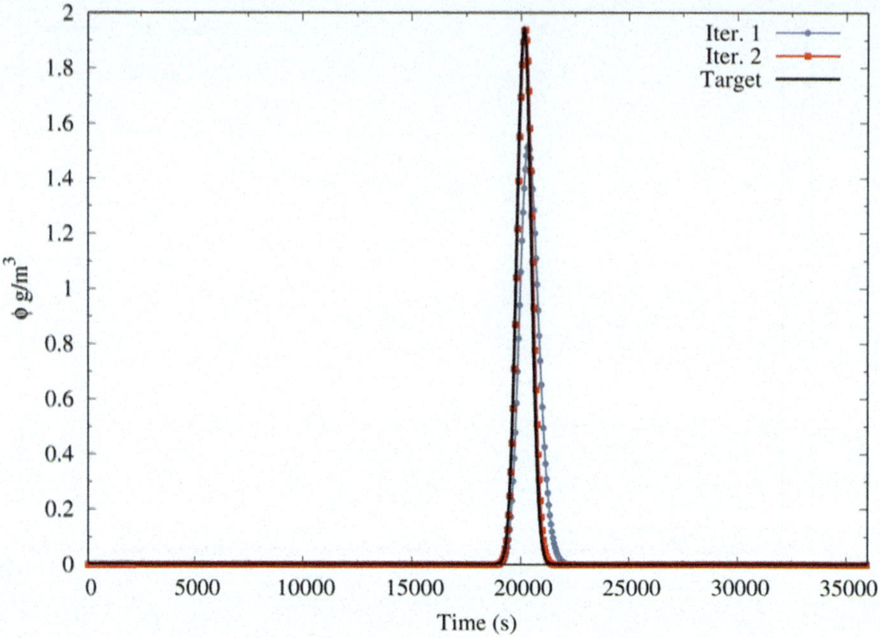

Fig. 2 Case 1. Evolution of the target

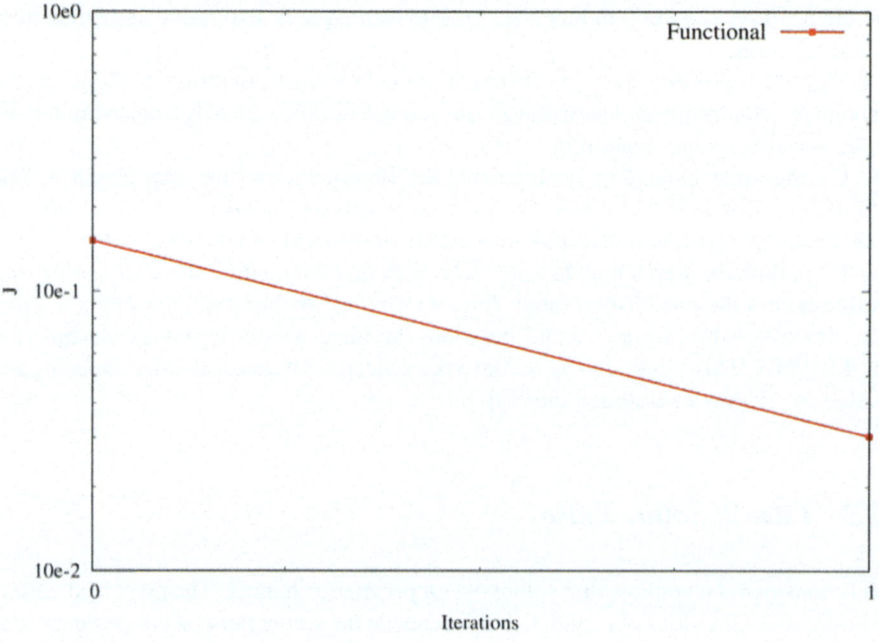

Fig. 3 Case 1. Evolution of the functional

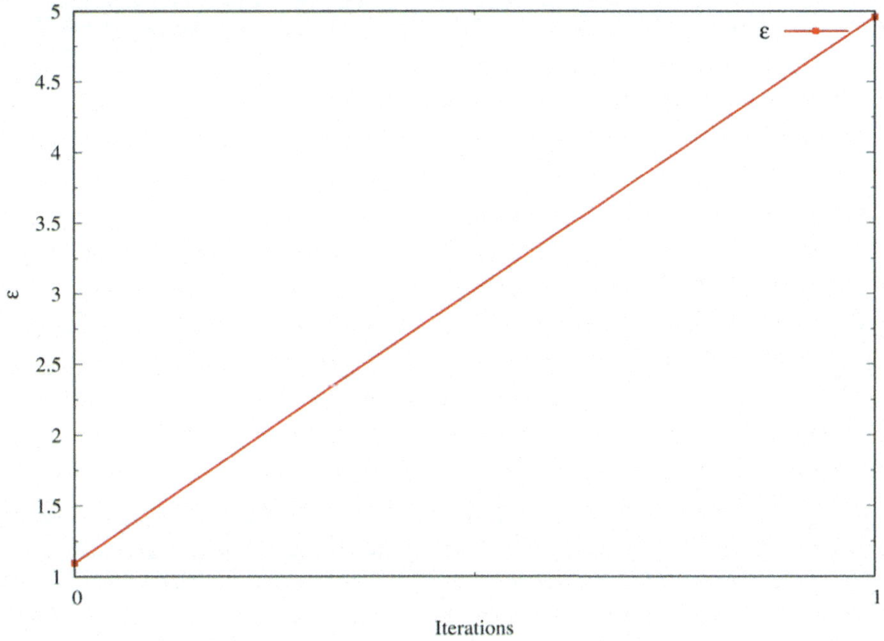

Fig. 4 Evolution of the step length

is 0.1 m/s, and the concentration of the boundary and initial condition is given by the expression (7). The diffusion coefficient in this case is 5 m^2/s, and the rate of reaction $k = 5 \times 10^{-4} s^{-1}$. The details of the case can be followed at Genuchten [20]

$$\phi(x,0) = \phi_E = 0 \quad and \quad \phi(0,t) = \phi_0 = \begin{cases} 1\ kg/m^3 & 0 < t < 1000s \\ 0\ kg/m^3 & t > 1000s \end{cases} \tag{7}$$

With the objective registered at $x = 1000$ m, proceed with the optimization process. The steps for this procedure are the same as in the previous case.

Figure 5 represents the reconstructions at each iteration of the boundary condition. The method can reach only five iterations (red line) of the theoretical boundary condition (black line). Both distributions are very close, making the proposed technique efficient and stable for the case presented.

Figure 6 similarly shows the recovery of the target in each iteration. The method tends to recover its objective. The concentration is above the target in the first iteration (grey line). Therefore, the technique automatically regulates the input closer to the target in the second iteration (blue line). This process is repeated until both the target and the last recovery (red line) are the same or until they are within the set tolerance.

The evolution of the functional begins at a value of 26 and reaches an approximate value of 8.6×10^{-3} (see Fig. 7). The value obtained in the last iteration is considered

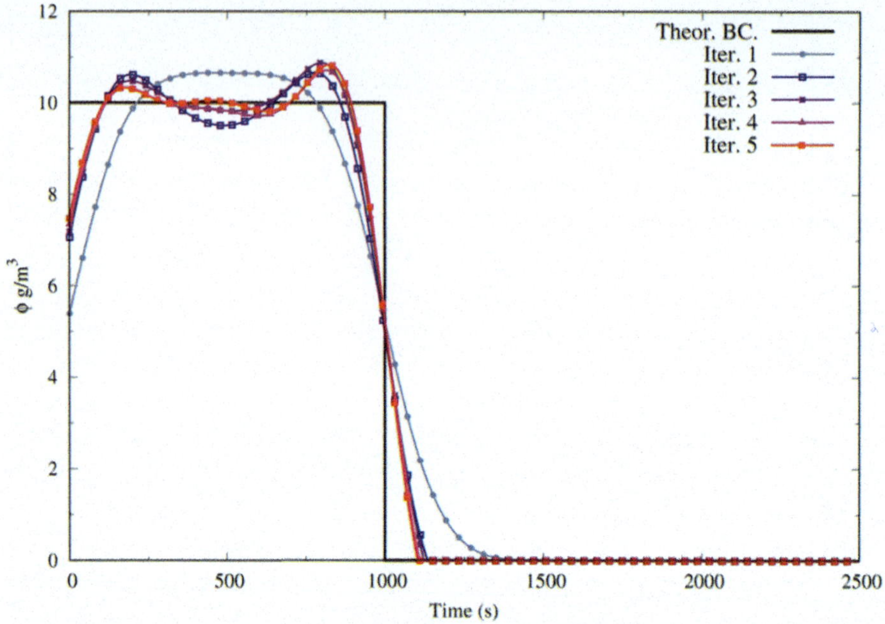

Fig. 5 Case 2. Evolution of the reconstruction of the boundary condition

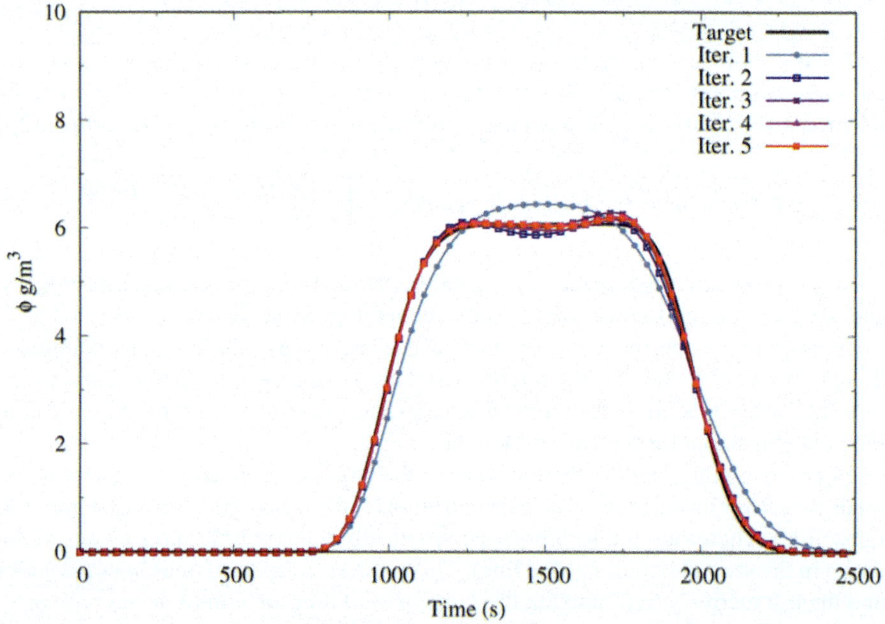

Fig. 6 Case 2. Evolution of the objective

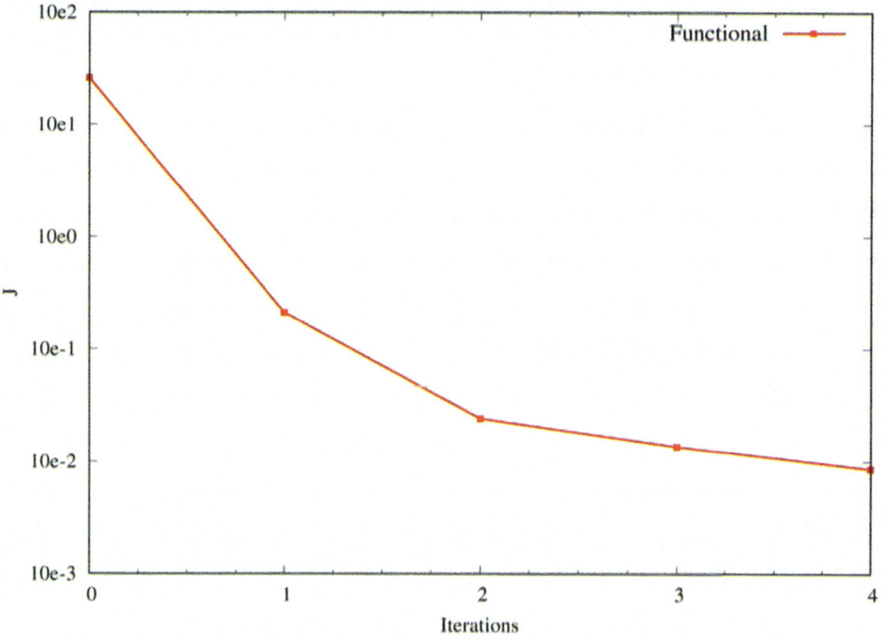

Fig. 7 Case 2. Evolution of the functional

valid because it allows for recovering both temporal (Fig. 5) and spatial distributions (Fig. 9).

Concerning the time step length ($\in$), its value varies in each iteration, achieving the objective outlined in this work. The first value chosen in a range from 0 to 1×10^5 (see Fig. 8), is 1.133, followed by 2.276 and ending in a $\in = 1.858$.

The method can also be checked by retrieving the spatial distribution of the solute. Its evolution is presented in Fig. 9. As before, the recovered distribution (red line) is very similar to the theoretical distribution shown in [20]. It is verifying again that the method retrieves all the distributions of solute transport over the channel.

4 Discussions

Some general considerations of the proposed method are presented below.

The adjoint transport equation can be solved with the same explicit first-order up-wind scheme. The method is easily implemented with almost no variations and simultaneously preserves computational efficiency.

The diffusive characteristic of the upwind method allows smoothing and damping of discontinuities in the solution. This effect could be distinguished in reconstructing the boundary condition of the cases presented (Fig. 2). However, when

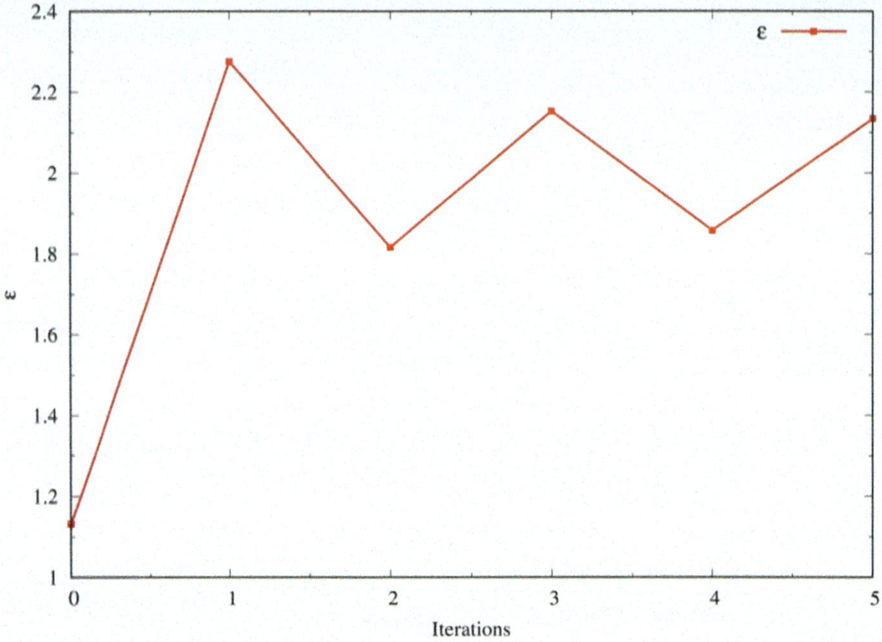

Fig. 8 Spatial evolution of the solute

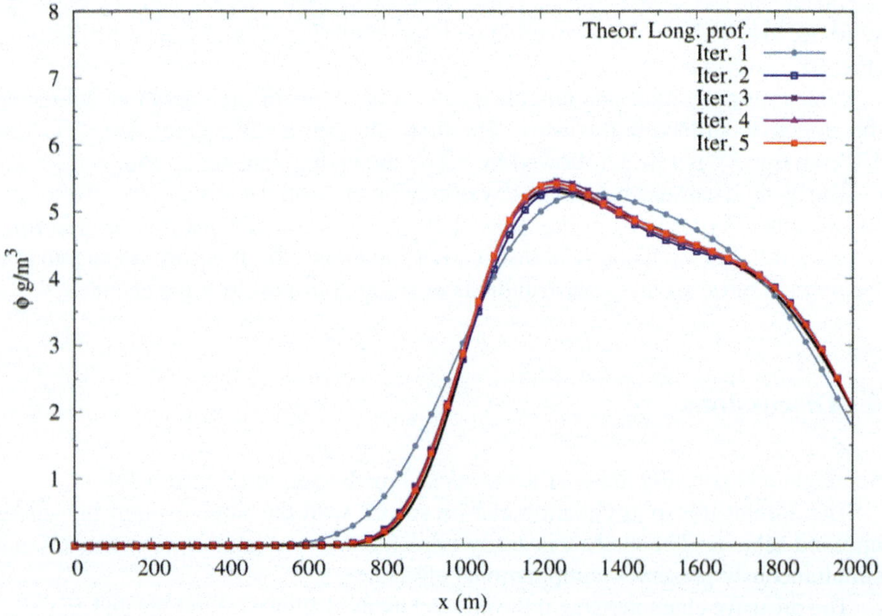

Fig. 9 Spatial evolution of the solute

the solute pulse is different than one step, the reconstructions improve remarkably, as demonstrated in Gordillo [2].

Analyzing the descending gradient method, it is known that it can converge to a local minimum. For this reason, the choice of the step length is automated using the golden section method to avoid resorting to different starting points. This procedure guaranteed the optimal selection of ϵ in each iteration (see Fig. 8). With the appropriate adjustment of ϵ, stable and rapid convergence is achieved in both cases.

Choice of learning rate: The learning rate is a critical hyperparameter in the gradient descent method, as it controls the magnitude of parameter changes at each iteration. If the learning rate is too large, the algorithm may not converge and oscillate unstablely. Conversely, if the learning rate is too small, the algorithm may converge slowly or get stuck in a local minimum. For this reason, it is essential to adjust the learning rate appropriately to achieve stable and fast convergence.

The adjoint equation requires fewer computational resources and takes advantage of most direct code [21]. Another advantage of the adjoint equation is that it allows for calculating all the variables at all the grid points with a single backward simulation. However, a new formulation is needed for each case [22, 23]. In addition, it requires significant memory space since all the information provided by the adjoint equation must be stored in each of the iterations.

5 Conclusions

The present work demonstrated that the optimal selection of the step length of the descending gradient method is achieved by implementing the golden section technique. The process determines the desirable value within a range established at the start of the simulation. Each iteration achieves a stable and fast convergence with the step length chosen by the method. The method can recover both the target and the longitudinal distribution and reconstruct the boundary condition with little computational effort. The method's main disadvantage is determined in the space required to store all the adjoint information in each iteration. The technique showed that the shorter and more discontinuous the boundary condition, the more difficult it will be to rebuild. Finally, using the same number schema requires the same computational mesh when solving the problem forwards (physical problem) and backward (adjunct problem). This feature allows for more efficient, stable, and robust solutions.

References

1. Ambrose, R., Wool, T., Martin, J.: The water quality analysis simulation program, WASP5, part a: model documentation. Environmental Research Laboratory. US Environmental Protection Agency, Athens (1993)
2. Gordillo, G., Morales, M., García, P.: A gradient-descent adjoint method for the reconstruction of boundary conditions in a river flow nitrification model. Environ. Sci. Process Impacts **22**(2), 381–397 (2020)
3. Van Griensven, A., Francos, A., Bauwens, W.: Sensitivity analysis and auto-calibration of an integral dynamic model for river water quality. Water Sci. Technol. **45**(9), 325–332 (2002)
4. Afshar, A., Kazemi, H., Saadatpour, M.: Particle swarm optimization for automatic calibration of large-scale water quality model (ce-qual-w2): application to karkheh reservoir. Iran. Water Resour. Manag. **25**, 2613–2632 (2011)
5. Muttil, N., Chau, K.: Neural network and genetic programming for modeling coastal algal blooms. Int. J. Environ. Pollut. **28**(3–4), 223–238 (2006)
6. Jeong, K., Joo, G., Kim, H., Ha, K., Recknagel, F.: Prediction and elucidation of phytoplankton dynamics in the nakdong river (korea) by means of a recurrent artificial neural network. Ecol. Model. **146**(1–3), 115–129 (2001)
7. Hameed, M., et al.: Application of artificial intelligence (AI) techniques in water quality index prediction: a case study in tropical region, Malaysia. Neural Comput. Appl. **28**, 893–905 (2017)
8. Masrur, A.: Prediction of dissolved oxygen in surma river by biochemical oxygen demand and chemical oxygen demand using the artificial neural networks (ANNs). J. King Saud Univ. Eng. Sci. **29**(2), 151–158 (2017)
9. Gebler, D., Szoszkiewicz, K., Pietruczuk, K.: Modeling of the river ecological status with macrophytes using artificial neural networks. Limnologica **65**, 46–54 (2017)
10. Csábrági, A., et al.: Application of artificial neural networks to the forecasting of dissolved oxygen content in the hungarian section of the river Danube. Ecol. Eng. **100**, 63–72 (2017)
11. Villas, M., Olivera, F., Azevedo, J.: Assessment of the water quality monitoring network of the piabanha river experimental watersheds in Rio de Janeiro, Brazil, using autoassociative neural networks. Environ. Monit. Assess. **189**, 1–15 (2017)
12. Nhantumbo, C., et al.: Applicability of a processes-based model and artificial neural networks to estimate the concentration of major ions in rivers. J. Geochem. Explor. **193**, 32–40 (2018)
13. Sun, S.: Simultaneous reconstruction of thermal boundary condition and physical properties of participating medium. Int. J. Therm. Sci. **163**, 106853 (2021)
14. Gordillo, G., Morales, M., García, P.: Solute transport control at channel junctions using adjoint sensitivity. Mathematics **10**(1), 93 (2022)
15. Batu, T., et al.: Testing closeness of discrete distributions. J. ACM (JACM) **60**(1), 1–25 (2013)
16. Gordillo, G., Morales, M., García, P.: Finite volume model for the simulation of 1D unsteady river flow and water quality based on the wasp. J. Hydroinf. **22**(2), 327–345 (2019)
17. Murillo, J., Navas, A.: A comprehensive explanation and exercise of the source terms in hyperbolic systems using roe type solutions. Application to the 1D-2D shallow water equations. Adv. Water Resour. **98**, 70–96 (2016)
18. Sun, W., Yuan, Y.: Optimization theory and methods: nonlinear programming, vol. 1. Springer Science & Business Media (2006)
19. Chapra, S.: Surface Water-Quality Modeling. Waveland press, Long Grove (2008)
20. Genuchten, M., et al.: Exact analytical solutions for contaminant transport in rivers 1. The equilibrium advection-dispersion equation. J. Hydrol. Hydromech. **61**(2),146–160 (2013)
21. Lacasta, A., et al.: Application of an adjoint-based optimization procedure for the optimal control of internal boundary conditions in the shallow water equations. J. Hydraul. Res. **56**(1), 111–123 (2018)
22. Ding, Y., Wang, S.: Optimal control of open-channel flow using adjoint sensitivity analysis. J. Hydraul. Eng. **132**(11), 1215–1228 (2006)
23. Piasecki, M.: Optimal wasteload allocation procedure for achieving dissolved oxygen water quality objectives. I: Sensitivity analysis. J. Environ. Eng. **130**(11), 1322–1334 (2004)

Assessing Carbon Capture, Transport, and Storage Processes: A Critical Review

Brayan E. Tarazona-Romero, Jenny P. Quizena-Fernandez, Nilson Y. Castillo-Leon, and Omar Lengerke-Pérez

Abstract Climate change is one of the most pressing challenges facing our planet today. To abort this problem, the importance of reducing greenhouse gas emissions, particularly carbon dioxide (CO2), has been considered. Applying carbon capture, transport, and storage systems is vital to the mitigation process in this context. However, evaluating and selecting the most influential and appropriate technologies for a specific environment can be a complex challenge due to the many variables involved. In this regard, a critical review of carbon capture, transport, and storage technologies is presented, applying the formal concept analysis (FCA) method to evaluate the compliance of a series of attributes or characteristics with the technologies to some extent. The review process is subdivided into CO2 capture, CO2 transport, and CO2 storage. For the treatment of the information, the free access software Concept Explorer is implemented to obtain a series of diagrams that simplify the treatment of the information and facilitate the understanding of the results. Finally, among the results, it is highlighted that traditional carbon sequestration technologies have partially overcome some limitations presented in decades: efficient application at a low scale. At the same time, traditional carbon transport methods prevail over emerging alternatives. Finally, traditional technologies stand out notoriously for storage at large-scale levels concerning the new proposals. However, there are emerging applications with potential storage capacity for low-scale systems.

Keywords Carbon capture · Carbon transport · Carbon storage · Formal concept analysis

B. E. Tarazona-Romero (✉) · J. P. Quizena-Fernandez · N. Y. Castillo-Leon · O. Lengerke-Pérez
College of Natural Sciences and Engineering, Unidades Tecnológicas de Santander, Bucara-Manga, Colombia
e-mail: btarazona@correo.uts.edu.com

G. F. Olmedo Cifuentes et al. (eds.), *Emerging Research in Intelligent Systems*, Lecture Notes in Networks and Systems 902, https://doi.org/10.1007/978-3-031-52255-0_8

1 Introduction

Anthropogenic activities have increased global energy consumption, increasing greenhouse gas emissions (GHG) [1]. As a result, climate change and global warming have become severe challenges [2] to face [3]. Globally, a growing number of countries are concerned about GHG emissions. According to statistics, CO2 represents 70% of the greenhouse gases emitted into the atmosphere, and the other 30% contains nitrous oxide [4], methane and some fluorinated gases [5]. Over the last five decades, CO2 emissions have intensified [6], resulting in a series of environmental impacts, such as melting glaciers, rising sea levels, and acidification of the oceans, causing irreparable damage to the ecological environment [7].

Consequently, two options have been considered to effectively address the problem of CO2 emissions: energy transition and CO2 capture. The former is subdivided into three processes: renewable energies, thermal storage, and green hydrogen production, addressing different research fields [8]. On the other hand, the CO2 capture process does not present subdivisions and offers effective results once implemented [9]. Currently, CO2 capture and storage (CCS) technology commercialization efforts are demonstrating [10] their applicability in industrial, commercial, and residential sectors [11].

CCS technology involves separating the CO2 directly from the emission source, transporting it to the storage site [12], depositing it, and keeping it isolated for long periods [13]. This technology can significantly reduce CO2 emissions from the atmosphere on a large scale and eliminate them directly from the source, which is a promising method for mitigating the greenhouse effect [14]. However, CCS has several limitations [15], mainly the high energy consumption to perform the processes of separation, transport, and storage of CO2 [16], which increases the process cost and the risk of leakage from geological storage [17], which directly affects its large-scale development [18].

In general, the document is subdivided into five sections. The current section presents the introduction. Subsequently, the review methodology used is described. The critical review of the technologies is carried out in Sect. 3. The formal analysis of the concepts is presented in Sect. 4, followed by a discussion of the review results. Finally, the most relevant results are presented in Sect. 5

2 Methodology

The evaluation of the information is based on the Formal Concept Analysis (FCA) method, based on the mathematical theory of order and the theory of lattices presented by Rudolf Wille in 1984. FCA is based on the triple relation H = (L, A, B), where L is a set of objects, A is a set of attributes, and B is a binary relation, $P \subseteq L*A$, which relates the object to the feature it fulfills, i.e., the references of L and A are presented through a selection matrix, where attributes occupy the columns and object the rows so that a Boolean value in cell (x, y) represent the fulfillment of an attribute for an object. The formal concept analysis (FCA) method does not involve a specific scale in quantitative measurements or numerical values. Instead, FCA analyzes qualitative relationships between concepts, objects, and attributes [19].

Consequently, this work presents the variables to be analyzed and established using a matrix. To classify the information related to CO_2 capture, transport, and storage, different attributes associated with the final objective of the work are evaluated in terms of compliance or non-compliance. Therefore, the following set of attributes is defined.

1. A1. Technological Maturity (MATURE): The higher the TRL, the higher the level of maturity or traditionality of the technology.
    - Research: TRL 1 to 3
    - Development: TRL 4 to 6
    - Demonstration: TRL 7 to 9
2. A2.Operational simplicity (EAS_USE)
3. A3. Low environmental impact (LOW_ENVIR)
4. A4. Performance at low scale (SMALL)
5. A5. Materials with high availability (AVAIL)
6. A6. Low investment needs (CHEAP)
7. A7. Low energy consumption (PASSIV).

Subsequently, the final visualization of the information is done using the open-access software "Concept Explorer" to analyze the relationships between the attributes and the objects, representing them graphically. It is important to emphasize that FCA is a qualitative method, i.e., it does not define whether an attribute is fulfilled.

3 Review

3.1 Carbon Capture Subsystem

Currently, different technologies for CO_2 capture are identified: industrial separation, post-combustion, pre-combustion, oxy-combustion, chemical loop combustion (CLC), and direct air capture (DAC) systems.

Industrial separation (IS). The industrial sector is perhaps the largest source of CO_2 emissions worldwide. Industrial processes can be adapted to capture CO_2 using meth-ods similar to post, pre, and oxy-combustion [20]. Additionally, it allows taking advantage of process feedstocks to produce natural gas. However, implementations are being developed in the steel, cement, and aluminum production sectors [21]. Its main advantage is its low cost of application in industrial processes with high CO_2 concentration levels in the range of 34.80 to 60.90 US$/t CO_2, as well as its high CO_2

removal efficiency of about 90% of the volume. However, its energy consumption and the CO_2 capture process are higher than the other techniques [22].

Capture post-combustion. Post-combustion is the capture when the combustion gases are released, that is, captured in the final phase of the process. Due to its simplicity and adaptability, this method is widely applied in thermoelectric plants and plants that use fossil fuels. The process captures the CO_2 at the exit of the combustion gases and passes through a separation process using the appropriate technology [23].

Post-combustion capture generally involves implementing separation technologies after capturing CO_2 from the combustion gases produced by fossil fuels [24]. Technologically, it is the most mature technology compared to the other alternatives and can be easily adapted to existing plants. However, it requires high energy demands, incurring high operating costs. In addition, its performance is affected by low CO_2 concentrations and partial pressures. Post-combustion capture achieves removal efficiencies of up to 90% of the volume, and its cost is lower than IS and DAC technology. Nevertheless, its price is higher than that of other alternative technologies. This technology's application costs around 46 to 74 US$/t CO_2 [22].

Capture pre-combustion. Pre-combustion is a process that seeks to produce sinter gas (composed mainly of hydrogen and carbon monoxide). Thus, it is a gasification process. This method captures CO_2 before combustion. Generally, syngases are transformed into CO_2 and H_2. CO_2 is captured using Selexol (a glycol-based solvent), while H_2 is used for electricity production using combined cycles [25].

It is a fully developed technology commercially deployed in some industries (adaptable to existing plants). It can capture high CO_2 concentrations of up to 90% CO_2 volume removal, improving its performance. However, the temperature differentials produced by heat transfer directly affect its performance. In addition, it consumes a lot of energy due to its solvent regeneration process. This incurs a high operating cost of around 34 to 63 US$/t CO_2.

Capture oxy-combustion. Oxy combustion is an alternative to gas combustion, mainly applied in thermal power plants. The method uses pure oxygen instead of air to burn carbon, eliminating nitrogen. The CO_2 is captured after water vapor and other contaminants are removed through the cooling and compression process. The result is a pure CO_2 stream suitable for storage and use. This method does not require a CO_2 capture system. However, an air separation unit is indispensable for oxygen production [26]. It is a mature technology that can operate in small processes. However, the higher the CO_2 concentration, the higher the performance, reaching removal efficiencies of over 90% of the CO_2 volume. The process has a high energy consumption, increasing operating costs to around 52 US$/t CO_2 [27].

Chemical looping combustion (CLC). The chemical loop combustion (CLC) process consists of capturing and separating CO_2 from N_2 and occurs without loss of combustion energy. It is an indirect combustion process. In other words, the fuel is burned without direct contact with the air. Consequently, the combustion products

Table 1 Attribute matrix for the carbon capture technology

	MATURE	EAS_USE	LOW_ENVIR	SMALL	AVAIL	CHEAP	PASSIV
IS						X	
Post	X	X			X	X	
Pre	X	X		X	X		
Oxy	X			X	X		
CLC		X	X		X	X	X
DAC		X	X				

contain only carbon dioxide (CO2) and water vapor. Therefore, high quantities of pure CO2 can be obtained by condensing the water [28].

CLC presents a significant projection to achieve a competitive concept of CO2 capture technology, achieving total cost reduction (around < 59.2 US$/t CO2) in future power systems. Pilot plants have been developed with different fuels in recent years, and experimental studies have found high combustion efficiencies and CO2 capture rates with most fuels. CLC is an emerging CO2 capture technology and differs in many ways from other carbon technologies, with the ability to adapt perfectly to electricity generation equipment and grid power [29].

Finally, CLC is applied at laboratory scales and requires that all technology components be tested at much larger scales, being economically feasible for their current size. In addition, they must improve their ability to withstand high pressures, various materials [25], and beds, among others. Additionally, industrial applications should be sought to evaluate their capacity. However, it can overcome the current limitations, reach a competitive position with mature technologies, and be integrated into pre and post-combustion [29].

Direct air capture (DAC). The DAC capture process is performed directly from the atmosphere and is an emerging technology in the demonstration stage (TRL 7). Commercially, there are two options: liquid solvents and solid sorbents. It is a highly efficient technique, reaching CO2 removal volumes of 93%. However, it is the most energy-consuming technology, costing around 140 to 340 US$/t CO2 [30].

Finally, the attributes described in Sect. 2 are evaluated, and the matrix shown in Table 1 is obtained. The features that fulfill each object to a certain extent are identified in the matrix with an X.

3.2 Transport Subsystem

CO2 transport is essential in the CO2 capture, transportation, and storage chain, connecting the CO2 sources to the storage site by different means. Table 2 presents the four existing means of CO2 transport and the type of phase in which it is transported. CO2 compression capture requires a dense phase at temperatures above >13 °C and pressures above >74 bar. [20]. It is important to note that there is H2O in the captured CO2. Therefore, dehydration must occur before transport to eliminate H2O and avoid corrosion or acid formation in pipes or storage vessels [31].

Pipeline transportation (PT). The most common way of transporting CO2 is by pipeline, which allows for moving large quantities and is positioned as a centralized technology (applicable for decentralized systems) (TRL9). Its primary application has been in power plants. It is performed onshore and offshore: for the first case, the pipelines have more than 2,500 km of implementation, and it is estimated that in the ocean, it reaches depths of up to 2.2 km [22].

Commercially, the pipelines operate between 86 and 152 bar to eliminate condensate (avoiding two phases inside). The high pressures simplify the process and reduce transportation costs. For temperature ranges, it can be transported between 13 °C and 41 °C, preferably above ambient temperature, to reduce the lining or insulation of the pipeline, lowering the transportation cost [32].

Its application can be done on land or at sea. However, transporting pipelines at sea is usually more expensive, about 1.7 times higher than transporting by land (pipelines) and 5.6 times when the pipeline section exceeds 180 km.

On the other hand, the capacity and distance depend on the project. Although pipeline transport is more economical than sea pipelines, its initial implementation cost is around 10.69 to 24.48 US$/t CO2 [20]. However, it allows the movement of large volumes and flows of CO2 [33]. The larger the size of the system, the shorter the payback time, i.e., although it applies to low-scale systems, its cost would be very high [22].

Rail transportation (RT). Rail transport is a mature technology and can be clarified as a TRL 7 to 9, applied with optimal performance and costs to systems handling low amounts of CO2 (4 tons to 100 tons) over distances not exceeding 1609 km [22] [34]. However, for large-scale applications with high volumes of CO2 and long distances (kilometers), its costs are very high [35], and its viability is low compared to ships and pipelines [36].

Truck transportation (TT). CO2 transport by truck is a mature process (TRL 8 to 9) that allows moving small volumes over distances of less than 322 km [22]. Truck transport is the least scalable option if large volumes are to be transported. It is an economical option for small-scale systems compared to the other methods. Trucks handle pressures of 20 bar and temperatures of –20 °C in their storage system [36].

Table 2 Classification of CO2 transport methods

Transport	Phase	Temperature (°C)	Pressure (Bar)
Ship		−40	17
Truck	Liquid	−20	20
Rail		N/A	N/A
Pipeline	Compression-dense	>13	>74

Table 3 Attribute matrix for the carbon Transport technology

	MATURE	EAS_USE	LOW_ENVIR	SMALL	AVAIL	CHEAP	PASSIV
PT	X		X	X			
RT	X		X	X		X	
TT	X		X	X	X	X	
ST	X		X	X		X	

Finally, it is a good option if integrated with sea transport, moving small quantities of CO_2 from the port to the place of use [31].

Ship transportation. Transporting CO_2 by ship is similar to ocean freight shipments of food or beverages, i.e., it is a mature technology (TRL 9) [22]. Its application started around three decades ago and was efficient for handling small volumes. Nowadays, the possibility of moving large-scale volumes over long distances at an economical cost. It could be complemented well with pipeline systems, as they allow mobility at offshore loading and unloading points. However, compared to pipeline transport, the ship provides for the phased supply of different storage or utilization sites [22]. In addition, your investment is around 9.43 to 18.51 US$/t CO_2.

Finally, the attributes described in Sect. 2 are evaluated, and the matrix shown in Table 3 is obtained. The features that fulfill each object to a certain extent are identified in the matrix with an X.

3.3 Storage Subsystem

CO_2 storage is currently performed at pressures above 74 bar in a supercritical or liquefied form in oil and gas fields, rocks, salt formations, coal seams, or deep ocean [20]. Each of the carbon storage technologies is described below.

Storage in depleted oil and gas reservoirs. When oil and gas fields deplete their reserves, they are no longer economically profitable for production. However, they can profitably store CO_2 capture. The CO_2 must be miscible with the oil. Otherwise, the CO_2 will not be distributed homogeneously and will migrate to the top of the reservoir, triggering possible CO_2 leakage into the atmosphere [37]. When storing CO_2 in gas reservoirs, there is usually a temperature drop due to the pressure difference between the pipeline and the reservoir. If the pressure drop is abrupt, it can reach levels of freezing of residual water in the reservoir, saturation due to thermal stress, and even hydrate formation. It is a viable and mature way to store CO_2. However, the cost varies depending on the reservoir capacity. The price would increase for storage in several small reservoirs concerning a single large-capacity reservoir [38].

Storage in Saline Formations. CO_2 storage in salt formations is a widely developed technique (TRL9), onshore and offshore. They are generally located close to

CO2 emission sources, facilitating their integration with CO2 separation and capture processes [39]. The process is simple: CO2 is stored at depths of up to 1000 m in porous rocks saturated with brackish, salt, or brine water [40].

There are several methods for storage in saline formations: (i) Hydrodynamic mechanisms: CO2 is introduced into the porous rock and is trapped in a gas–liquid phase. If not applied in rocks with low porosity, this method may allow the release of CO2 from the reservoir, (ii) residual trapping: to perform this method, the rocks' viscosity-gravity-heterogeneity force relationship must be considered. This method allows storing CO2 in small quantities, (iii) solubility trapping: CO2 is dissolved in an equilibrated formation fluid. However, this process can take years to complete, and (iv) mineral trapping is a prolonged process and depends on the type of rock in which the CO2 is stored since CO2 reacts with the minerals present in the rocks and forms a stable compound (solid).

In general, applying any of the methods described above, the cost of storage exceeds the costs of storage in depleted oil and gas fields.

CO2-EOR storage. CO2-EOR (enhanced oil recovery) is a widely developed technology (TRL 9) with applications in the oil and gas industry for the last five decades [20]. The storage process is simple: CO2 is injected into depleting oil and gas reservoirs to mix it and reduce the viscosity of the crude oil so that it can be released to the production area, i.e., it enhances the extraction process. It is estimated that EOR improves conventional extraction processes by up to 60%. However, the process allows the release of CO2 into the atmosphere, thus affecting the environment. Based on this, procedures have been developed that recapture CO2, reducing the environmental impact. This storage method allows handling of small, medium, and large volumes of CO2 because the storage takes place underground. It is currently the most cost-effective form of CO2 storage [41].

Storage in basaltic and ultramafic rocks. CO2 storage in basaltic and ultramafic rocks is an emerging technology under investigation (TRL 3) [22]. The process consists of mineralizing the CO2 that is injected into the rocks, i.e., the captured CO2 reacts with the selected mineral in a series of operations. This chemical reaction is exothermic, i.e., it releases heat and transforms the CO2 into solid carbonates. The mineralization process can take years or decades. There is a potential in basaltic and ultramafic rocks in various terrestrial and oceanic regions worldwide. However, CO2 injection has drawbacks due to the low permeability of the rocks [42].

One of the biggest problems of this CO2 storage technology is silica formations, a small layer that comes out at the surface of the rocks and reduces their porosity, reducing permeability. The lower the permeability of the rocks, the lower the flow of CO2 that can be injected. Several research processes are being carried out at the laboratory level, and two pilot plants based on mineralization have been implemented on a low scale [43]. However, the results show low injection rates in the rocks, and the technology for monitoring conditions and operational variables of conventional storage methods do not apply to this process. Further technological developments and operating cost analysis are required [44].

Table 4 Attribute matrix for the carbon storage technology

	MATURE	EAS_USE	LOW_ ENVIR	SMALL	AVAIL	CHEAP	PASSIV
B&UR				X	X		
SF	X			X			
DO&GR	X	X		X			
EOR	X	X		X		X	
ECBM sales				X			
Deep water	X	X					

Storage in coal seams through CO2-ECBM. This CO2 storage technology, based on enhanced coal bed methane recovery (ECBM), injects CO2 into coal seams. The coal seams or porous coal absorbs the injected CO2 to allow the release of natural gas into the atmosphere [45]. The storage capacity varies, depending on the coal seam. It is implemented at low, medium, and large scales. A limitation is the permeability of the carbon during the CO2 injection process, decreasing the size of the veins. Currently, the technology is being considered at the research level (TRL 3), and cost analyses performed by some authors reveal that the technology is costly compared to conventional ones [38].

Deep-water storage. The deep-water storage process is carried out in the oceans and can be done in two ways: pipeline and water column injection. The first process uses a pipeline to release CO2 to a depth of 2 km, while the second involves transporting the fluid to an offshore platform and injecting it into a vertical column in the sea. Pipeline drilling is commercial, while transport to the platform requires cargo tankers, increasing costs and limiting marketability [22]. Both ways involve environmental damages that have not been studied in depth. However, there are reports of the pH of the seawater being affected and some marine species being affected. Finally, cost studies should be carried out to overcome the current limitations and bring the technology to a commercial environment [46].

Finally, the attributes described in Sect. 2 are evaluated, and the matrix shown in Table 4 is obtained. The features that fulfill each object to a certain extent are identified in the matrix with an X.

4 Discussion

Figure 1 presents the flow diagram provided by the Concept Explorer software based on the attributes matrix in Table 1, which was obtained by applying the FCA method corresponding to the CO2 capture subsystem. This diagram relates 6 CO2 capture technology options to the seven attributes presented in Sect. 2. The chart has a series of nodes that connect and signify: (i) Blank nodes that act only as branch connectors;

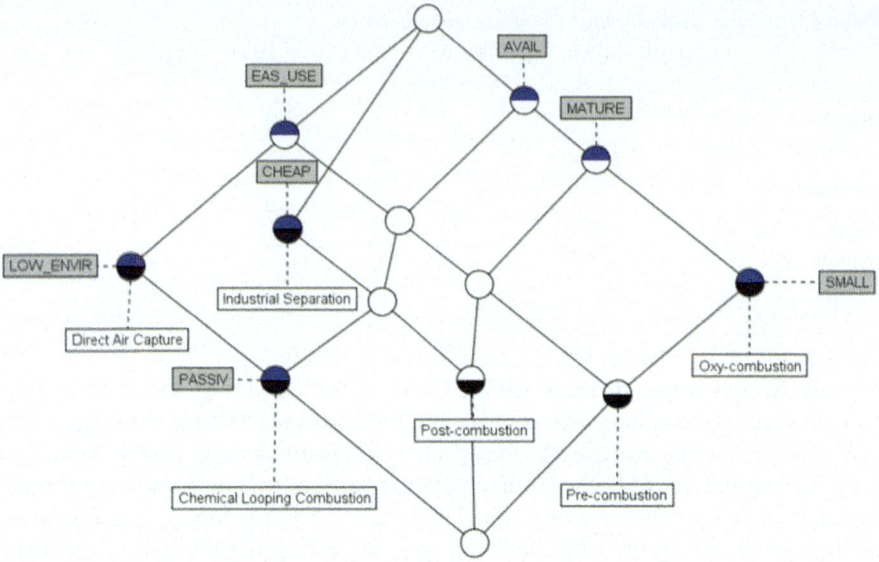

Fig. 1 FCA carbon capture technologies

(ii) Blue-black nodes that act as containers of CO2 capture technologies related to the proposed attributes; (iii) White-black nodes that act as containers of only CO2 capture technologies and (iv) White-Blue, are nodes that act as condemners of only attributes.

With this in mind, the analysis of the FCA diagram presented in Fig. 1 is straightforward. Each technology option is related to the attributes that connect up-wards with it, up to the origin node. Thus, the Chemical Looping Combustion technology represents the technology option that connects with the most significant number of attributes based on the analysis performed in this document since it relates five attributes and connects individually with 1 of them. Specifically, Chemical Looping Combustion is the only option with low energy consumption (PASSIV). Moreover, Chemical Looping Combustion shares additional attributes with Industrial Separation and Post-Combustion, showing low-cost (CHEAP) implementation in centralized systems. In turn, it meets features that express simplicity in the use of the technology (EAC_USE), wide availability of materials and/or equipment (AVAIL), and low environmental impact (LOW_ENVIR), shared with technologies such as direct air capture, post-combustion, pre-combustion and oxy-combustion. On the other side of the diagram, there are matured or traditional technologies (MATURE) with low-scale applications (SMALL).

This analysis shows that traditional (mature) carbon capture technologies have sought to simplify their application at the industrial level and generate low-scale systems, reducing costs. However, they are still developing to meet these objectives. As an alternative to this approach, industrial separation capture technology has been successfully applied in centralized natural gas processes. However, in low-scale systems, its efficiency and energy consumption are low. On the other hand, direct air capture (DAC), despite showing high efficiency in CO2 elimination, lacks application at low concentration levels, generating increased energy consumption costs. Finally, the chemical loop combustion for the current analysis shows that despite being an emerging technology, problems of excess energy consumption have been overcome, reducing operating costs. In addition, the materials used in the process are low-cost.

Analogically, Fig. 2a presents the FCA diagram of the CO2 transport methods, relating the different attributes given in the methodology and the four forms of CO2 transport. Thus, the characteristics of manual use (EAS_USE) and low energy consumption (PASSIV) are not met by any of the objects (CO2 transport methods). The CO2 transport method that connects with the most attributes is transported by truck (5 features). Specifically, it is the only way with available materials (AVAIL) and the most remarkable transport ease anywhere. In turn, it shares the characteristic of being economical on a low scale (CHEAP) with the method of transport by rail and ship. Finally, it has factors such as small-scale applicability (SMALL), technological maturity (MATURE), and low environmental impact (LOW_EVIR) compared to the other three transportation methods.

For this analysis, it follows that the CO2 trucking method is the simplest way to transport small volumes over short distances (kilometers) at a reasonable cost. In addition, it can supply CO2 volume movements of other methods. On the other hand, the process of CO2 transport by ship is very good at low scale and economical. However, its application is limited to coastal areas. On the other hand, traditional transport allows moving high volumes at low costs. However, the pipe stretch implemented at a standard scale is very high and cannot compete with the other

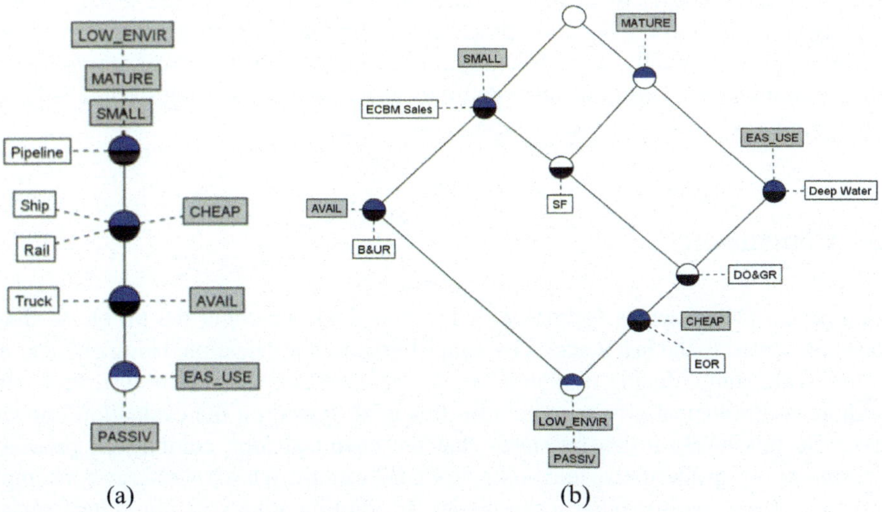

Fig. 2 **a** FCA carbon transports. **b** FCA carbon storage

methods. Finally, none of the CO2 transport methods is easy to use and has low energy consumption. Each one requires a series of processes to move the volumes of CO2, as well as auxiliary equipment that consumes energy to reach the equilibrium point that allows the safe transport of CO2.

On the other hand, Fig. 2b presents the FCA diagram of the CO2 storage technologies related to the attributes described in the methodology section. The diagram is simple to interpret. None of the CO2 storage technologies meet the characteristics of low environmental impact (LOW_ENVIR) and low energy consumption (PASSIV). On the other hand, the EOR technology meets the highest number of attributes (4 attributes), being the only one with economically viable applications (CHEAP). At the same time, it shares application characteristics in small-scale systems (SMALL) with technologies such as B&UR, SF, DO&GR, and ECBM sales. In addition, EOR is a mature technology (MATURE), as are SF, DO&GR, and Deep-Water technologies. Also, EOR can be used simply as DO&GR and Dees Water technologies. Finally, there is the B&UR technology that stands out from the others for being the only one with available materials for its application (AVAIL) in the Project environment.

This analysis affirms that existing CO2 storage technologies have high energy consumption and must be monitored to measure storage volumes and identify losses to take corrective actions. Losses generate release into the atmosphere or marine sources, generating environmental damage. EOR, SF, DO&GR, and Deep-Water technologies have high maturity levels (TRL 9). Emerging technologies, ECBM salts, and B&UR are alternatives with low-scale applicability. However, further technological developments and research projects are needed to overcome the laboratory experimentation phase.

Finally, based on the characteristics outlined in the methodology section, the EOR technology currently offers a more compact process for CO2 storage. However, it must face problems such as Environmental damage due to CO2 leakage, high energy consumption for injection processes (a problem present in all CO2 storage technologies), simplicity in its application, and improvement in the cost-storage ratio, i.e., improving the current balance that indicates that the lower the storage volume applying EOR, the higher the operating cost of the process.

5 Conclusions

Traditional CO2 capture technologies have tried to overcome the problems they have presented in the last decades by simplifying their application, being efficient at a low scale, and reducing implementation and operation costs. However, they still require technological efforts to achieve this goal. Based on the evaluation criteria taken in this work, it can be stated that the chemical loop combustion process. Despite being an alternative technology for CO2 capture, it has low energy consumption levels compared to other technologies. In addition, it presents more inadequate implementation and operation costs.

The traditional methods of CO2 transport are viable on a large scale since they allow for the movement of high volumes of CO2 with tall lengths. However, when scaling it to a small system, its cost becomes relatively high. In turn, the method of transport by ship has been implemented on a low scale. However, the operational costs and application limitations to maritime areas restrict its application level. On the other hand, truck transport adapts to other traditional methods. Truck transport allows small volumes to be moved and is economically viable over short distances. Despite the different individual advantages of the evaluated transport methods, improving the application processes and reducing the energy consumption intensity is necessary to reach an equilibrium point that allows safely moving CO2.

Traditionally, CO2 storage systems are viable for large storage volumes despite the possible environmental impacts of the process if storage sites are not adequately monitored. When looking for recent alternatives, ECBM and B&UR technologies can be applied to small volumes. However, since they are in the process of experimentation and lack technological maturity. The EOR technology presents the most significant benefits based on the results obtained by the methodology used. However, further studies are required to find an economically viable storage system at any scale that guarantees the stored fluid's total hermeticity.

References

1. McLaughlin, H., et al.: Carbon capture utilization and storage in review: sociotechnical implications for a carbon reliant world. Renew. Sustain. Energy Rev. **177**, 113215 (2023). https://doi.org/10.1016/j.rser.2023.113215
2. Tarazona-Romero, B.E., Campos-Celador, A., Maldonado-Muñoz, y Y.A.: Can solar desalination be small and beautiful? A critical review of existing technology under the appropriate technology paradigm. Energy Res Soc Sci **88**, 102510 (2022). https://doi.org/10.1016/j.erss.2022.102510
3. Gizer, S.G., Polat, O., Ram, M.K., Sahiner, y N.: Recent developments in CO2 capture, utilization, related materials, and challenges. Int. J. Energy Res. **46**(12), 16241–16263 (2022). https://doi.org/10.1002/er.8347
4. Lau, H.C., Tsai, y S.C.: A decarbonization roadmap for Taiwan and its energy policy implications. Sustain. Switz. **14**(14) (2022). https://doi.org/10.3390/su14148425
5. Godin, J., Liu, W., Ren, S., Xu, y C.C.: Advances in recovery and utilization of carbon dioxide: a brief review. J. Environ. Chem. Eng. **9**(4), 105644 (2021). https://doi.org/10.1016/j.jece.2021.105644
6. Tan, Z., Zeng, X., Lin, y B.: How do multiple policy incentives influence investors' decisions on biomass cofiring combined with carbon capture and storage retrofit projects for coal-fired power plants? Energy **278**, 127822 (2023). https://doi.org/10.1016/j.energy.2023.127822
7. Fu, L. et al.: Research progress on CO2 capture and utilization technology. J. CO2 Util. **66**, 102260 (2022). https://doi.org/10.1016/j.jcou.2022.102260
8. Li, K., Tan, X., Yan, Y., Jiang, D., Qi, y S.: Directing energy transition toward decarbonization: the China story. Energy **261**, 124934 (2022). https://doi.org/10.1016/j.energy.2022.124934
9. Fawzy, S., Osman, A.I., Doran, J., Rooney, y D.W.: Strategies for mitigation of climate change: a review. Environ. Chem. Lett. **18**(6), 2069–2094 (2020). https://doi.org/10.1007/s10311-020-01059-w
10. Tackie-Otoo, B.N., Haq, y M.B.: A comprehensive review on geostorage of H2 in salt caverns: prospect and research advances. Fuel **356** (2024). https://doi.org/10.1016/j.fuel.2023.129609

11. Khosroabadi, F., Aslani, A., Bekhrad, K., Zolfaghari, y Z.: Analysis of carbon dioxide capturing technologies and their technology developments. Clean. Eng. Technol. **5**, 100279 (2021). https://doi.org/10.1016/j.clet.2021.100279
12. Hua, W., Sha, Y., Zhang, X., Cao, y H.: Research progress of carbon capture and storage (CCS) technology based on the shipping industry. Ocean Eng. **281**, 114929 (2023). https://doi.org/10.1016/j.oceaneng.2023.114929
13. Rogieri Pelissari, M., San Martín Cañas, S., Oliveira Barbosa, M., Gaeta Tassinari, y C.C.: Decarbonizing coal-fired power plants: Carbon capture and storage applied to a thermoelectric complex in Brazil. Results Eng. **19**, 101249 (2023). https://doi.org/10.1016/j.rineng.2023.101249
14. Heuberger, C.F., Staffell, I., Shah, N., Mac Dowell, y N.: Quantifying the value of CCS for the future electricity system. Energy Environ. Sci. **9**(8), 2497–2510 (2016). https://doi.org/10.1039/c6ee01120a
15. Al-Shafi, M., Massarweh, O., Abushaikha, A.S., Bicer, y Y.: A review on underground gas storage systems: natural gas, hydrogen and carbon sequestration. Energy Rep. **9**, 6251–6266 (2023). https://doi.org/10.1016/j.egyr.2023.05.236
16. Shu, D.Y., Deutz, S., Winter, B.A., Baumgärtner, N., Leenders, L., Bardow, y A.: The role of carbon capture and storage to achieve net-zero energy systems: trade-offs between economics and the environment. Renew. Sustain. Energy Rev. **178**, 113246 (2023). https://doi.org/10.1016/j.rser.2023.113246
17. Al-Breiki, M., Bicer, y Y.: Liquified hydrogen vs. liquified renewable methane: evaluating energy consumption and infrastructure for sustainable fuels. Fuel **350** (2023). https://doi.org/10.1016/j.fuel.2023.128779
18. Xiao, H., Mandic, y D.P.: Variational embedding multiscale sample entropy: a tool for complexity analysis of multichannel systems. Entropy **24**(1) (2022). https://doi.org/10.3390/e24010026
19. Tarazona-Romero, B.E., Campos-Celador, A., Lengerke-Perez, O., Cas-tillo-Leon, N.Y., Rincon-Quintero, y A.D.: Appropriate technology for solar ther-mal desalination by concentration applied the humidification-dehumidification method. en Applied Technologies, vol. 1756. In: Botto-Tobar, M., Zambrano Vizuete, M., Montes León, S., Torres-Carrión, P., Durakovic, y B. (eds.) en Communications in Computer and Information Science, vol. 1756 (415–428). Springer Nature Switzerland, Cham (2023). https://doi.org/10.1007/978-3-031-24971-6_30.
20. Kearns, D.D., Liu, H., Consoli, y C.: Technology Readiness and Costs of CCS
21. Gomilšek, R., Čuček, L., Homšak, M., Tan, R.R., Kravanja, y Z.: Carbon emissions constrained energy planning for aluminum products. Energies **13**(11), Art no 11 (2020). https://doi.org/10.3390/en13112753
22. Hong, W.Y.: A techno-economic review on carbon capture, utilisation and storage systems for achieving a net-zero CO2 emissions future. Carbon Capture Sci. Technol. **3**, 100044 (2022). https://doi.org/10.1016/j.ccst.2022.100044
23. Wilcox, J.: Introduction to Carbon Capture. en Carbon Capture. Wilcox, J., (ed.), pp. 1–34. Springer, New York, NY (2012). https://doi.org/10.1007/978-1-4614-2215-0_1
24. Wang, X., Song, y C.: Carbon capture from flue gas and the atmosphere: a perspective. Front. Energy Res., **8** (2020). Accedido: 22 de febrero de 2023. [En línea]. Disponible en: https://www.frontiersin.org/articles/https://doi.org/10.3389/fenrg.2020.560849
25. Mohammad, M., Isaifan, R.J., Weldu, Y.W., Rahman, M.A., Al-Ghamdi, y S.G.: Progress on carbon dioxide capture, storage and utilization. Int J Glob Warm. **20**(2), 124–144 (2020). https://doi.org/10.1504/IJGW.2020.105386
26. Seif El Nasr, A., Nelson, T., Kataria, A., Abu-Zahra, y M.R.M.: Benchmarking of a novel solid sorbent CO2 capture process for NGCC power generation. Int J Greenh Gas Control **42**, 583–592 (2015). https://doi.org/10.1016/j.ijggc.2015.09.014
27. Rubin, E.S., Davison, J.E., Herzog, y H.J.: The cost of CO2 capture and storage. Int. J. Greenh. Gas Control **40**, 378–400 (2015). https://doi.org/10.1016/j.ijggc.2015.05.018

28. Schnellmann, M.A., Heuberger, C.F., Scott, S.A., Dennis, J.S., Mac Dowell, y N.: Quantifying the role and value of chemical looping combustion in future electricity systems via a retrosynthetic approach. Int. J. Greenh. Gas Control **73**, 1–15 (2018). https://doi.org/10.1016/j.ijggc.2018.03.016
29. Lyngfelt, A.: Chemical looping combustion: status and development challenges. Energy Fuels **34**(8), 9077–9093 (2020). https://doi.org/10.1021/acs.energyfuels.0c01454
30. Adánez, J., Abad, y A.: Chemical-looping combustion: status and research needs. Proc. Combust. Inst. **37**(4), 4303–4317 (2019). https://doi.org/10.1016/j.proci.2018.09.002
31. Emodi, N.V., Inekwe, J.N., Zakari, y A.: Transport infrastructure, CO2 emissions, mortality, and life expectancy in the Global South. Transp. Policy **128**, 243–253 (2022). https://doi.org/10.1016/j.tranpol.2022.09.025
32. Li, Y., Wang, W., Chen, Z.-F., Li, y Y.-X.: Integrity assessment of supercritical CO2 transport pipelines. Geoenergy Sci. Eng. **221**, 211355 (2023). https://doi.org/10.1016/j.geoen.2022.211355
33. Martynov, S.B., Talemi, R.H., Brown, S., Mahgerefteh, y H.: Assessment of fracture propagation in pipelines transporting impure CO2 streams. Energy Procedia **114**, 6685–6697 (2017). https://doi.org/10.1016/j.egypro.2017.03.1797
34. IPCC, I.P. on C. Change (IPCC), y I.P. on C.C.W.G. III Carbon Dioxide Capture and Storage: Special Report of the Intergovernmental Panel on Climate Change. Cambridge University Press (2005)
35. Ben Jebli, M., Belloumi, y M.: Investigation of the causal relationships between combustible renewables and waste consumption and CO2 emissions in the case of Tunisian maritime and rail transport. Renew. Sustain. Energy Rev. 71, 820–829 (2017). https://doi.org/10.1016/j.rser.2016.12.108
36. Orchard, K. et al.: The Status and Challenges of CO 2 Shipping Infrastructures. Rochester, NY, 17 de marzo de (2021). https://doi.org/10.2139/ssrn.3820877
37. Twerda, A., Belfroid, S., Neele, y F.: CO2 Injection In Low Pressure Depleted Reservoirs. presentado en Fifth CO2 Geological Storage Workshop, European Association of Geoscientists & Engineers, pp. 1–4 (2018). https://doi.org/10.3997/2214-4609.201802976
38. Bock, B., Rhudy, R., Herzog, H., Klett, M., Davison, J., Simbeck, y D.: Economic evaluation of Co2 storage and sink enhancement options. Tennessee Valley Authority (US) (2003). https://doi.org/10.2172/826435
39. Orlic, B.: Geomechanical effects of CO2 storage in depleted gas reservoirs in the Netherlands: Inferences from feasibility studies and comparison with aquifer storage. J. Rock Mech. Geotech. Eng. **8**(6), 846–859 (2016). https://doi.org/10.1016/j.jrmge.2016.07.003
40. Zhang, D., Song, y J.: Mechanisms for geological carbon sequestration. Procedia IUTAM **10**, 319–327 (2014). https://doi.org/10.1016/j.piutam.2014.01.027
41. Hill, B., Hovorka, S., Melzer, y S.: Geologic carbon storage through enhanced oil recovery. Energy Proced. **37**, 6808–6830 (2013). https://doi.org/10.1016/j.egypro.2013.06.614
42. Kelemen, P., Benson, S.M., Pilorgé, H., Psarras, P., Wilcox, y J.: An overview of the status and challenges of CO2 storage in minerals and geological formations. Front. Clim. **1** (2019). Accedido: 14 de marzo de 2023. [En línea]. Disponible en: https://www.frontiersin.org/articles/https://doi.org/10.3389/fclim.2019.00009
43. Bachu, S.: Carbon dioxide storage capacity in uneconomic coal beds in Alberta, Canada: methodology, potential and site identification. Int. J. Greenh. Gas Control **1**(3), 374–385 (2007). https://doi.org/10.1016/S1750-5836(07)00070-9
44. Xing, T., Zhu, W., Fusseis, F., Lisabeth, y H.: Generating porosity during olivine carbonation via dissolution channels and expansion cracks. Solid Earth **9**(4), 879–896 (2018). https://doi.org/10.5194/se-9-879-2018
45. Bernstein, L., Lee, A., Crookshank, y S.: Carbon dioxide capture and storage: a status report. Clim. Policy **6**(2), 241–246 (2006). https://doi.org/10.1080/14693062.2006.9685598
46. Rae, J.W.B. et al.: CO2 storage and release in the deep Southern Ocean on millennial to centennial timescales. Nature **562**(7728), Art. n.o 7728 (2018). https://doi.org/10.1038/s41586-018-0614-0

Antenna Densification in Non-uniform Linear and Planar Arrays

Fernando Lara, Daniel Altamirano, Jaime Cepeda, and Felipe Grijalva

Abstract Multiple-input multiple-output (MIMO) systems rely on expanding the number of antennas to minimize user interference. The expansion in antennas requires additional physical space for antenna arrays. This article proposes to optimize the positioning of elements in non-uniform linear and non-uniform planar arrays, aiming to minimize the interference reduction factor and reduce the array's physical footprint. The findings indicate a potential reduction of 29% in length for linear arrays and 38% in area for planar arrays.

Keywords Multiple-input multiple-output · Uniform linear array · Uniform planar array · Interference reduction factor

1 Introduction

New telecommunications systems such as 5G use millimeter wave (mmWave) frequency bands [1, 2]. At these frequencies, the attenuation phenomena become critical; therefore, new technologies are used to overcome channel losses, such as Multiple-Input Multiple-Output (MIMO) [3] systems.

MIMO systems, which involve equipping the base station (BS) with a substantial number of antennas and the user equipment (UE) with one or more antennas [1], are

---

F. Lara (✉)
Departamento de Electrónica, Telecomunicaciones y Redes de Información (DETRI), Escuela Politécnica Nacional, Quito 170525, Ecuador
e-mail: marco.lara@epn.edu.ec

D. Altamirano
Departamento de Ingeniería Eléctrica, Electrónica y Telecomunicaciones, Universidad de las Fuerzas Armadas ESPE, Sangolqui 171103, Ecuador

J. Cepeda
Departamento de Energía Eléctrica (DEE), Escuela Politécnica Nacional, Quito 170525, Ecuador

F. Grijalva
Colegio de Ciencias e Ingenierías "El Politécnico", Universidad San Francisco de Quito USFQ, Quito 170157, Ecuador

G. F. Olmedo Cifuentes et al. (eds.), *Emerging Research in Intelligent Systems*, Lecture Notes in Networks and Systems 902, https://doi.org/10.1007/978-3-031-52255-0_9

employed to overcome channel losses in mmWave [4]. The utilization of antenna arrays offers various advantages, including electronic beam control, beamforming [5], and enhanced interference reduction factor (IRF) [6, 7], among others. The commonly employed antenna array types are the uniform linear array (ULA) and the uniform planar array (UPA) [8].

Increasing the number of antennas generates narrower and more directional lobes; however, this requires a larger physical space for the antenna array [1].

This effect is reflected in the IRF, which quantifies the normalized reduction in user interference [6, 7].

In the publication by Peng et al. [9], Non-Uniform Linear Arrays (NULA) are introduced as arrays that enhance channel capacity compared to ULA by positioning the antenna elements at Fekete points. Altamirano et al. [6] present the increase in IRF in UPA by augmenting the number of antennas and reducing the separation between them within a limited physical space. Liu et al. [10] propose non-uniformly spaced rectangular arrays to improve spectral efficiency. Furthermore, in the study by Liu et al. [11], non-uniform planar arrays are employed to estimate the received signal's angle of arrival (AoA). Non-uniform antenna densification by optimizing IRF to improve multi-user capability in MIMO systems is not explored in the presented papers.

This paper proposes non-uniform linear array (NULA) and non-uniform planar array (NUPA) configurations designed to optimize the placement of isotropic elements using heuristic optimization algorithms [12]. The utilization of NULA and NUPA enables the densification of antennas, allowing for an increased number of antennas within a given physical space. This study focuses on a 2D MIMO scenario with line of sight (LoS), explicitly considering the horizontal lobe and azimuthal angle [6, 8].

The remaining sections of this paper are organized as follows. Section 2 overviews the fundamental concepts related to MIMO channels and steering vectors. In Sect. 3, the methodology employed to optimize the position of the elements is described, along with the formulation of the optimization problem. The results related to the reduction of IRF and antenna densification are presented in Sect. 4. Lastly, Sect. 5 presents the conclusions drawn from this study.

2 Theoretical Background

The steering vector is defined with the position and configuration of the array of elements [6, 7, 13]. If an environment is considered where the BS has a ULA of N antennas and a link is established with k users with LoS, in this scenario, the steering vector for each user is specified as:

$$\mathbf{h}_{k(x,y,z)} = \left[1, \exp\left(-i\Phi_{k,(x,y,z)}\right), \cdots, \exp\left(-i\left(N_{x,y,z} - 1\right)\Phi_{k,(x,y,z)}\right)\right] \quad (1)$$

where Φ_k is the phase di fference between two adjacent antennas considering isotropic elements, Φ_k depends on the axis in which the adjoining antenna is aligned. In (2) [7], Φ_k for each axis is presented as follows:

$$\begin{aligned}\Phi_{k,x} &= 2\pi\,\Delta_x \cos(\phi_k)\sin(\theta_k)\\ \Phi_{k,y} &= 2\pi\,\Delta_y \sin(\phi_k)\sin(\theta_k)\\ \Phi_{k,z} &= 2\pi\,\Delta_z \cos(\phi_k)\end{aligned} \tag{2}$$

where Δ is the separation between elements, ϕ_k is the elevation angle, and θ_k is the azimuthal signal arrival angle according to the user's position.

The Kronecker product defines the steering vector for a BS with a UPA array [14]. This is because UPA is a two-dimensional array with two axes. In (3), the steering vector for UPA is introduced.

$$\begin{aligned}\mathbf{h}_{k,xy} &= \mathbf{h}_{k,x} \otimes \mathbf{h}_{k,y}\\ \mathbf{h}_{k,xz} &= \mathbf{h}_{k,x} \otimes \mathbf{h}_{k,z}\\ \mathbf{h}_{k,yz} &= \mathbf{h}_{k,y} \otimes \mathbf{h}_{k,z}\end{aligned} \tag{3}$$

In an antenna array, the direction of the main lobe can be controlled [15] with the array phase. This directional control is depicted in Fig. 1, where the elevation angle is constant ($\phi_k = \pi$), and the azimuth angle can be modified (θ). With nine elements ($N = 9$), it is observed how the main lobe goes to the location of the user k.

The radiation pattern of the MIMO array is defined as:

$$R = \frac{\left|h_j^H h_k\right|}{N} \tag{4}$$

Increasing the number of antennas reduces the azimuthal aperture of the main lobe, which diminishes the IRF. This is depicted in Fig. 2, where the radiation pattern of two UPA is presented, with $N_1 = 9$ and $N_2 = 25$. In Fig. 2, observe that the main lobe of the UPA (25 elements) is narrower concerning UPA (9 elements).

The IRF is defined in [7] as:

$$\gamma = E\{R^2\} \tag{5}$$

In [7], the IRF of a ULA is given by:

$$\gamma_x = \frac{1}{N_x} + \frac{2}{N_x^2}\sum_{m=1}^{N_x-1}(N_x - m)J_0^2(2\pi\,\Delta_x m) \tag{6}$$

And the IRF of UPA is:

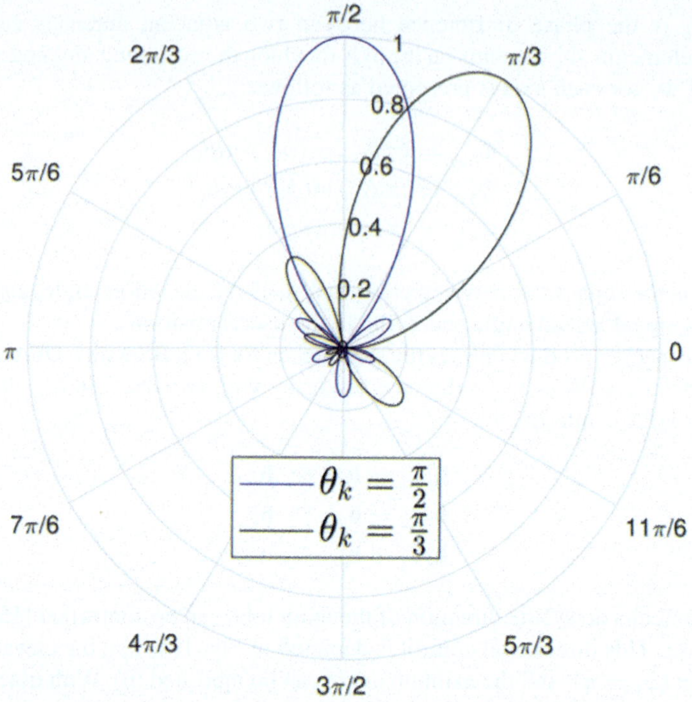

Fig. 1 9-Antennas arrays (UPA) with different azimuth angles

$$\begin{aligned}\gamma_{xy} &= \frac{1}{N_x N_y} + \frac{2N_y}{N_x^2 N_y^2} \sum_{m=1}^{N_x-1} (N_x - m) J_0^2(2\pi \Delta_x m) \\ &+ \frac{2N_x}{N_x^2 N_y^2} \sum_{n=1}^{N_y-1} (N_y - n) J_0^2(2\pi \Delta_y n) \\ &+ \frac{4}{N_x^2 N_y^2} \sum_{m=1}^{N_x-1} \sum_{n=1}^{N_y-1} \left[\begin{array}{l} (N_x - m)(N_y - n) \\ \times J_0\left(2\pi \Delta_x \sqrt{m^2 + n^2}\right) J_0\left(2\pi \Delta_y \sqrt{m^2 + n^2}\right) \end{array} \right] \end{aligned} \tag{7}$$

where $J_0(x)$ corresponds to the Bessel function of order 0, N_x is the antenna's number on the x–axis, and N_y is the antenna's number on the y–axis.

The number of antennas must be increased to reduce the IRF, which increases the array's physical size, a fundamental problem in MIMO systems [16]. On the other hand, if the distance between antennas Δ is reduced, the IRF worsens since there are more antennas in the same physical space. This work proposes heuristic optimization algorithms [17] to find the optimal Δ values.

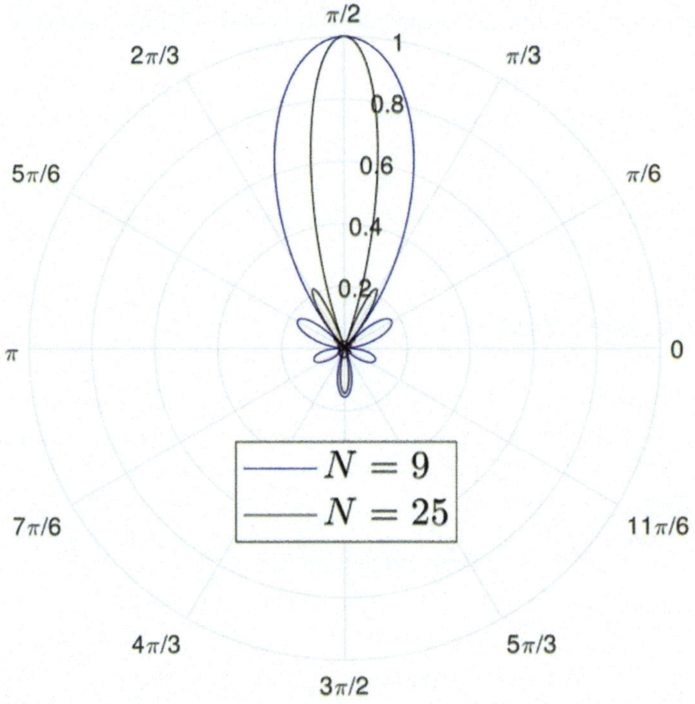

Fig. 2 UPA with a different number of elements

3 Methodology

3.1 ULA Densification

A modification of the Δ_x, i.e., separation between ULA elements, is proposed in this paper. This means generating an array with non-uniformly spaced elements, i.e., NULA. For this, in (6), the $\Delta_x = 0.5\lambda$ value is modified. In ULA Δ, the value is a scalar, and NULA Δ is a vector. By modifying Δ, an array $\mathbf{\Delta}_x = [\Delta_{x1}, \Delta_{x2}, \Delta_{x3}, \ldots, \Delta_{xNx}]^{\mathrm{T}}$ is defined with different separations between each element. This is a NULA array.

With this modification, the optimization $\mathbf{\Delta}_x$ is proposed to increment γ^{-1} and reduce the physical size of the array.

$$\begin{aligned} & Minimize : \gamma_x(\mathbf{\Delta}_x) \\ & subject\ by : \\ & 0.25\lambda < \mathbf{\Delta}_x < 0.5\lambda \end{aligned} \tag{8}$$

Two heuristic algorithms are tested: Particle Swarm Optimization (PSO) [18, 19] and Mean–Variance Mapping Optimization (MVMO) [20].

These optimization processes determine the vector $\mathbf{\Delta}_x$. As part of the study, a test is carried out for different numbers of antennas and changes in the radiation pattern to be verified.

A comparison between a ULA, considering (6), and Monte Carlo simulations with the ULA and NULA array are performed to verify the changes in IRF.

3.2 UPA Densification

For UPA, the optimization of (7) generates values that do not agree with the simulations. Therefore, the optimization function is modified to find the IRF from Monte Carlo simulations [21] and perform an optimization process based on stochastic programming.

$$\begin{gathered} Minimize: \gamma_x(\Delta_x, \Delta_y) \\ subject\ by: \\ 0.25\lambda < \Delta_x < 0.5\lambda \\ 0.25\lambda < \Delta_y < 0.5\lambda \end{gathered} \tag{9}$$

The optimization problem is formulated in (9), where γ_{xy} is associated with (5) and (4). In this context, the vector h_j is generated using uniformly distributed random angles θ ranging from $-\pi$ to π. To minimize the IRF, the optimal values for Δ_x and Δ_y are sought using the MVMO algorithm [20].

In the case of the NUPA array, it is assumed that the distances between elements along the X and Y axes are uniform. However, it is considered that Δ_x is different from Δ_y ($\Delta_x \neq \Delta_y$).

The changes in IRF between ULA and NULA are compared. This comparison is made through theoretical Eqs. (7) and the Montecarlo simulation.

Algorithm 1 shows the minimization of γ_{xy}, where $N_{Montecarlo}$ corresponds to the number of Monte Carlo simulations.

Algorithm 1 Montecarlo-based γ_{xy} function for stochastic programming of NUPA

for $i = 0$ to $N_{Montecarlo}$ do
  $\theta_k \leftarrow U(-\pi, \pi)$ ▷ One number
  $\theta_j \leftarrow U(-\pi, \pi)$ ▷ One number
  $\mathbf{h}_{k1} \leftarrow \exp(-j\mathbf{N}\Phi_x(\phi, \theta_k, \Delta_x))$
  $\mathbf{h}_{k2} \leftarrow \exp(-j\mathbf{N}\Phi_y(\phi, \theta_k, \Delta_x))$
  $\mathbf{h}_k = \mathbf{h}_{k1} \otimes \mathbf{h}_{k2}$
  $\mathbf{h}_{j1} \leftarrow \exp(-j\mathbf{N}\Phi_x(\phi, \theta_j, \Delta_y))$
  $\mathbf{h}_{j2} \leftarrow \exp(-j\mathbf{N}\Phi_y(\phi, \theta_j, \Delta_y))$
  $\mathbf{h}_j = \mathbf{h}_{j1} \otimes \mathbf{h}_{j2}$
  $R[i] = |\mathbf{h}_k \times \mathbf{h}_j^T|^2$
end for

$$\lambda_{xy} = \frac{\overline{R}}{N^2}$$

4 Results and Discussion

4.1 Distance Reduction

In Fig. 3, the position of the antennas in ULA ($\Delta_x = 0.5$) and NULA ($\boldsymbol{\Delta}_x$ obtained with MVMO) is presented. For both scenarios, the configuration of 20 antennas is shown. In Fig. 3, the NULA array has a concentration of antennas at the extremes, and there is a reduction in physical space from the ULA of length 9.5 λ to the NULA of length 7.1λ.

For the example shown in Fig. 3, the ULA has γ^{-1} of 13.1, and the NULA has γ^{-1} of 12.5. These values show an IRF reduction from ULA to NULA, even with the array length reduction. Figure 4 shows how the main lobes of the NULA (20 elements) and ULA (20 elements) are similar.

Table 1 shows the optimization results for a different number of antennas. It is observed that there is an average length reduction of 71%, and γ^{-1} remains equivalent between the ULA and NULA.

The proposed approach allows for a more significant number of antennas to be accommodated within the same length D. Figure 5 shows the radiation pattern of a ULA that can accommodate 15 antennas within D = 7[λ], whereas a NULA that can accommodate 20 antennas within the same length (as shown in Table 1). This increase in antennas leads to an improvement in the γ^{-1} value. Specifically, the ULA array achieves $\gamma^{-1} = 7.2$, while the NUPA array achieves $\gamma^{-1} = 12.5$, representing a 40% increase in γ^{-1}. Furthermore, as depicted in Fig. 5, the NULA array exhibits a narrower main lobe than the ULA array.

In another scenario, the UPA and NUPA arrays are considered, both consisting of 25 antennas, as shown in Fig. 6. To achieve $\Delta_Y = 0.36\lambda$ and $\Delta_X = 0.46\lambda$, the function γ_{xy} is constructed using Algorithm 1 and optimized using MVMO. It is worth noting that the NUPA array exhibits a rectangular shape as it reduces its area

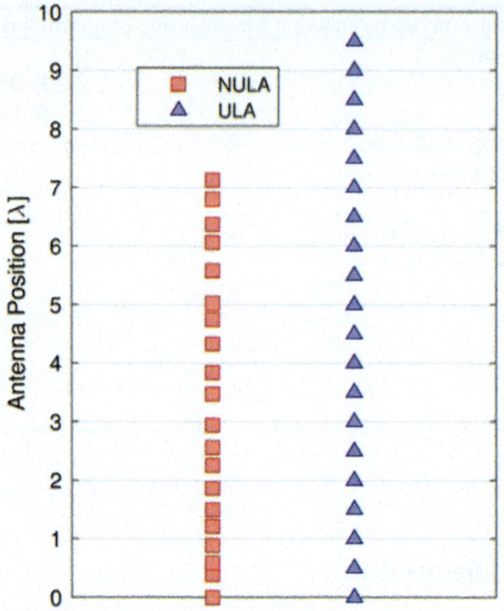

Fig. 3 20-Antennas arrays ($N = 20$): NULA and ULA

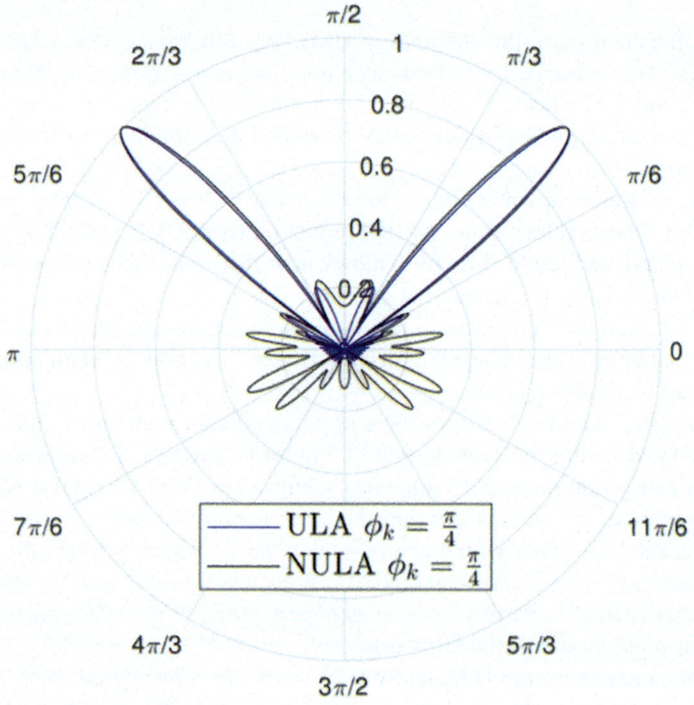

Fig. 4 ULA and NULA radiation pattern

Table 1 ULA and NULA antenna arrays with optimal distance

N	D_{ULA} [λ]	D_{NULA} [λ]	$\overline{\gamma}^{-1}$	$\frac{D_{NULA}}{D_{ULA}}$
20	9.5	7.1	12.5	0.75
50	24.5	17.3	25.6	0.71
80	39.5	28.0	38.5	0.71
100	49.5	35.8	53.5	0.72
150	74.5	53.1	67.1	0.71
200	99.5	71.0	84.0	0.71

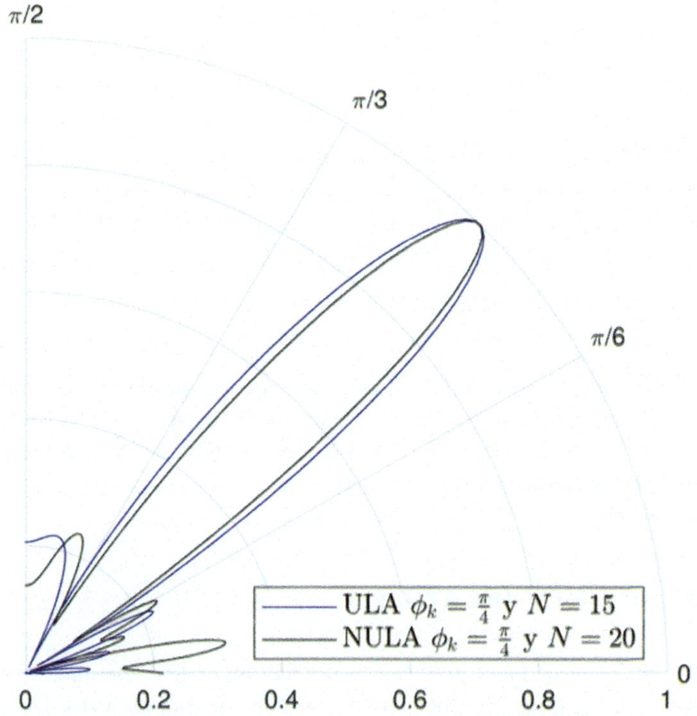

Fig. 5 ULA and NULA main lobe

while accommodating the 25 antennas. Simulation results indicate that the UPA array achieves $\gamma_{xy}^{-1} = 10.4$, while the NUPA array achieves $\gamma_{xy}^{-1} = 11.6$. These results are further supported by the radiation patterns shown in Fig. 7. Thus, it is possible to conclude that densification does not compromise the reduction of the IRF in the NUPA array.

This study employs the area as a metric for the arrays to quantify the reduction in physical space. For instance, the UPA array consisting of 25 elements occupies an area of $4.0\lambda^2$, while the NUPA array occupies a smaller area of $2.7\lambda^2$.

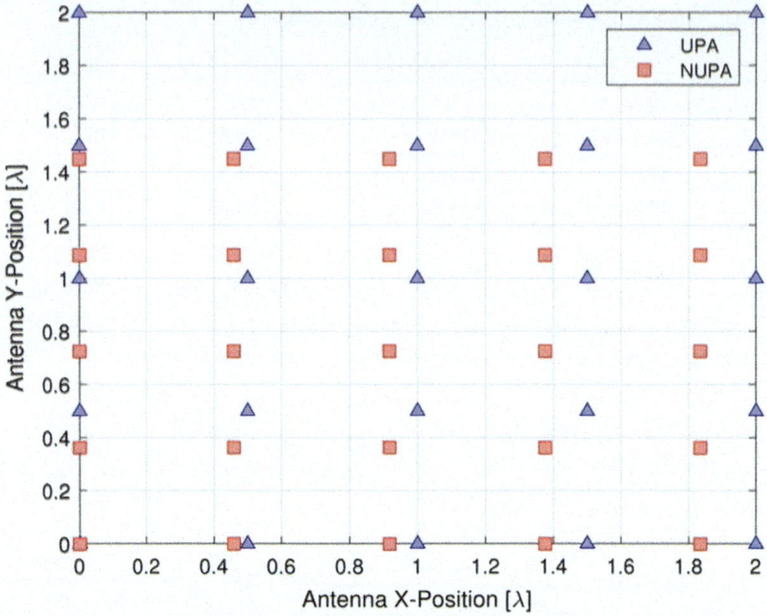

Fig. 6. 25-Antennas arrays ($N = 25$): NUPA and UPA

Table 2 presents the values of γ^{-1} and the corresponding area reduction for the NUPA arrays. The values of Δ_x and Δ_y are determined using the MVMO algorithm for each array with different values of N. Besides, Δ_x remains constant at 0.46λ, and Δ_y remains constant at 0.36λ for all arrays.

4.2 *IRF for NUPA and NULA*

To illustrate the improvement in γ^{-1} or the reduction of IRF, the maximum number of antennas within a given distance (for ULA or NUPA) or area (for UP or NUPA) is considered.

Figure 8 demonstrates the increase in γ^{-1} by comparing the number of antennas in the ULA and NULA arrays at a fixed distance. The simulation values used in this comparison are consistent with those presented in Table 1. It is observed that the NULA array allows for 29% more antennas compared to ULA within the same distance.

In the case of the UPA and NUPA arrays, which are two-dimensional, the increase in IRF relative to the occupied area is examined. Figure 9 shows γ^{-1} as a function of the array area, where it can be observed that the NUPA allows for 34% more antennas within the exact location than UPA. The simulated values of UPA and NUPA used in this analysis correspond to those provided in Table 2.

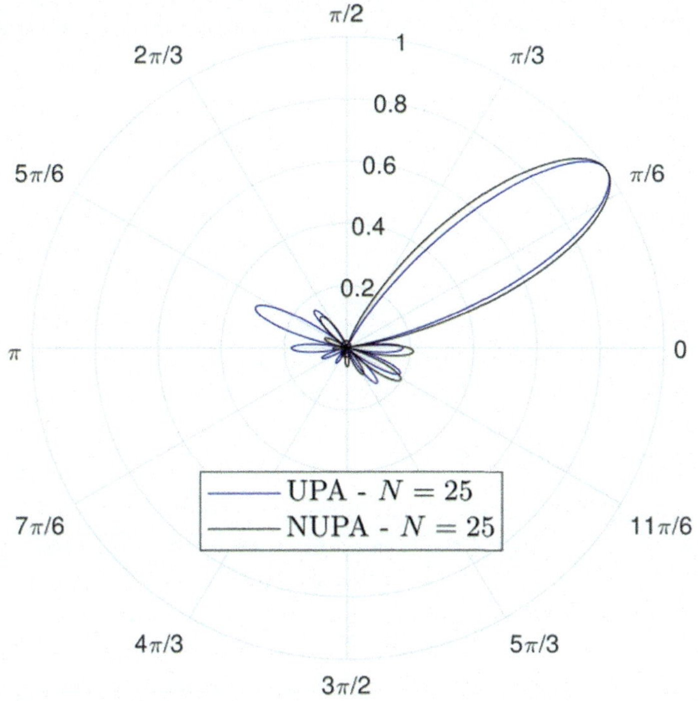

Fig. 7 UPA and NUPA radiation pattern

Table 2 UPA and NUPA antenna arrays with optimal distance

N	A_{ULA} [λ^2]	A_{NULA} [λ^2]	$\overline{\gamma}^{-1}$	$\frac{A_{NULA}}{A_{ULA}}$
25	4.0	2.7	10.4	0.66
49	9.0	6.0	16.2	0.66
81	16.0	10.6	21.6	0.66
100	20.3	13.5	24.3	0.66
144	30.3	20.1	30.0	0.66
196	42.3	28.1	35.5	0.66

The theoretical equations for NUPA and NULA, depicted in Fig. 8 and Fig. 9, are derived from the relationships between distances (see Table 1, D_{NULA}/D_{ULA}) and area (see Table 2, D_{NUPA}/D_{UPA}), considering the theoretical equations of UPA and ULA (6) and (7), respectively.

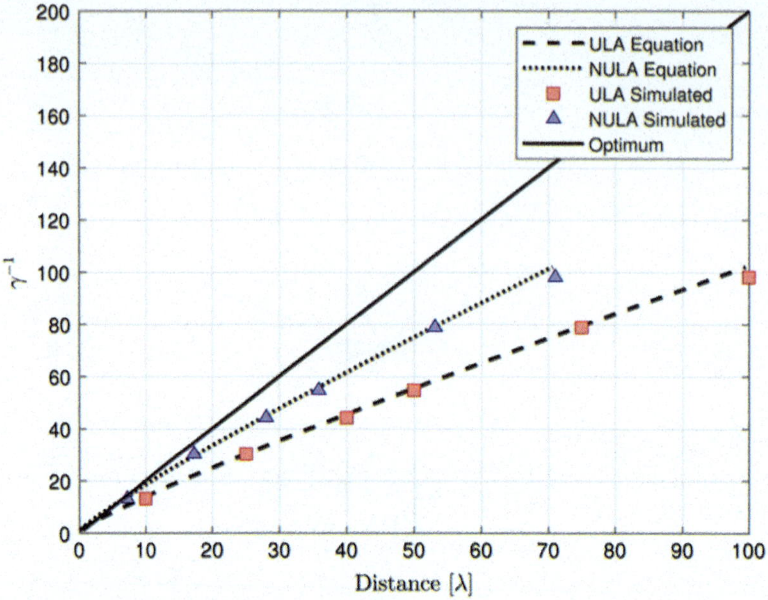

Fig. 8 Inverse IRF for ULA and NULA by distance

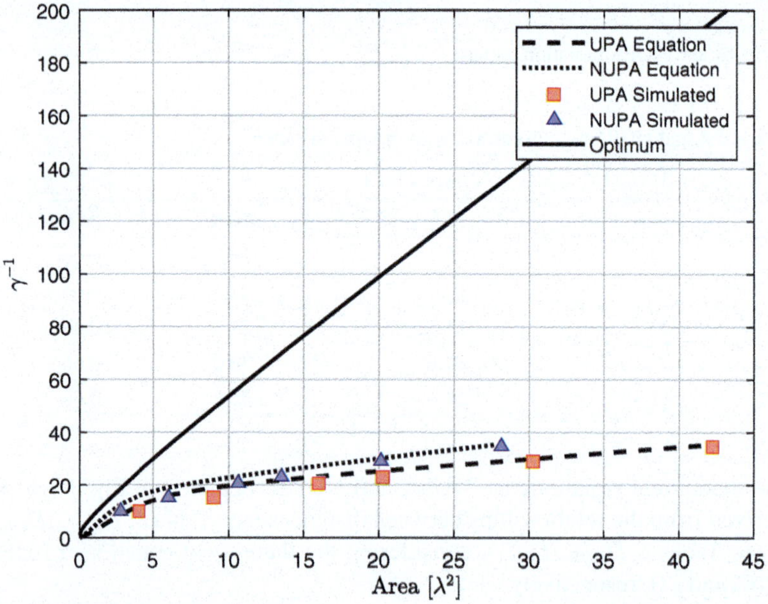

Fig. 9 Inverse IRF for UPA and NUPA by area

5 Conclusions

Modifying the distance between elements increases the Interference Reduction Factor (IRF), as evidenced by the ability to accommodate more antennas within the same space compared to a uniform arrangement. Compared to other ULA densification proposals, such as [7], a similar γ^{-1} value is obtained. However, the developed proposal requires less physical space with NULA densification. It has been shown that the uniform densification approach reduces physical space for ULAs, with an average distance reduction of 29%. At the same time, for Uniform Planar Arrays (UPAs), there is an average area reduction of 34%. With massive MIMO [22], it is expected to have antennas on buildings and other structures, so in the future, it is expected to have different geometries in the antenna arrays such as circular arrays [23] or 3D array (cylinder array, cube array, among others) [23, 24]. The optimal location for these array configurations will be determined in future works.

References

1. Lopez, A., Chervyakov, A., Chance, G., Verma, S., Tang, Y.: Opportunities and challenges of mmwave nr. IEEE Wirel. Commun. **26**(2), 4–6 (2019)
2. Niu, Y., Li, Y., Jin, D., Su, L., Vasilakos, A.: A survey of millimeter wave communications (mmwave) for 5g: opportunities and challenges. Wireless Netw. **21**(8), 2657–2676 (2015)
3. Biglieri, E., Calderbank, R., Constantinides, A., Goldsmith, A., Paulraj, S., Poor, H.V.: MIMO wireless communications. Cambridge university press (2007)
4. Akoum, S., El Ayach, O., Heath, R.W.: Coverage and capacity in mmwave cellular systems. In: 2012 Conference Record of the Forty Sixth Asilomar Conference on Signals, Systems and Computers (ASILOMAR), pp. 688–692 (2012)
5. Vook, F.W., Ghosh, A., Thomas, T.A.: Mimo and beamforming solutions for 5g technology. In 2014 IEEE MTT-S International Microwave Symposium (IMS2014), pp. 1–4 (2014)
6. Altamirano, C.D., de Almeida, C.: Inter-user interference reduction factor for 3-d massive mimo systems. In 2016 8th IEEE Latin-American Conference on Communications (LATINCOM). IEEE, pp. 1–5 (2016)
7. Altamirano, C., de Almeida, C.: Inter-user interference reduction in massive mimo for linear and planar arrays. Int. J. Comm. Antenna Propagat. (IRECAP) **9**(1), 30 (2019)
8. Sohal, R.S., Grewal, V., Kaur, J.: Analysis of different antenna array configurations in massive mimo cellular system for line of sight. Wirel. Pers. Commun. **120**(3), 2029–2041 (2021)
9. Wang, P., Li, Y., Peng, Y., Liew, S.C., Vucetic, B.: Non-uniform linear antenna array design and optimization for millimeter-wave communications. IEEE Trans. Wirel. Commun. **15**(11), 7343–7356 (2016)
10. Liu, W., Wang, Z., Sun, C., Chen, S., Hanzo, L.: Structured non-uniformly spaced rectangular antenna array design for fd-mimo systems. IEEE Trans. Wireless Commun. **16**(5), 3252–3266 (2017)
11. Liu, C., Yang, B., Zhang, P., Wang, H., Wang, C.-X., You, X.: Multiple angles of arrival estimation using broadband signals and a non-uniform planar array. IEEE Trans. Commun. **70**(6), 4093–4106 (2022)
12. Bugnolo, D.: A quasi-" isotropic" antenna in the microwave spectrum. IRE Trans. Antennas Propag. **10**(4), 377–383 (1962)
13. Marzetta, T.L., Yang, H.: Fundamentals of massive MIMO. Cambridge University Press (2016)
14. Schacke, K.: On the kronecker product. Master's thesis, University of Waterloo (2004)

15. Balanis, C.A.: Modern antenna handbook. John Wiley & Sons (2011)
16. Marzetta, T.L.: Noncooperative cellular wireless with unlimited numbers of base station antennas. IEEE Trans. Wirel. Commun. **9**(11), 3590–3600 (2010)
17. Lee, K.Y., El-Sharkawi, M.A.: Modern heuristic optimization techniques: theory and applications to power systems, vol. 39. John Wiley & Sons (2008)
18. Miranda, V., Fonseca, N.: Epso-evolutionary particle swarm optimization, a new algorithm with applications in power systems. In IEEE/PES Transmission and Distribution Conference and Exhibition, vol. 2, pp. 745–750. IEEE (2002)
19. Shi, Y., Eberhart, R.: A modified particle swarm optimizer. In 1998 IEEE International Conference on Evolutionary Computation Proceedings. IEEE World Congress on Computational Intelligence (Cat. No. 98TH8360), pp. 69–73. IEEE (1998)
20. Rueda, J.L., Erlich, I.: Evaluation of emerging metaheuristic strategies on opimal transmission pricing. In 2014 IEEE PES General Meeting, pp. 1–5 (2014)
21. Harrison, R.L.: Introduction to monte carlo simulation. In AIP Conference Proceedings, vol. 1204, no. 1, pp. 17–21. American Institute of Physics (2010)
22. Marzetta, T.L.: Massive mimo: an introduction. Bell Labs Tech. J. **20**, 11–22 (2015)
23. Yaacoub, E., Husseini, M., Ghaziri, H.: An overview of research topics and challenges for 5g massive mimo antennas. In 2016 IEEE Middle East Conference on Antennas and Propagation (MECAP), pp. 1–4. IEEE (2016)
24. Tashiro, K., Hoshino, K., Nagate, A.: Cylindrical massive mimo system for haps: capacity enhancement and coverage extension. In 2021 IEEE 93rd Vehicular Technology Conference (VTC2021-Spring), pp. 1–6. IEEE (2021)

Greinacher Rectifier Evaluation for WiFi Band Energy Harvesting

Javier Rojas, Nataly Valencia Pavón, David Atupaña, and José Campos

Abstract Wireless sensor networks (WSN) have significantly developed in recent years and are highly interested in acquiring data from remote systems. These networks require the availability of sensors with autonomy and low power consumption. Autonomous energization focuses on recycling and optimizing energy sources whose principle of use is based on specific characteristics of the environment and which are not conventionally used. One alternative, with great potential, is the recycling of electromagnetic energy, which is radiated by WiFi internet connection equipment currently found almost everywhere, capturing and using them as sources of electrical energy. The main element is the power converter that captures radio frequency energy radiated by access points of a WiFi network and generates electrical energy. This paper evaluates the potential of the Greinacher rectifier and three circuit variants. The rectifier is popular because of its low power consumption. The energy capture is performed with a commercial omnidirectional antenna tuned to 2.4 GHz. The rectifier is evaluated in two scenarios. Results verify that it is possible to obtain a voltage of up to 1.4 [V] and energize a 20 [mA] LED for 1.2 [s] after a charging time of approximately 24 [h], information that can be used for the development of autonomous energization circuits in WSN.

Keywords Energy harvesting · Greinacher rectifier · Rectenna · Wireless sensor networks

J. Rojas (✉) · N. V. Pavón
Escuela Superior Politécnica de Chimborazo, Riobamba, Ecuador
e-mail: javier.rojas@espoch.edu.ec

N. V. Pavón · D. Atupaña · J. Campos
Universidad Politécnica Salesiana, Quito, Ecuador

G. F. Olmedo Cifuentes et al. (eds.), *Emerging Research in Intelligent Systems*, Lecture Notes in Networks and Systems 902, https://doi.org/10.1007/978-3-031-52255-0_10

1 Introduction

Wireless Sensor Networks (WSNs) have gained widespread popularity, driving IoT and cloud computing development and representing a key component of Industry 4.0. Developing devices for WSNs typically involves a sensor for data acquisition, a processor for operation control, and a wireless transmission system. Research in this area primarily focuses on the ability to transmit acquired information to a data processing center while overcoming obstacles and distance-related issues and reducing latency and information loss. However, energization has not been prioritized, even though the operation or maintenance time depends on it, especially in mobile applications or remote areas [1–3].

WSN device's energization is based on rechargeable batteries, which reduces the operating time, so machines have a limited sampling period, data acquisition, and transmission frequency. Additionally, they require periodic maintenance for battery replacement. To address this issue, energy harvesting devices that use solar cells or kinetic mechanisms could generate electric energy to help maintain battery charge [4, 5].

As an energy harvester, electromagnetic radiation waves from WiFi devices can be captured by an antenna, tuned to a specific frequency, obtaining an AC voltage that electronic circuits can rectify for use in the energization of devices or storage in a battery. This set of elements is known as rectenna. The amplitude of the captured voltage is minimal, so it is necessary to consider electronic circuits with low power consumption. This study aims to evaluate the potential of the Greinacher rectifier to manage the captured voltage, taking advantage of the capability of acting as a voltage multiplier and improving the energization method [6–8].

It represents an attractive idea because WiFi networks are ubiquitous and can be found in homes, offices, public spaces, and even remote areas using satellite technology. WiFi radio frequency waves are a type of electromagnetic energy that is readily available and capable of penetrating walls and obstacles, but more importantly, it can provide constant power. Harnessing this energy for WSNs can keep devices operational without needing battery replacements.

Several research studies where the performance of the rectenna is optimized have been developed on this topic but are still subject to simulation and laboratory conditions. In Taiwan, at the National University of Ilan [9], a rectenna that collects RF energy from the WiFi frequency band is developed and simulated for powering medical devices. As a result, a rectenna with a maximum output voltage of 1.05 [V] is obtained, collecting RF waves in a radius of 1.5 [m] at a WiFi frequency of 2.45 [GHz], allowing specific medical devices, which need electrical power and are in an acceptable range, to be powered wirelessly without disturbing the medical process. Another study, which integrates rectennas, is the one developed in Chennai, India, at the Faculty of Engineering of the University of Sri Venkateswara [10], where the need to take a step towards the new generation of recycling of radio frequencies and electromagnetic energy that are scattered in the environment is seen, with the design of a simulated rectenna capable of feeding sensors installed in integrated devices with IoT technology, obtaining. As a result of a voltage of 7.9 [V] product of radio frequency waves that were collected at a frequency of 2.45 [GHz], the study takes ideal situations without taking into account the fact that there are disturbances in the signal, given by electromagnetic interference typical of a hostile industrial environment. Rob Matheson, a researcher at MIT [11], published research that focuses on designing a flexible rectenna only a few atoms thick, using a two-dimensional semiconductor that converts the AC signal from the environment into DC. The author mentions that the output voltage is unstable during experimental tests over time and proposes to continue developing new studies in this field. Currently, most of the studies related to rectennas obtain results from simulation software and designs oriented to the development of antennas without placing greater emphasis on the rectification stage and, even less, the application of this technology in actual conditions [12].

In contrast to the works mentioned above, this paper shows the rectifier circuit design based on Greinacher's rectification theory [13–15]. An omnidirectional antenna tuned to 2.4 [GHz] with 2dBi isotropic gain captures WiFi radio frequency waves. The rectifier is evaluated in three different schemes and two scenarios. First, a signal generator is configured with a peak-peak value and frequency approximate to those received by the antenna in an environment with radiating devices. Secondly, in a controlled WiFi environment without obstacles and external interferences.

This paper has four sections: the methodology and the rectifier's design are explained in Sect. 2, the analysis of results is shown in Sect. 3, and finally, in Sect. 4, the conclusions are mentioned.

2 Methodology

This work is developed based on the block diagram shown in Fig. 1, where the main parts of a system for rectifying radio frequency waves are presented. The energy collector is responsible for capturing the radio frequency waves in the WiFi bandwidth, and the rectifier is responsible for converting the radio frequency waves into DC voltage so that the frequency does not affect the system's operation. Subsequently, the storage stage accumulates the captured energy to improve the energization of the load.

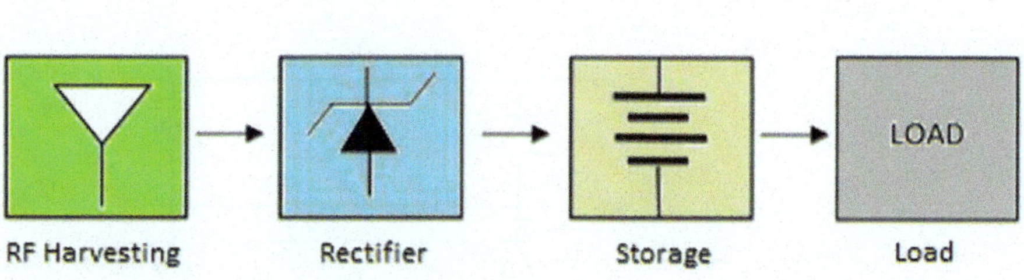

Fig. 1 Block scheme of the energy harvester

Table 1 Antenna technical characteristics

Frequency [GHz]	Gain [dBi]	Impedance [Ω]	VSWR	Polarization	Wavelength	Radiation
2.4–2.5	3.0	50	≤1.5	Vertical	0.25 dipole	Omni

2.1 RF Harvesting

This block is responsible for capturing the WiFi radiofrequency waves tuned to 2.4 [GHz] so that it operates in the industrial, scientific, and medical (ISM) radio band. It is implemented by a commercially available W1030 omnidirectional antenna manufactured by Pulse Electronics. It is used in WLAN devices for WiFi (802.11 b/g), Bluetooth, and ZigBee technology. Some critical technical values are shown in Table 1.

This passive antenna model focuses on omnidirectional point-to-multipoint RF signal pickup. In the 2.4 and 2.5 [GHz] bandwidth, it has a minimum of 2dBi gain and an average percentage of 87% pickup efficiency with a power of 100 mW. The antenna's range depends greatly on the manufacturer and where they are placed. Based on the manufacturer's data, the model used has an ideal range of 100 m that is subject to variations by external factors that make this distance of coverage less [16].

2.2 Rectifier

The rectifier allows the DC voltage to be obtained from the captured WiFi waves. The topology is based on Greinacher's rectification theory because it has the best performance when working with low voltages and high frequencies. The circuit is a full-wave rectifier of the AC signal at the input and doubles the DC voltage at the output depending on the number of phases or layers connected in a cascade [17]. The topology of the rectifier is shown in Fig. 2, where it can be seen that it is formed by four diodes and four capacitors [18, 19].

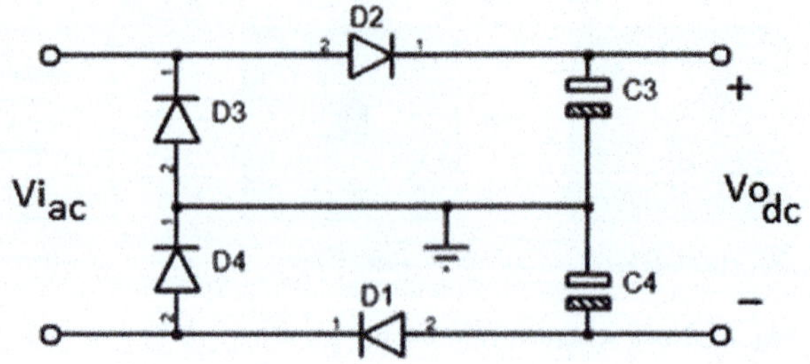

Fig. 2 Greinacher rectifier topology

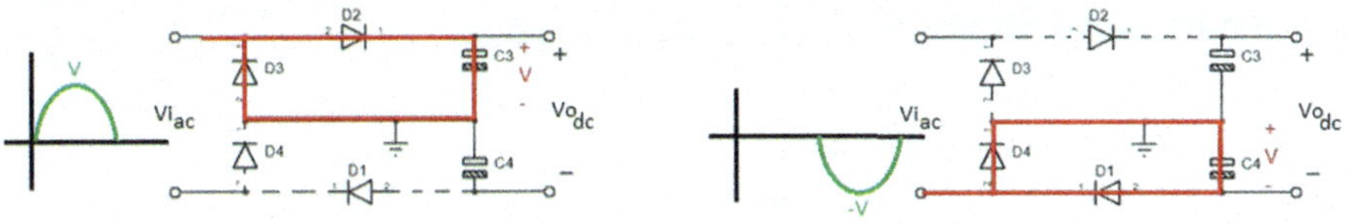

Fig. 3 Rectifier equivalent circuits

The equivalent circuits for each half-cycle can be observed in Fig. 3. During the positive half-cycle, diode D2 is directly biased, so C3 is charged up to the maximum input voltage, and diode D3 closes the current loop so that all the energy only flows in C3. in the negative half-cycle diode D1 is directly biased charging capacitor C4 and closing the loop with D4. Capacitors C3 and C4 are alternately charged and discharged, which helps to keep the output voltage constant and reach twice the input voltage.

The rectifier is implemented with low threshold voltage diodes to reduce conduction losses. Germanium diodes, such as the 1N4148, with a threshold voltage of 0.2 [V] and a reverse recovery time of 4 [ns], are used to allow operation in the WiFi frequency band. To maintain a constant DC voltage during each half cycle, 100 [uF] electrolytic capacitors are used. Their values are recommended by Carrod Electronics, a company specializing in producing and distributing electronic products [19] for circuits that work with radiofrequency signals to moderate voltage and current fluctuations in the rectified output. Additionally, 220 [nF] input ceramic capacitors are added to smooth the signal captured by the antenna. Modifications shown in Fig. 4a and b have been made to improve energy capture and output voltage level and are called rectifier two and rectifier three, respectively. In the first case, the connection of two rectifier circuits in a series powered by the same antenna is evaluated. In contrast, increasing the input voltage by connecting a second antenna is proposed in the second case.

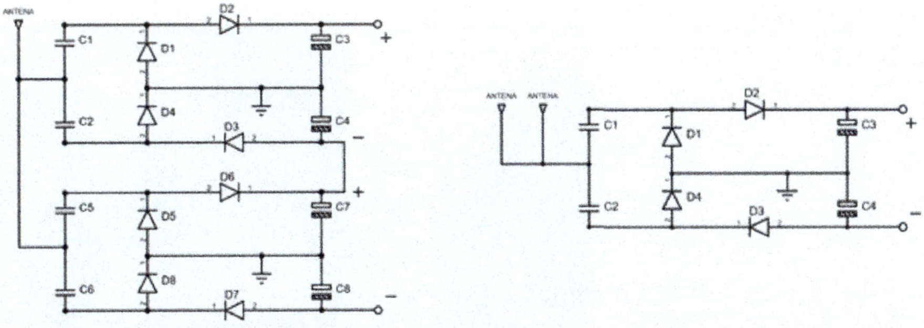

(a) (b)

Fig. 4 Rectifier equivalent circuits **a** series rectifier and **b** two inputs rectifier

3 Analysis of Results

Similar works obtain results only from simulations and are focused on antenna tuning. In this article, each topology is physically implemented and evaluated in a controlled environment with radio frequency waves in the WiFi band. The rectifiers use a controlled input signal from a function generator as a comparison scenario. An oscilloscope and a multimeter are used to visualize and measure the voltage obtained at the output. The test bench implemented for the two scenarios can be seen in the Fig. 5.

3.1 WiFi Electromagnetic Wave Test

This test is performed in a room of 4 × 7 m, where Wi-Fi routers are installed to avoid object interference. The room only has a table where the router is located and a table where the prototype under test is located along with the measuring instruments, and there is no circulation of people while the measurements are taken. In Fig. 6, a layout of the room is shown.

Fixed distance test. In this test, the rectifier is placed in the center of the room, and the distance between the WiFi router and the rectifier (d) is kept constant for the test duration to evaluate its performance. A total of 10 measurements are performed for 5 h at equal time intervals. Figure 7 shows the results for Rectifier 1. It can be noticed that the output voltage has a minimum variation of around 280 [mV] for each sample. The signal delivered by the antenna shows considerable noise with peak-to-peak values of 36 [mV] and an approximate frequency of 1.3 [MHz].

The results for the other rectifiers are similar in distribution. Still, a lower output voltage for rectifier two and a higher output voltage for rectifier three show that it is possible to amplify the output voltage, and rectifier 3 is the one that allows obtaining

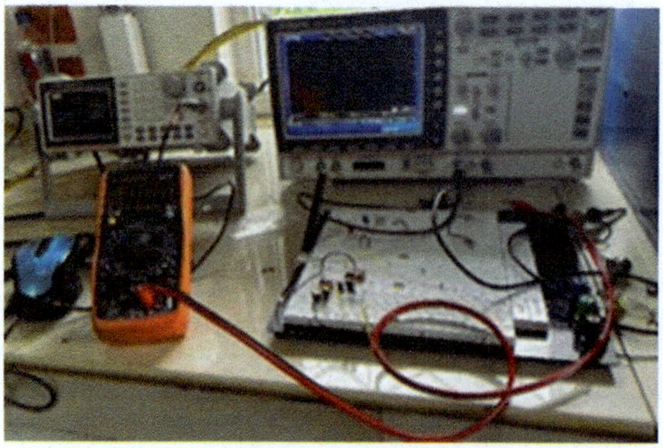

Fig. 5 Prototype and AC voltage captured

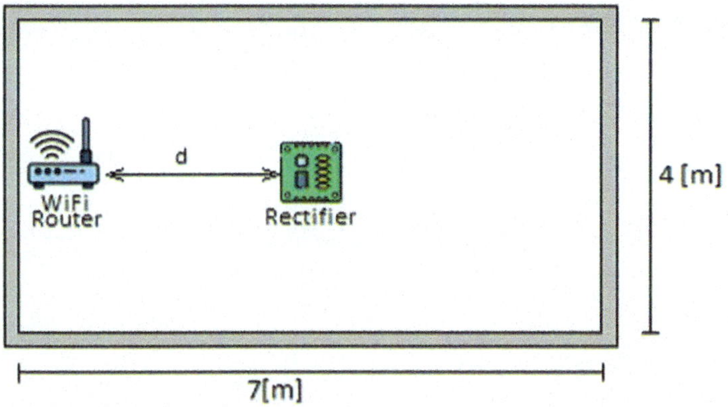

Fig. 6 Test room layout

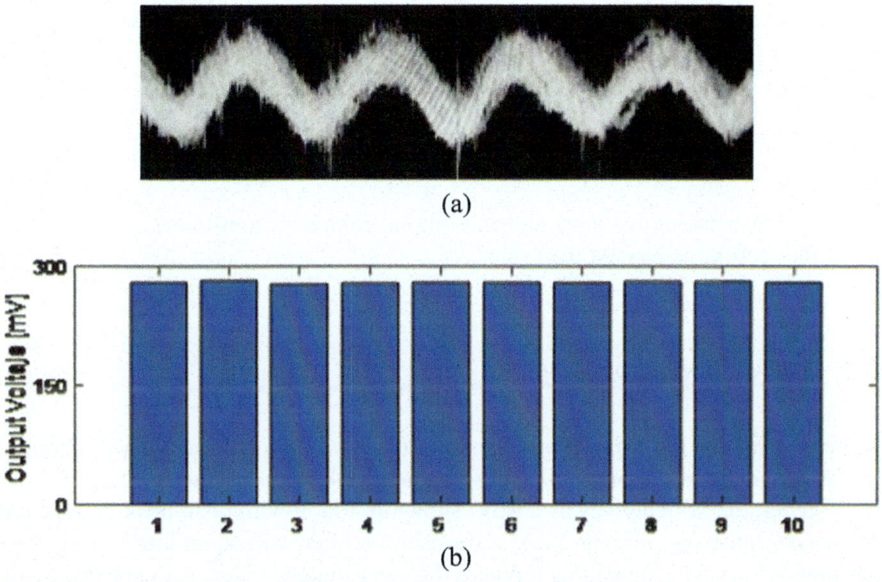

Fig. 7 Fixed distance test results **a** input signal and **b** output voltage

the highest voltage. The average results are presented in Table 2, and it can be seen that there is also an increase in the input voltage delivered by the antenna.

Variable distance test. In this test, the rectifier is placed 1m from the router, and measurements are taken the same way as in the test described in Sect. 3.1. The rectifier is moved, increasing the separation distance (d) in 1m steps until it reaches the end of the room. Figure 8 shows the results for rectifier 1. It can be noticed that the output voltage has a maximum when (d) is 1m and a minimum when (d) is 6m.

Table 2 Data from the WiFi waves test

Rectifier in study	$V_{i,pk-pk}$ [mV]	f_i [Hz]	$V_{o,DC}$ [mV]
Rectifier 1	192	3.48	280
Rectifier 2	156	1.47	227
Rectifier 3	480	8.39	528

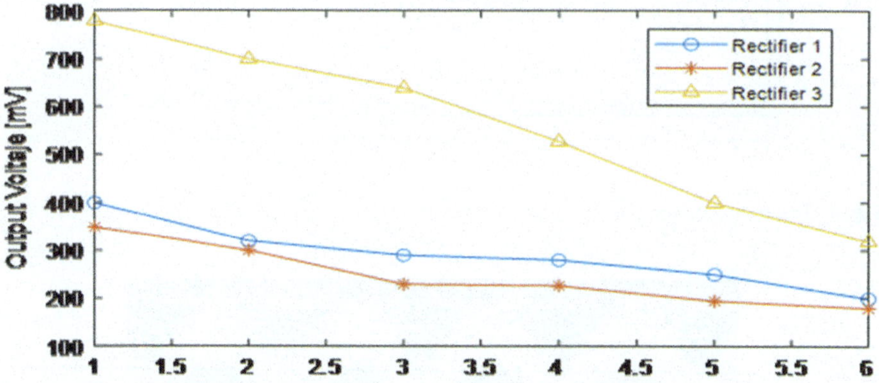

Fig. 8 Variable distance test results

The relation between distance and the output voltage is non-linear, so it is better when the rectifier is nearest the router.

3.2 *Wave Generator Test*

To better evaluate the rectifier performance, the adverse conditions of a WiFi signal in the environment are omitted. The antenna signal input is simulated using a sine wave generator, which gives us a wave shown in Fig. 9 with peak-peak voltage and frequency values detailed in Table 3 for each rectifier, corresponding to values for the simulation of the signal that replaces the electromagnetic energy captured by the antenna and allows us to optimize the rectifiers performance evaluation designed by stabilizing the values of frequency and peak-peak voltage at the input in 10 min for each rectifier.

With this, tests and measurements of the peak-peak voltage and frequency values at the input and output of the rectifiers are performed. In the first case, a maximum load value of 0.289 [V] is obtained at approximately 9 [min]. Rectifier 2 is the one that has less charging voltage and the one that takes longer to reach the maximum voltage point, with a value of 0.235 [V] in 9.5 [min]. This is because when entering an AC signal in parallel, this is weakened and decreases the amount of electrons flowing into the rectifier circuit. Rectifier 3 doubles the value of the previous cases, with a maximum load voltage of 0.537 [V] in a time of approximately 4.5 [min]

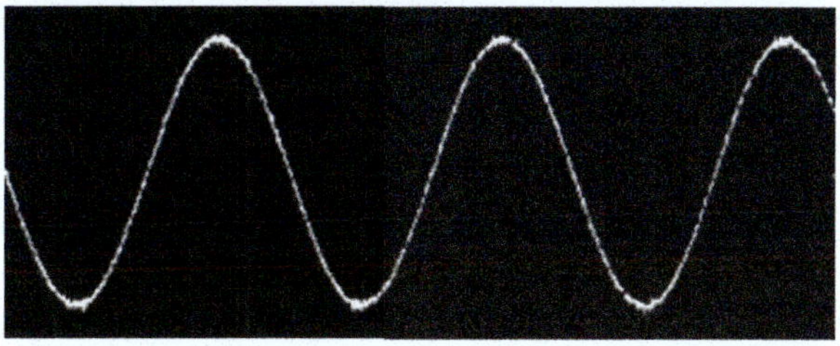

Fig. 9 AC voltage obtained from the function generator

Table 3 Data obtained from the wave generator test

Rectifier in study	$V_{i,pk-pk}$ [mV]	f_i [Hz]	$V_{o,DC}$ [mV]
Rectifier 1	300	5.00	289
Rectifier 2	276	1.47	235
Rectifier 3	370	8.39	537

for its stabilization, being the one with the highest performance, both in the output voltage and in the time it takes to reach this value, the data obtained are synthesized in Table 3.

3.3 Capacitors Charging Time Test

For this test, a capacitor is added at the output as a storage element to analyze the time it takes for the rectifiers to reach their maximum load. After several tests, it was obtained that in 10 min, the output voltage remains constant to visualize the load values taken at an interval of 30 s, resulting in a total of 20 tests per circuit. The mean results are shown in Fig. 10. The maximum voltage obtained is 1.24 [V] DC at the output. This voltage could turn on an LED slightly for less than a second and immediately discharge from the circuit, confirming that the circuit can capture and store energy. However, it does not represent an appropriate or close to the expected performance with RF wave recycling.

Figure 10 shows the charging curves concerning the voltage stored in the capacitor array of the three rectifiers. With one rectifier and a simulated RF signal, the first case performs slightly better than the second one, obtaining a curve stabilizing at 9 min of charging. The second case, with the two rectifiers modified with the simulated RF signal from an antenna, presents the lowest performance, the product of locating the electromagnetic energy source in parallel that reduces the amount of incoming wave flux, stabilizing at 9.5 min of load. The design of the third case, which has a higher flow of radio frequency wave reception, with the simulated signal corresponding to two antennas as electromagnetic energy sources, allows an increase in performance of 50% more than the other two designs, doubling the final voltage and reducing the charging time by half with 4.5 min for its stabilization.

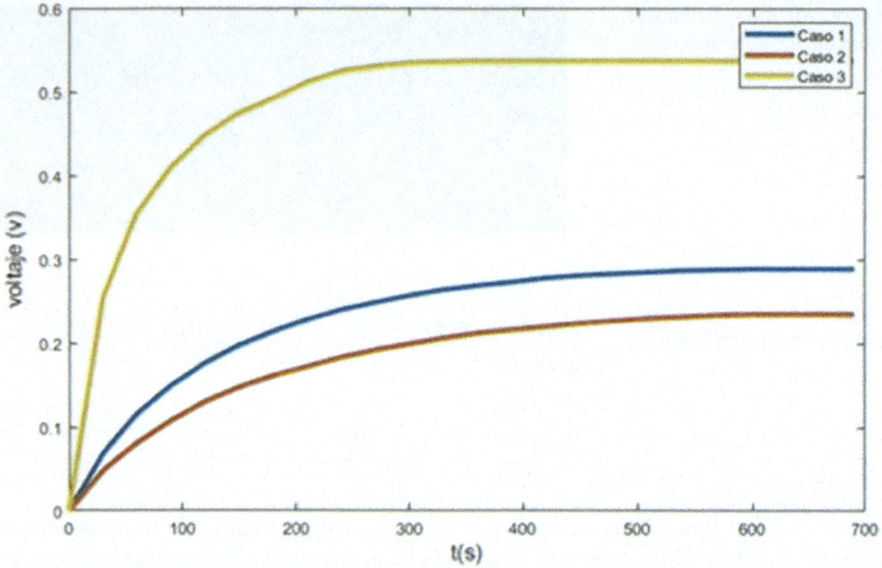

Fig. 10 Charge curves of the output voltage

3.4 *Capacitor Discharge Time Test*

After knowing the load curves, the discharge analysis of each of the designed rectifiers is performed, with the simulated RF signal, to evaluate the behavior of the stored voltage against a load that consumes DC energy and to obtain the curves of the duration time of the stored load. For this test, a 220 [Ω] resistive load is used in parallel to each rectifier's output point of the capacitor array.

Figure 11 shows the time it takes for the stored voltage of the three rectifiers to be consumed. The behavior of each one is practically similar in terms of the time it takes for the circuits to reach a voltage of 0 [V] by the load because the current flowing through the rectifiers does not exceed 20 [mA] due to the minimum voltage drops present in the diodes and capacitive components that, in front of output voltages in millivolts, has a significant impact. The approximate discharge time of the three rectifier designs analyzed is 0.2 s, demonstrating that the designs are in the optimization process and subject to future studies to be used as power sources for sensors and low-power devices.

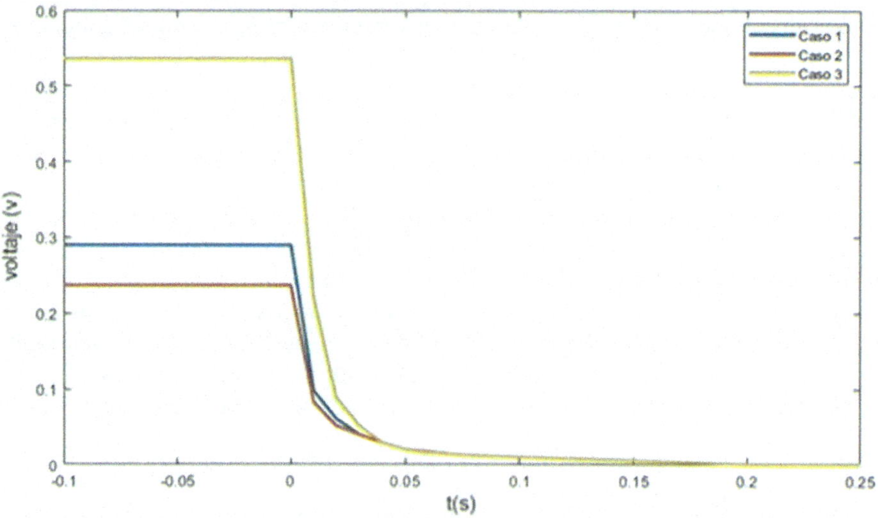

Fig. 11 Discharge curves of the output voltage

4 Conclusions

Greinacher's theory is efficient for high frequencies and low electromagnetic signal voltages, such as WiFi waves at 2.4 [GHz] captured by the omnidirectional antenna implemented within an optimized space to improve its capture capacity. However, these waves were unstable and irregular due to several factors, evident in the number of tests performed for each circuit designed, with a total of 60 tests, with considerably stable voltage values obtained. For the validation of the same, it is necessary to resort to the simulation of the electromagnetic signal of the antenna, taking as a reference the load voltage data obtained in the circuits.

The performance and output DC voltage of the three cases of rectifiers analyzed are directly proportional to the amount of RF waves captured. Rectifier 1, with the signal from an antenna to a rectifier, gives a maximum load of 0.289 [mV]. Rectifier 2, with an antenna as an RF source for two circuits with their outputs in series, delivers a whole load of 0.235 [mv]. In rectifier 3, the signal from two antennas to a rectifier circuit provides a maximum load of 0.537 [mV]. With this, the relationship between the output DC voltage and the number of captured waves is verified, with rectifier 3 having the best performance. Under normal conditions, obtaining a storage voltage of 1.24 [V] in a load time of 24 [h], which fed a low-intensity LED, representing the load necessary to feed low-power sensors, is possible. However, the discharge time is approximately 0.2 [s], as in the rectifiers with the simulated antenna signal. This is insufficient for a constant supply of electronic elements that need much more stable voltage and current. However, this rectifier can be improved with a storage system for long charge duration.

Results obtained can't be compared with similar works because only simulations are performed in most cases. The results refer to the antenna capacity and not to the captured energy, and the working scenarios are of unknown conditions, so the results are not equivalent. However, an experimental panorama promotes experimentation in this knowledge area.

The RF signal reception, delivery to the circuit, and rectifier design must be optimized to improve signal stability and increase the final load voltage to enhance performance. With the mV values obtained, electronic components such as low-power sensors do not have enough energy to consume constantly. Despite not using impedance couplings or antennas with higher range and receiving power, the energy transmitted by WiFi radio frequency at 2.4 [GHz] capable of delivering an acceptable voltage consumed in a short period was recycled, allowing the results of this research to be used as a reference to improve the MEH or Microwave Energy Recycling.

References

1. Gajanan, P., Arora, B., Goel, T., Patnaik, A.: A dual-band microwave Energy harvesting rectenna system for WIFI sources. In: 2019 IEEE Indian Conference on Antennas and Propogation (InCAP) (2019)
2. Lin, C.-H., Chiu, C.-W., Gong, J.-Y.: A wearable rectenna to harvest low-power RF energy for wireless healthcare applications. In: 2018 11th International Congress on Image and Signal Processing, BioMedical Engineering and Informatics (CISP-BMEI) (2018)
3. Okba, A., Takacs, A., Aubert, H.: Compact flat dipole rectenna for energy harvesting orwireless power transmission applications. In: 2018 IEEE International Symposium on Antennas and Propagation & USNC/URSI National Radio Science Meeting (2018)
4. Benyetho, T., Zbitou, J., El Abdellaoui, L.: Fractal multiband planar antenna for wireless power transmission. In: 2014 International Renewable and Sustainable Energy Conference (IRSEC) (2014)
5. Isobe, H., Xu, S.-Y., Fu, L.: High-frequency rectification via chiral Bloch electrons. Science Advances (2020)
6. Raghavandaar, M., Prathiksh, M., Revathy, S., Habiba, H.U.: Energy harvesting using 2.45ghz Rectenna for powering sensors in IOT devices. In: 2020 International Conference on Electronics, Information, and Communication (ICEIC) (2020)
7. Montesdeoca-Mendez, P.A., Gallegos-Segovia, P.L., Leon-Paredes, G.A.: Implementation of fog computing for data pre-processing in an IOT and cloud computing scenario. In: 2019 IEEE Colombian Conference on Communications and Computing (COLCOM) (2019)
8. Alejandro, C.V., Cristian, R.P., Oscar, R.G.: Mejora de la Eficiencia en la transferenciaInalámbrica de Energía del Enlace Inductivo de 3 Bobinas de Tipo SCMR. 2020 IEEE Congreso Bienal de Argentina (ARGENCON) (2020)
9. Song, C., Huang, Y., Zhou, J.: A high-efficiency broadband rectenna for ambient wirelessenergy harvesting. IEEE Trans. Antennas Propag. **63**, 3486–3495 (2015)
10. Gozel, M.A., Kahriman, M., Kasar, O.: Design of an efficiency-enhanced greinacher rectifier operating in the GSM 1800 band by using Rat-race coupler for RF energy harvesting applications. Int. J. RF Microwave Comput. Aided Eng. (2018)
11. Wang, C., Shinohara, N., Mitani, T.: Study on 5.8 GHz band Rectenna rectifying circuit for internal wireless system of satellite. In: 2016 IEEE Wireless Power Transfer Conference (WPTC) (2016)

12. Sathiyapriya, T., Gurunathan, V., Vimala, T., et al.: Voltage doubler design for RFEnergy harvesting system. In: 2020 7th International Conference on Smart Structures and Systems (ICSSS) (2020)
13. Aboualalaa, M., Elsadek, H.: Rectenna systems for RF energy harvesting and wireless power transfer. Recent Wireless Power Transfer Technologies (2020)
14. Carrod (n.d.).: Capacitor electrolítico 100 F varios voltajes. In: Carrod. https://www.carrod.mx/products/capacitor-electrolitico-100-fvarios-voltajes. Accessed 29 July 2022
15. Sedeek, A., Tammam, E., Hasaneen, E.-S.: High efficiency 2.45 GHz low power hybrid junction rectifier for RF energy harvesting. In: 2018 International Japan Africa Conference on Electronics, Communications and Computations (JAC-ECC), Alexandria, Egypt, pp. 147–150 (2018). https://doi.org/10.1109/JEC-ECC.2018.8679546.
16. Alneyadi, F., Alkaabi, M., Alketbi, S., Hajraf, S., Ramzan, R.: 2.4GHz WLANRF energy harvester for passive indoor sensor nodes. In: 2014 IEEE International Conference on Semiconductor Electronics (ICSE2014), Kuala Lumpur, Malaysia, pp. 471–474 (2014). https://doi.org/10.1109/SMELEC.2014.6920900
17. Fan, S., et al.: A novel rectenna array with RBR for long-distance WLAN energy harvesting system. In: 2018 IEEE Wireless Power Transfer Conference (WPTC), Montreal, QC, Canada, pp. 1–3 (2018).https://doi.org/10.1109/WPT.2018.8639266
18. Mansour, M.M., Kanaya, H.: Compact RF rectifier circuit for ambient energy harvesting. In: 2017 IEEE International Symposium on Radio-Frequency Integration Technology (RFIT), Seoul, Korea (South), pp. 220–222 (2017).https://doi.org/10.1109/RFIT.2017.8048256
19. Hosain, M.K., Kouzani, A.Z.: Design and analysis of efficient rectifiers for wireless power harvesting in DBS devices. In: 2013 IEEE 8th Conference on Industrial Electronics and Applications (ICIEA), Melbourne, VIC, Australia, pp. 651655 (2013). https://doi.org/10.1109/ICIEA.2013.6566448

Experimental Measurement of Potential Energy in Stairways with Hydraulic Damping Systems

Miguel A. Duran-Sarmiento, Brayan E. Tarazona-Romero, Nilson Y. Castillo-Leon, and Javier G. Ascanio-Villabona

Abstract This article outlines the creation of a model designed to capture energy from staircases. The setup revolves around a hydraulic mechanism meant to retain the energy produced by individuals of varying weights (ranging from 72 to 85 kg) as they step on an tilting step. The design process considered the system's operational modes, followed by experimental assessments aiming to measure the potential energy by vertically moving a 12 kg load. Findings indicate variations in the device's performance concerning alterations in the angles of hydraulic actuators and shifts in the contact point with the stairs, thereby emphasizing the dependence on the contact area. The average values for inclination angles and contact area approximately yield 0.082 J per step. Challenges in subsequent phases include adjusting the angle across more than three positions and accurately quantifying pressure and volumetric flow within the circuit.

Keywords Energy harvesting · Hydraulic circuit · Staircase

1 Introduction

In the search for sustainable and clean energy solutions, using renewable energy sources has gained significant attention worldwide. The history of energy generation by humans dates back to ancient times when physical effort was the primary energy source. Over time, various sources and methods of energy production were discovered, leading to increased demand and the development of tools and machines. Nowadays, renewable energy sources such as wind [1], water [2], organic matter, heat [3], and motion [4] are considered valuable resources as long as the energy generated exceeds the losses incurred during processing and transportation.

A process that captures and stores energy from the environment is known as "energy harvesting" and typically involves converting it into electrical energy. Energy

M. A. Duran-Sarmiento (✉) · B. E. Tarazona-Romero · N. Y. Castillo-Leon · J. G. Ascanio-Villabona
Unidades Tecnológicas de Santander, 680005 Bucaramanga, Colombia
e-mail: aduran@correo.uts.edu.co

G. F. Olmedo Cifuentes et al. (eds.), *Emerging Research in Intelligent Systems*, Lecture Notes in Networks and Systems 902, https://doi.org/10.1007/978-3-031-52255-0_11

harvesting has become essential to social development, aiming to reduce environmental impact and energy consumption gradually. Various energy sources, including light, wind, and even human motion, have been explored, such as piezoelectric energy [5] generated from footsteps. Transforming these environmental energies into electrical power can contribute to a sustainable and positive impact, particularly in the face of global warming.

In this context, this article focuses on implementing energy harvesting systems and explores the utilization of human motion, specifically through an experimental staircase prototype. Places with high pedestrian traffic, such as shopping malls, schools, and universities, offer ideal conditions for generating electricity through people's movement. By harnessing the gravitational potential energy generated by each step on the prototype staircase, the power is transferred to a hydraulic system. This system activates hydraulic cylinders to transfer the oil to a constant pressure accumulator, effectively storing the residual energy generated by people's movements.

Integrating renewable energy sources in buildings presents numerous advantages, such as reducing greenhouse gas emissions, sustainability, and cost-effectiveness for building owners and tenants. It also contributes to the global effort to mitigate climate change and decrease dependence on fossil fuels. Harnessing the energy from daily human activities and structures through various mechanisms, such as fluid movement, vibration, and impact, poses a challenge that requires innovative strategies to combine energy sources and types of excitations effectively.

This article explores the importance of renewable energy sources, the concept of energy harvesting, and the utilization of human motion as an alternative method of energy generation. The focus on an experimental staircase prototype highlights the potential for sustainable energy production in high pedestrian traffic areas. Understanding and harnessing ambient energies makes it possible to pave the way toward a greener future and contribute to the global goal of sustainable development.

2 Energy Harvesting

Energy harvesting refers to the technology that allows for the capture and utilization of energy from various environmental sources, such as solar light, movement, residual heat, and electromagnetic waves, to convert it into usable energy. This approach has become a promising alternative for addressing energy generation and supply challenges.

One of the most common ways to collect energy is by capturing solar energy using photovoltaic cells [6]. These cells convert solar light into electricity that power low-consumption electronic devices, such as sensors or portable devices.

In addition to solar energy, other environmental energy sources are also being explored. For example, piezoelectric devices can convert mechanical energy from movement or vibrations into electricity [7]. This technology has been used in applications such as battery recharging in portable devices through human body movement.

Residual heat is also considered a potential source of exploitable energy. Thermoelectric materials can convert temperature differences into electricity through the Seebeck effect [8]. This could be used, for instance, in heat recovery in industrial processes or power generation from human body heat.

Another area of research focuses on capturing energy from electromagnetic waves, such as radio signals or Wi-Fi waves. Electromagnetic energy harvesting devices can convert these signals into electricity [9], enabling the power supply to low-consumption electronic devices in environments where a conventional power source is unavailable.

These energy harvesting strategies can be classified based on their energy source, the type of excitation, and the conversion mechanism [10]. The energy sources can be classified as shown in Fig. 1.

Energy harvesting offers promising alternatives for sustainable energy generation and autonomous powering of electronic devices. As technology advances, energy harvesting is expected to play an increasingly significant role in pursuing alternative energy sources and reducing dependence on conventional energy sources.

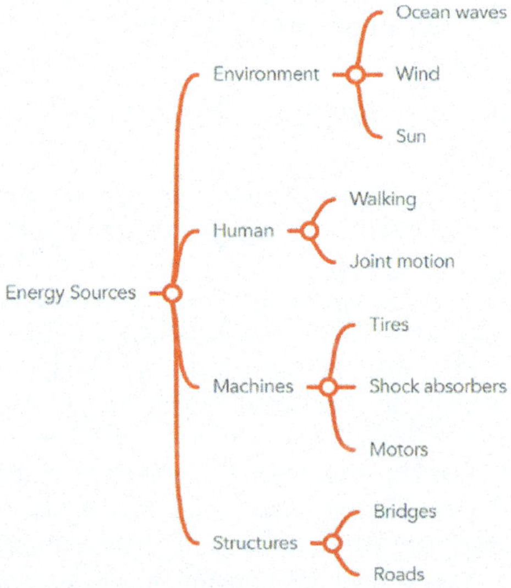

Fig. 1 Energy sources used in energy harvesting [10]

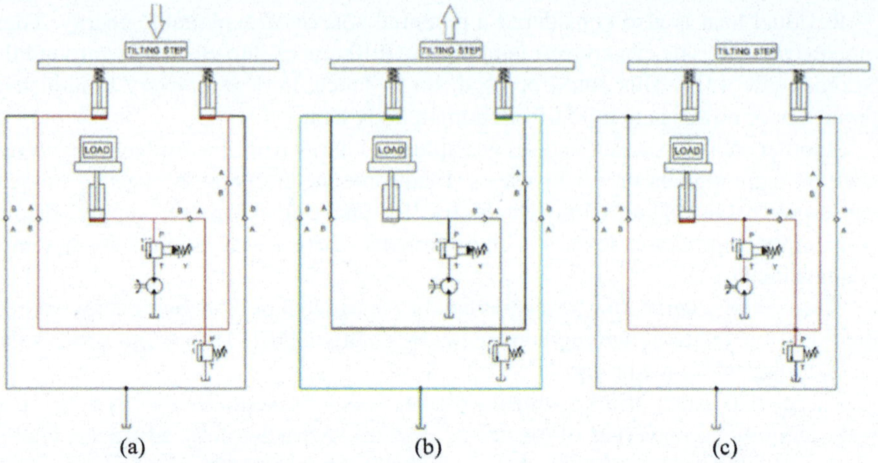

Fig. 2 Hydraulic circuit **a** Loading mode, **b** Suction mode, **c** Transformation mode [11]

3 Materials and Methods

3.1 Design of the Hydraulic Circuit

The prototype incorporates a hydraulic suspension system comprising two actuators and two pairs of springs designed to absorb and restore energy to the initial position. This motion prompts the actuators to displace oil towards a third actuator (a constant pressure accumulator) vertically, consequently shifting a load in the same direction.

The hydraulic circuit's design delineates three operational modes: loading, suction, and transformation. The loading mode, illustrated in Fig. 2a, initiates when an individual traverses the step, propelling oil from the reservoir to the vertical actuator (functioning as a constant pressure accumulator). The suction mode, depicted in Fig. 2b, arises when the contact force dissipates, enabling the springs to facilitate the return of the tilting step to its original position. This action prompts the cylinder chamber to draw oil from the reservoir until the load returns.

Lastly, the transformation mode, as seen in Fig. 2c, facilitates the conveyance of stored oil within the actuator, driven by gravity, through the hydraulic motor. The energy assessment at this phase will estimate the potential energy variance of the mass within the actuator.

3.2 Conceptual Design

The design methodology for a staircase prototype that recovers energy through hydraulic circuits involves several fundamental steps. Firstly, an analysis of the

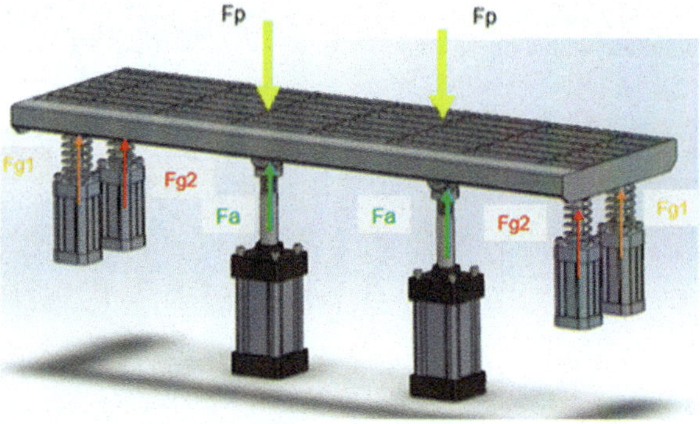

Fig. 3 Working principle of the ladder

requirements and specifications of the model is conducted. This includes determining the maximum load expected to be lifted by the staircase and the desired height for potential energy accumulation, as well as the arrangement of the system's cylinders and springs, as shown in Figs. 2 and 3. Factors such as energy efficiency, system size, and weight are also considered.

Once the requirements have been established, the conceptual design of the staircase is implemented. The actuators and springs suitable for the system are selected, considering their load capacity and the force required to lift the mass. The hydraulic circuit is then designed, consisting of valves and hydraulic cylinders, which allow the flow and control of the fluid to lift the load.

In the design stage, drawings are prepared, and exact dimensions and materials are defined for each staircase component. Calculations and simulations are performed to ensure the structural integrity and efficiency of the system. In addition, safety is considered, including protection mechanisms against overloads and potential component failures (Fig. 4).

Once the design is completed, the prototype construction begins. All components are assembled, and actuators, springs, and the closed hydraulic circuit are installed. Subsequently, tests and adjustments are conducted to verify its proper functioning and efficiency in potential energy recovery when lifting the mass. Design iterations and adjustments are made to optimize the system's performance if necessary.

Fig. 4. 3D modeling of the prototype

Fig. 5 Prototype of a staircase with a tilting step

The prototype features a structural steel profile, and the steps are constructed using industrial-grade steel mesh. Each step measures 23 cm by 53 cm, with a vertical distance of 21 cm between consecutive steps.

3.3 Tests Performed on the Prototype

The experimental tests are carried out using a constant load applied to the staircase to determine the optimal configuration of the prototype (see Fig. 5). This included determining the tilt angle of the energy-absorbing hydraulic actuators and the appropriate contact points on the tiltable step. Once the best configuration is established, additional tests are conducted to analyze the transit of people with varying weights on the staircase.

These tests aimed to estimate the amount of energy recovered per unit of mass. The contact areas with the tiltable step, as shown in Fig. 6, are positioned between the hydraulic actuators. Zones 1 and 5 are located above the actuators, while Zone 3 represents the central area of the staircase. Zones 2 and 4 are positioned between the central and lateral ends of the stairs.

Tests were conducted using masses of 80, 77, 72, and 85 kg, distributed across five contact zones, while varying the inclination angle of the hydraulic actuators. These tests were performed at angles of 90, 70, and 60°.

The measured output variable was the height attained by a 12 kg load in the vertical actuator, serving as an estimate for potential energy. Contact areas were marked using adhesive tape to ensure tests were conducted with consistent input data. Tests were conducted with foot contact on the step, simulating the usual action when ascending stairs.

From its conception, the methodology for testing the prototype involved defining the energy accumulation mechanism and conducting preliminary tests to verify the

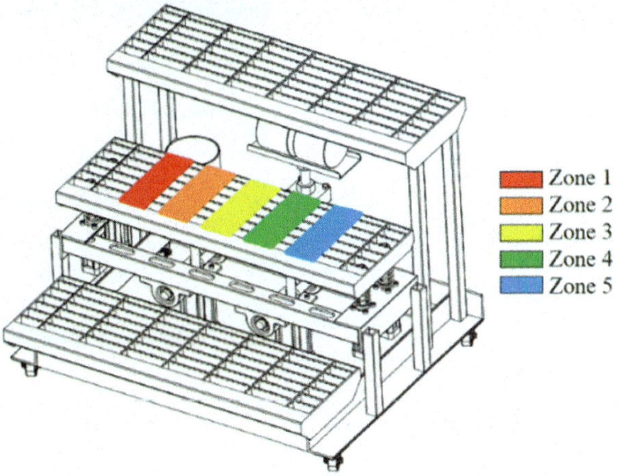

Fig. 6 Contact zones in the prototype [11]

correct functioning of the hydraulic system in loading and suction modes. Subsequently, efforts were made to ensure that tests were consistently executed under similar conditions.

4 Results

The outcomes derived from the examinations provide insights into the prospective contact areas that convert the greatest volume of energy, along with the most advantageous actuator setup, specifically, the inclination angle. The energy computed within the prototype denotes potential energy, while the actuator operates as a constant-pressure hydraulic reservoir, sustaining and elevating a 12 kg load. Measurement of the height is facilitated by employing a linear potentiometer, as illustrated in Fig. 8. This potential energy, contingent upon mass, height, and gravitational acceleration, triggers the propulsion of oil back through the hydraulic circuit, utilizing this momentum to produce power via a hydraulic motor generator. The inclination angle configuration trend of the actuator is depicted in Fig. 7, where the 90° angle resulted in superior outcomes compared to the 70 and 60° angles.

The results obtained are deemed plausible due to the correlation between the actual displacement of the inclined step and the compression displacement of the hydraulic actuators. For the 70 and 60° angles, the ratio of displacement is contingent upon the vertical segment of each angle. The resultant energy metrics from these examinations are displayed in Fig. 9. Moreover, the contact surface area on the tilting step influences the ultimate potential energy metrics, demonstrating that areas 1 and 5 (correlating to the placement of the hydraulic actuators) exhibit superior outcomes in comparison to areas 2, 3, and 4.

The effect of the person's mass passing over the step is analyzed by measuring the pressure in the line using an oil pressure sensor as shown in Fig. 10.

Fig. 7 Actuator tilt angle [11]

Fig. 8 Linear potentiometer installed

This measurement is carried out after calibrating the sensor with a pressure standard in the laboratory (see Fig. 11), the calibration data can be found in Table 1.

The pressure measurement in each test shows no significant pressure variation concerning the person's weight since the pressure in the line depends on the mass in the accumulator (vertical actuator) as shown in Fig. 12.

Additionally, the contact area in the inclination stroke impacts the final potential energy values (see Table 3), where it is observed that zones 1 and 5 (areas that coincide with the location of the hydraulic actuators) have better results compared to zones 2, 3, and 4, as shown in Table 2.

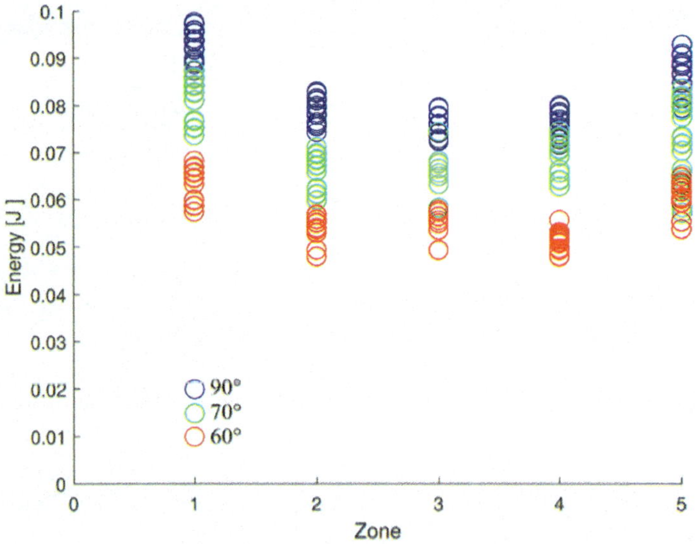

Fig. 9 Accumulated energy as a function of contact area and tilt angle [11]

Fig. 10 Oil pressure sensor

5 Conclusions

During its prototype phase, this system facilitates the gathering and utilization of sustainable and renewable energy derived from natural sources, consequently reducing reliance on non-renewable energy resources. Initial experimental data suggests that an actuator tilt angle of 90° yields the highest energy acquisition, averaging 0.082 J/step. Furthermore, the tests illustrate that the system captures

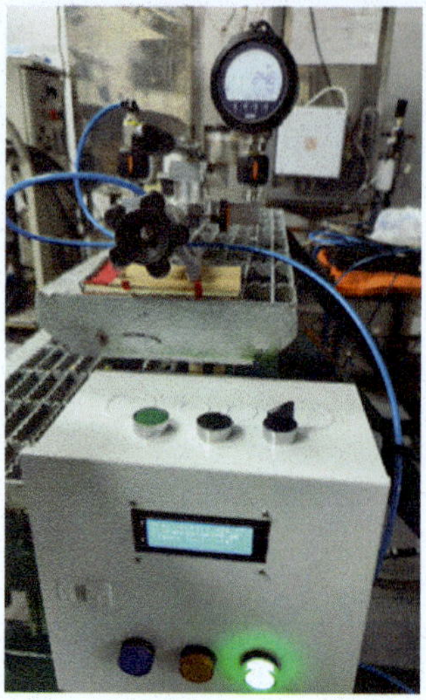

Fig. 11 Pressure sensor calibration

Table 1 Pilot tests on the pressure sensor

Pattern (psi)	Sensor reading (psi)	Error rate (%)
35.1	35.44	−1
32.7	32.27	1
29.5	29.19	1
29.5	29.13	1
28.4	28.67	−1
27.5	27.09	1
26	25.76	1
27	26.97	0
23	22.97	0
19.9	19.6	2
16.4	16.66	−2
12	11.59	3
10.9	10.32	5
6.8	6.496	4
4.2	4.076	3

greater energy when the step is positioned in alignment with the actuator. Implementing a mechanism that enables the complete lowering of the inclined step to a horizontal position is imperative to ascertain if the contact area continues to impact the conclusive outcomes.

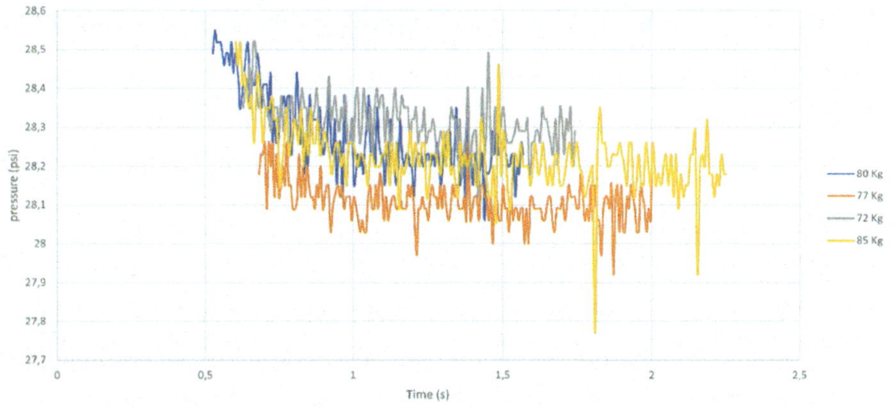

Fig. 12 Pressure measured in the system during the person's passage

Table 2 Average energy per zone [11]

Average energy per zone [J]				
Zone 1	Zone 2	Zone 3	Zone 4	Zone 5
0.079	0.066	0.064	0.065	0.074

Table 3 Average energy per angle [11]

Average energy per zone [J]		
90°	70°	60°
0.082	0.071	0.056

References

1. Wang, J., Zhou, S., Zhang, Z., Yurchenko, D.: High-performance piezoelectric wind energy harvester with Y-shaped attachments. Energy Convers. Manage. **181**(September 2018), 645–652 (2019). https://doi.org/10.1016/j.enconman.2018.12.034
2. Jiang, T., Yao, Y., Xu, L., Zhang, L., Xiao, T., Wang, Z.L.: Spring-assisted triboelectric nanogenerator for efficiently harvesting water wave energy. Nano Energy **31**, 560–567 (2017). https://doi.org/10.1016/j.nanoen.2016.12.004
3. Thielen, M., Sigrist, L., Magno, M., Hierold, C., Benini, L.: Human body heat for powering wearable devices: from thermal energy to application. Energy Convers. Manage. **131**, 44–54 (2017). https://doi.org/10.1016/j.enconman.2016.11.005
4. Wang, W., Cao, J., Zhang, N., Lin, J., Liao, W.H.: Magnetic-spring based energy harvesting from human motions: design, modeling and experiments. Energy Convers. Manage. **132**, 189–197 (2017). https://doi.org/10.1016/j.enconman.2016.11.026
5. Hwang, G.T., Byun, M., Jeong, C.K., Lee, K.J.: Flexible piezoelectric thin-film energy harvesters and nanosensors for biomedical applications. Adv. Healthc. Mater. **4**(5), 646–658 (2015). https://doi.org/10.1002/adhm.201400642
6. Chang, S.Y., Cheng, P., Li, G., Yang, Y.: Transparent polymer photovoltaics for solar energy harvesting and beyond. Joule **2**(6), 1039–1054 (2018). https://doi.org/10.1016/j.joule.2018.04.005

7. Mohanty, A., Parida, S., Behera, R.K., Roy, T.: Vibration energy harvesting: a review. J. Adv. Dielectr. **9**(4) (2019). https://doi.org/10.1142/S2010135X19300019
8. Zhu, T., Fu, C., Xie, H., Liu, Y., Zhao, X.: High efficiency half-Heusler thermoelectric materials for energy harvesting. Adv. Energy Mater. **5**(19), 1–13 (2015). https://doi.org/10.1002/aenm.201500588
9. Huynh, N.D., Lin, Z.H., Choi, D.: Dynamic balanced hybridization of TENG and EMG via Tesla turbine for effectively harvesting broadband mechanical pressure. Nano Energy **85**(February), 105983 (2021). https://doi.org/10.1016/j.nanoen.2021.105983
10. Duran-Sarmiento, M.A., Borras-Pinilla, C., Del Portillo-Valdes, L.A.: Energy harvesting: energy sources, excitation type and conversion mechanisms. Commun. Comput. Inf. Sci. 1756(CCIS), 355–369 (2023). https://doi.org/10.1007/978-3-031-24971-6_26
11. Duran Sarmiento, M.A., Del Portillo Valdés, L.A., Pinilla, C.B.: Analysis of the energetic potential in a staircase with hydraulic circuit: an experimental energy tests. In International Conference on Electrical, Computer, Communications and Mechatronics Engineering (ICECCME 2023), no. July, pp. 19–20 (2023). https://doi.org/10.1109/ICECCME57830.2023.10253332

Vulnerability Characteristics of Wetlands in the Páramo of Guaranda, Central Ecuador—Implications of Long-Term Sustainability

Eva Gavilanes-Betancourt, María Vallejo-Ilijama, Grey Barragán-Aroca, José Paucar-Camacho, and Theofilos Toulkeridis

Abstract Wetlands are transcendental ecosystems of great ecological importance that offer diverse goods and services to the nearby populations. Unfortunately, during the last years, they have suffered affectations and variations in their physical, biological, and chemical characteristics, as well as the decrease or loss of their flora and fauna. The current research obtained knowledge of vulnerabilities through a scientific-technical, socio-cultural, socio-economic, and socio-environmental exercise closely linked to its administrative dimension, which has been adjusted to descriptive trans-sectional designs and correlations-causals in a determined space of time. For the identification of the vulnerability factors of the wetlands, their equilibrium as an ecosystem, Biocenosis-Biotype, is considered. The system equilibrium calculations are conducted by using a variable categorization matrix, the operationalization of the variables, with the ecosystem equilibrium equation (EECO), calculating the result of the correspondence between the variables defined as Biotope, Edaphotope, Biotype, Biocenosis, Biocides, Limiting factors, using the mathematical relation. Finally, the value obtained has led to determining the equilibrium using the table corresponding to the interrelation between ecosystem variables and categories. The results indicate that the balance of the wetland ecosystem maintains a tendency of risk of disappearance with values from 0.33 to 2.00. Its vulnerability is medium, with a tendency to increase. Among the impacts are the decrease in native plant and animal species, the degree of water contamination due to the dynamics of occupation, as well as changes in the geographical space of the wetlands.

Keywords Wetlands · Biotype · Biocenosis · Vulnerability · Impact · Ecuador

E. Gavilanes-Betancourt · M. Vallejo-Ilijama · G. Barragán-Aroca · J. Paucar-Camacho
Facultad de Ciencias de la Salud y del Ser Humano, Universidad Estatal de Bolívar UEB, Av. Ernesto Che Guevara SN y Av. Gabriel Secaira, Guaranda 020102, Ecuador

T. Toulkeridis (✉)
Departmento de Ciencias de la Tierra y Construcción, Universidad de las Fuerzas Armadas ESPE, General Rumiñahui SN y Ambato, Sangolquí 171103, Ecuador
e-mail: ttoulkeridis@espe.edu.ec

G. F. Olmedo Cifuentes et al. (eds.), *Emerging Research in Intelligent Systems*, Lecture Notes in Networks and Systems 902, https://doi.org/10.1007/978-3-031-52255-0_12

1 Introduction

Lake wetlands are ecosystems of world importance but are threatened and affected by many factors, such as pollution, biological resource use, natural system modification, agriculture and aquaculture, and climate change [1, 2]. Wetlands appear in various forms, such as tidal wetlands, inland freshwater wetlands, and riparian wetlands, amongst many others, hosting several fundamental amounts of the local biodiversity [3, 4]. As a result, these ecosystems are also responsible for carbon sequestration, the fishing industry, and several more ecological and socio-economic aspects [5, 6].

In the Andes, the wetlands are unique to even extreme environments of high altitude and often influenced by active volcanism [7–10]. The high Andean wetlands of Ecuador are an integral part of the paramo ecosystem. This region is located above 3,500 m above sea level, with a territorial extension of 1,250,000 ha, representing about 6% of the country's total area. These wetlands are ecosystems with extreme temperature, humidity, porous volcanic soils, unique biodiversity adapted to these conditions, and limitations for using natural resources, including soils in productive agricultural or livestock activities [11]. From the different studies conducted regarding the paramo and its wetlands, it is known that they were considered in the context of the worldview of the original peoples of present-day Ecuador as sacred places, using these spaces as sanctuaries. However, the conquest and colonization made the indigenous population flee the colonial system and seek refuge in wild places [12]. Subsequently, the agrarian reforms in the country (from the 60 s to the 90 s) motivated the expansion of the agricultural frontier in these ecosystems.

By handing over the paramo lands to various indigenous communities and landowners, rural settlements were created or increased, whose territorial spatial expression reflects the adaptations to extreme environmental conditions, executing, in most cases, activities incompatible with the ecological characteristics, paramo environment, and consequently of its wetlands [13]. The investigations indicate that the use of these spaces dates back to pre-Inca times, Inca times, and the Spanish conquest [14, 15]. These were periods of occupation with successive transformations or modifications of the physical environment. During the conquest, the data on introducing sheep and other species was reported during the mills' time. For example, around the year 1630, there were a million sheep grazing in the Chimborazo paramo [16].

At present, activities of overgrazing, breeding of fighting and dairy cattle, exploitation of stone materials, burning of grasslands, the introduction of exotic tree species, tourism, construction of communication routes, provision of electrical energy, construction of irrigation channels are reported as well as rupture of water management with lagoon drying practices, significant public and private investments, a sum of inappropriate practices with the environment, which together with wind and water erosion, have generated a situation of growing negative impact, threat to the ecosystem and for the quality of life of the people who live on it. This also counts indirect users located at lower levels, which depend on the environmental services of the paramo and its wetlands [17].

In the current political and legal context, proper management and protection of the paramo, as well as wetlands, acquired importance since it is the first time in the republican history of Ecuador that the rights of nature are incorporated into the Constitution in the Second Section Healthy Environment Art.14 [18]. In addition, among indigenous organizations, the potential of environmental services produced by these ecosystems is beginning to be valued [19] as a possibility of improving the quality of life of the new ruralities settled in fragile ecosystems.

In the paramo, the wetland is an ecosystem typical of the northern sector of the Andes. It begins at approximately 3,500 m above sea level (presence of permafrost) and goes to the limit with perpetual snow. Its flora and fauna are unique worldwide because these elements have adapted to inhospitable climatic conditions, increasing their resilience [20]. Wetlands constitute one of the most important sources of fresh surface water on the Earth [21, 22] due to their ability to retain water and regulators of the water regime, especially the high Andean wetlands. However, those that are in the territory of Bolívar are being increasingly attacked environmentally, despite the ecological-functional importance, for being the basis for the formation of hydrographic micro-basins, which in turn generate three sub-basins, Chimbo, Zapotal, and Catarama. These are direct tributaries of the great basin of the Guayas River [23] and state that of the 36,573m^3/year of water received by the flow of the Guayas River, between 30 and 40% of the water comes from the water systems of the province of Bolívar. The functionality dynamics of wetlands, with their vulnerabilities, need to be determined by attributes or factors, whether physical [24] or social, related to the dynamics of surrounding human settlements within the framework of complex historical processes, whose socio-cultural configuration, economic and political of the territory [25] is the result of the implementation of various development models.

Based on the context above, the purpose of the current study is highly justified, which is the identification of the vulnerability characteristics of wetlands by determining the biotic and abiotic factors and the social-productive dynamics of the surrounding communities (see Fig. 1).

2 Methodology

Identifying vulnerability characteristics of the paramo wetlands of Guaranda parish is developed in the territories of the Guanujo, Salinas, and Simiatug parishes and comprises 1456.56 ha. These are twelve wetlands called Moya Grande, Chauchivi, Capadia, Cocha Colorada, Jolopogyo, Laigua, Peñas Blancas, Rincón de los Andes, Pachancho, Yurak Uscha, Galo Cocha (see Fig. 1).

Knowledge of the vulnerabilities of wetlands becomes a scientific-technical, socio-cultural, socio-economic, and socio-environmental exercise closely linked to its political-administrative dimension and based on a methodological design that represents how to answer the purpose of the present study. For this case, in particular, the descriptive and correlational-causal trans-sectional methods are adjusted to the extent that they analyze the occurrence, impact, and the values in which one

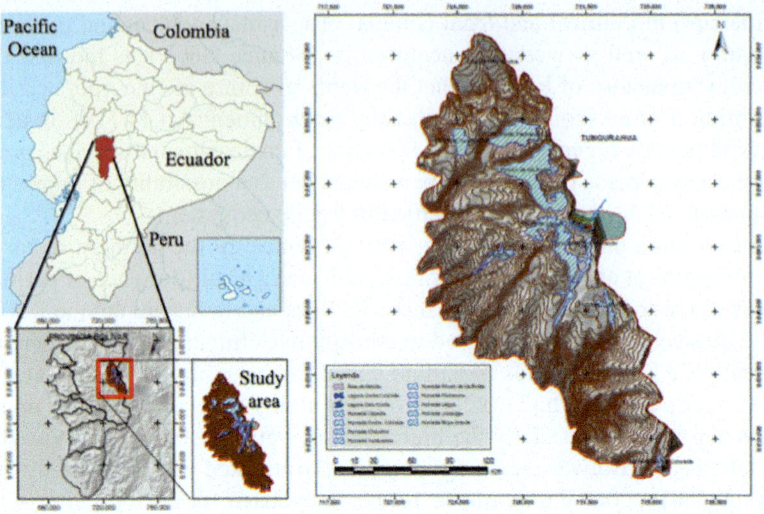

Fig. 1 Map of the geographical location of the wetlands of the paramo of the Guaranda parish [26]

or more variables are manifested (quantitative approach) or locate, categorize, and provide a vision of a community, an event, a context, a phenomenon or a situation (qualitative approach). On the other hand, they describe relationships between two or more categories, concepts, or variables in a given space of time. It is assumed that the causes and effects have already occurred [27], in reality, or are occurring during the development of the study, and the researcher observes and reports them.

Identifying the vulnerability factors of wetlands requires information in the first instance of its balance as a system. This balance occurs when the biotic community or Biocenosis (set of individuals of different species), with its different Biotopes (living environment) and a Biotope in particular, the flow of energy allows reciprocal interdependence between all species, without alterations in climatic conditions (Edaphotope). These can become limiting for the ecosystem's survival, or Biocides or substances harmful to living beings that deplete resources or limit the species' reproductive capacity [28], even more so the resilience of the ecosystem.

The system balance calculations are performed following the method of Aranda [28], in which the categorization matrix of the variables (Table 1) is used, followed by the operationalization of the variables, with the ecosystem balance equation (EECO), calculating the result of the correspondence between the variables defined as Biotope, Edaphotope, Biotype, Biocenosis, Biocides, Limiting factors, through the mathematical relationship (1). Finally, the value obtained led to the balance determination using the corresponding table (Tables 1 and 2).

$$\text{EECO} = \frac{(\text{Biotope} + \text{Edafotope}) \cdot \text{Biotype}}{(\text{Biocenosis} + \text{Biocidas}) \cdot \text{Limiting factors}} \tag{1}$$

Table 1 Interrelation between ecosystem variables and categories based on [28]

Variables	Categories			
	High 3	Medium 2	Low 1	None 0
Biotope	Excellent for the species and biotope	Normal for the species and biotope	Bad for the species and biotope	Void for the species and biotope
Edafotope	Excellent for the species and biotope	Normal for the species and biotope	Bad for the species and biotope	Void for the species and biotope
Biotype	Excellent for the species and biotope	Normal for the species and biotope	Bad for the species and biotope	Void for the species and biotope
Biocides	Excess pollution destroyed the system	Excessive toxic substances that put the ecosystem at risk	There are toxic substances that put the components of the ecosystem at risk	There are no toxic substances in the ecosystem
Biocenosis	They are populations with growth that destroy the balance of the ecosystem or it has become extinct	Excessive population for biotype and biotope or in low quantity due to ongoing species extinction	Normal population of each species that allows interdependence and little limits the resources for their survival	Ideal population of each species that allows interdependence and does not limit resources or put the ecosystem or the survival of the species at risk
Limiting factors	They have destroyed the ecosystem	Are already inhibiting the survival of the species or biotopic or biotypic conditions	Factors that put the survival of the ecosystem at high risk	There are no factors that are limiting the biotype and biocenosis and their interdependence

Table 2 Ecosystem balance

Correlation value between variables	Ecosystem balance
≥12	Balance
8–11.99	Imbalance risk
4–7.99	Imbalance
0.125–3.99	Risk of disappearance
0–0.1249	Disappearance

The previous analysis implies the study of each one of the components of the ecosystem, a valued definition of each one of the variables, the calculation of the EECO, and the value of the result leads us to establish the balance of the ecosystem, which allows to visualize the trend scenario of the wetland prospectively. The environmental and ecological vulnerability, or the limits that the environment establishes for the proper existence of living beings, is connected with the deterioration of the quality of air, water, and soil, exploitation of resources, exposure to toxic pollutants, loss of biodiversity, and the breakdown of the self-recovery of the ecological system [29], this contributes to increasing vulnerability.

VL equals low vulnerability; VM equals medium vulnerability; VH high vulnerability; VVH very high vulnerability.

To establish the level of vulnerability, Table 3 can be used, among other alternatives, with variables and characteristics according to the level of weakness in the site where the risk is estimated. The information required for the wetlands' vulnerability level is possible with data collection, field information gathering, and analysis of the information contained in the satellite images through a geographic information system (GIS). The final product expected is to improve knowledge of the processes that make them vulnerable at different levels.

Table 3 Environmental and ecological vulnerability [29]

Variable	VL	VM	VH	VVH
	Less 25%	25–51%	51–75%	76–100%
Weather conditions	Normal average temperature levels	Temperature levels slightly above the normal average	Temperature levels above normal average	Stable higher than normal average temperature levels
Air composition and quality	Without any degree of contamination	With a moderate level of contamination	High degree of contamination	Pollution level not suitable
Water composition and quality	Without any degree of contamination	With a moderate level of contamination	High degree of contamination	Pollution level not suitable
Ecological conditions	Conservation of natural resources, planned population growth, deforestation, and pollution are not practiced	Moderate level of exploitation of natural resources, slight population growth, and the level of pollution	High level of exploitation of natural resources, increase in population, and pollution	Indiscriminate exploitation of natural resources, population growth outside planning, deforestation, pollution

3 Results and Discussion

The balance of the wetland ecosystem in the study area maintains a trend of risk of disappearance, with values from 0.33 to 2.00. Severe cases correspond to the wetlands in the Simiatug area, with values less than 0.23, in danger of disappearing. The only sector wetland with a bio-restoration plan has reached equilibrium value 6 (Table 4).

The vulnerability of wetlands is at medium vulnerability levels with a tendency to increase (Table 5). The data obtained (Table 6) qualitatively and quantitatively indicate the impacts and sequelae derived from the anthropic intervention in these fragile wetlands. The surface of the wetlands, the current land use, and the surface covered with pads correspond with the economic activities carried out by the inhabitants of the surrounding human settlements or in the wetland space. Among the impacts on the ecosystem, the reduction of native plant and animal species in the place and the degree of water contamination stand out. Both the equilibrium and vulnerability level values allow for establishing the occupation dynamics and the changes in the geographic space of the wetlands. It is necessary to emphasize the importance of the pads (Plantago rigida, sp), whose function is to maintain and purify the wetlands' water reserves and sustain biodiversity. The felling of these species is catastrophic for all living beings that depend on the water from the Chimborazo paramo.

The Pearson coefficient r value is about -0.636. This value of r indicates a moderate negative relationship between the variables. Then, the verification of correlation between variables is applied using student's t analysis.

Table 4 Equilibrium values of the wetland ecosystem of the study area

Wetland	Edafotope	Biotype	Biocenosis	Biocidas	Limiting factors	EECO
Cocha Colorada	1	0.4	1	1	1	0.40
Moya Grande	2	1.5	2	2.5	2	0.67
Chauchivi	2	1.5	2	1	2	1.00
Peña Blanca	2	2	2	1	2	1.33
Capadia	2	1.5	2	1	1	2.00
Laigua	1	2	2	2	2	0.75
Pachancho	1	2	2	2	2.8	0.36
Yurack2	1	1	1	1.5	2.4	0.33
Yurack1	1	3	1	1	1	6.00
Jolopogyo	1	2	2	2	2	0.75
Galo Cocha	1	1	1	2	2.9	0.23
Cocha Colorada	1	1	2	2	2.9	0.17
Rincón de los Andes	1	2	2	2.5	2.7	0.33
Average						1.10
Standard deviation						1.56

Table 5 Level of vulnerability of the wetlands of the paramo of the Guaranda parish

Wetland	Weather	Air contamination	Water contamination	Exploitation of natural resources and anthropism	Value in %	Level of vulnerability
Cocha Colorada	25	27	70.58	32.74	38.83	VM
Moya Grande	25	27	70.58	38.26	40.21	VM
Chauchivi	25	30	55.88	38.1	37.25	VM
Peña Blanca	25	26	90	44.46	46.37	VM
Capadia	25	25	31.44	60.03	35.37	VM
Laigua	25	25	60.58	53.14	40.93	VM
Pachancho	27	30	95	53.14	51.29	VA
Yurack2	27	30	90.88	100	61.97	VA
Yurack1	20	15	26.47	53.3	28.69	VM
Jolopogyo	30	32	90	50.479	50.62	VM
Galo Cocha	28	27	97	100	63.00	VA
Cocha Colorada	35	37	100	100	68.00	VA
Rincón de los Andes	27	30	98	46.716	50.43	VM
Average					47.15	
Standard deviation					11.83	

It is possible to apply the linear regression model to estimate the relationships between the variables, balance, and vulnerability of wetlands since they are continuous. Once the trends are plotted in the dispersion diagram [30], the coefficient r is closer to the unit (see Fig. 2) when the model is potential. This regression model is an alternative when the linear model does not achieve an appropriate coefficient of determination or when the phenomenon under study has a behavior that can be considered potential or logarithmic [31], as is the case of wetland ecosystems, which may be confounding factors responsible for changes in the variables of the system, natural phenomena are highly uncertain and social ones. Sometimes, they are not predictive due to the human complexity [32–35].

Calculated t is less than the critical t in the left tail of the student's t plot in rejecting the null hypotheses. The Null Hypothesis is rejected if there is an inverse relationship between wetlands' balance and vulnerability variables. R^2 tends to 1. It could be interpreted as the existence of a mathematical relationship between the model variables, ruling out chance D damaged pad; M pad in good condition; S

Table 6 Data chosen for quantitative and qualitative calculations

Wetland	Limiting factors	Total surface	Surface covered with pads	Intervened surface, destruction of pads in %	Health status of the pads	Main economic activities	Total coliforms in wetland waters	Presence of Escherichia coli
Arenal								
Moya Grande	Onion crops, potatoes, blue grass, bleeding, induced burning, garbage, paved road	55.99	36.71	34.4	D	Grazing, agriculture	240	220
Capadia	Human settlement, induced burning, road construction	60.89	27.99	54	M	Grazing	67	204
Jolopogyo	Paved road, human settlement garbage, water, and wind erosion	51.66	28.19	45.4	D	Grazing agriculture, livestock, surrounding human settlement	140	TNTEC
Chauchiví	Bleeding, cultures onion, blue grass, induced burning, garbage, water and wind erosion, human settlement	223.28	146.71	34.3	D & M	Grazing, agriculture	190	206
Cocha Colorada	Construction of roads, houses, irrigation canal intake, erosion	19.34	13.64	29.5	M	Grazing, livestock, incipient tourism		
Salinas								
Peña Blanca		55.38	33.22	40	M	Moderate grazing	TNTCT	TNTEC
Rincón de Los Andes	Human settlement in the wetland, latrines, trails, garlic cultivation, induced burning, bleeding, garbage	448.93	260.18	42	D	Livestock, agriculture	TNTCT	479

(continued)

Table 6 (continued)

Wetland	Limiting factors	Total surface	Surface covered with pads	Intervened surface, destruction of pads in %	Health status of the pads	Main economic activities	Total coliforms in wetland waters	Presence of Escherichia coli
Pachancho	Human settlement in the wetland, latrines, trails, blue grass, water pollution, induced burning, bleeding, garbage	105.58	55.09	47.8	D	Livestock, microbusiness, agriculture	178	TNTEC
Yurakuscha 1	Protected by the community	86.92	26.85	69.1	S	Biorestoration	90	
Yurakuscha 2	Human settlement, latrines, trails, blue grass, erosion	62.66	0	100	D	Grazing, livestock, microenterprise	309	TNTEC
Laigua		105.58	55.09	47.8	D	Grazing, agriculture	206	104
Simiatug								
Cocha Colorada	Human settlement in the wetland, use of fertilizers, pesticides, agriculture, livestock. Pine forest	4.34	0	100	D	Agriculture, livestock, exploitation of the pine forest	340	TNTEC
Galo Cocha	Human settlement in the wetland, use of fertilizers, pesticides, agriculture, livestock. Pine forest, dam construction	26.98	1,9	93	D	Exploitation of pine forest, livestock, grazing, agriculture	330	TNTEC

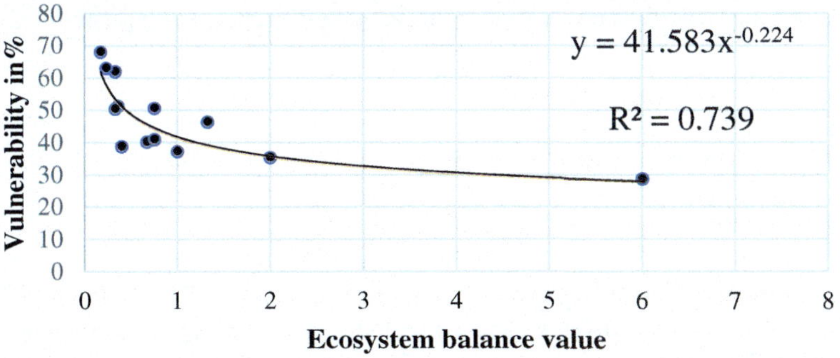

Fig. 2 Potential regression model for the wetlands of the paramo of the Guaranda parish

healthy pad TNTCT countless colonies of Total Coliforms; TNTEC uncountable colonies of Escherichia coli vulnerability (Tables 7 and 8).

The linearized equation is the following:

$$\log y = \log(41.58) - 0.224 \cdot \log x \tag{2}$$

where Ho there is no correlation between variables, Hi if there is a significant correlation between variables and t computed is less than the critical t, on the left tail. The

Table 7 Test the existence of a correlation between the variables

X	Y					
EECO	Vulnerability %	X − Xm	$(x - xm)^2$	y − ym	$(y - ym)^2$	(x − xm) (y − ym)
0.4	38.83	−0.702	0.492	−8.321	69.235	5.837
0.67	40.21	−0.432	0.186	−6.941	48.174	2.995
1	37.25	−0.102	0.010	−9.901	98.025	1.005
1.33	46.37	0.228	0.052	−0.781	0.610	−0.178
2	35.37	0.898	0.807	-11.781	138.787	−10.585
0.75	40.93	−0.352	0.124	−6.221	38.698	2.187
0.36	51.29	−0.742	0.550	4.139	17.133	−3.069
0.33	61.97	−0.772	0.595	14.819	219.610	−11.434
6	28.69	4.898	23.995	-18.461	340.800	−90.429
0.75	50.62	−0.352	0.124	3.469	12.036	−1.220
0.23	63.00	−0.872	0.760	15.849	251.198	−13.813
0.17	68.00	−0.932	0.868	20.849	434.690	−19.422
0.33	50.43	−0.772	0.595	3.279	10.753	−2.530
14.32	612.96		29.158		1679.749	−140.655
Xavg	*Yavg*		Root		Root	
1.102	47.151		5.400		40.985	

Table 8 Potential regression model for the wetlands of the Guaranda parish paramo

r	r^2	$1 - r^2$	n^{-2}	$(1 - r^2)/n^{-2}$	t calculated	Critical t positive or negative
−0.63556	0.40393	0.59607	11	0.05419	−2.7303	2.2010
				0.23278		

model is a potential linear regression with a negative trend. That is, the relationship between the variables is inverse. According to the potential regression mathematical model, the more significant the balance of the ecosystem, the lower the impacts that have significantly altered its balance and potentially increased its vulnerability.

4 Conclusions

The wetlands continue to maintain one of their primary functions as freshwater reserves, coming from the glaciers of the Chimborazo volcano, groundwater, precipitation, and mist due to the type of Andosol soils, young volcanic, porous, covered by padded vegetation that meet for this purpose.

The contamination of the water exceeds the hygienic-sanitary values, allowing for the health of ecosystems, according to international and national standards, whose average values are: Total coliforms maximum of 50UFC/100 ml; Escherichia coli 200UFC/100 ml, in turn, constitute an alert to the presence of other microorganisms or viruses, pathogens, that may be present, requiring a new collection of water samples for more detailed studies in which heavy elements are included due to the use of chemicals in agricultural activities—the water intake for human consumption irrigation. Recreation is found in the paramo, serving the provinces of Tungurahua, Chimborazo, Bolívar, Los Ríos, and Guayas. The vulnerability characteristics are due to geographic factors, a decrease in the surface of wetlands, and a decrease in the surface covered with cushions expressed in the indicator change in land use, climatic variability, extreme rainfall or drought, and biodiversity loss.

Another factor is the social dynamics: demographics as the population increases with its needs and demands for survival, which is why they engage in various economic activities that are, in most cases, incompatible with these ecosystems. The location of surrounding human settlements or wetland spaces reflects the policy of development models based on extractivism, which is in force in Ecuador—construction of infrastructure works without regard to the fragility of these ecosystems.

In short, protecting wetlands is the responsibility of the governors and the governed at different levels. It is a problem of science and conscience. The parish GAD has exclusive powers to execute the land use plan, categorizing the soils of fragile ecosystems. This is urgent.

References

1. Xu, T., Weng, B., Yan, D., Wang, K., Li, X., Bi, W., Li, M., Cheng, X., Liu, Y.: Wetlands of international importance: status, threats, and future protection. Int. J. Environ. Res. Public Health **16**(10), 1818 (2019)
2. Toulkeridis, T., Tamayo, E., Simón-Baile, D., Merizalde-Mora, M.J., Reyes–Yunga, D.F., Viera-Torres, M., Heredia, M.: Climate change according to Ecuadorian academics–Perceptions versus facts. LA GRANJA. Revista de Ciencias de la Vida **31**(1), 21–46 (2020)
3. Lázaro, W.L., Oliveira-Júnior, E.S., Silva, C.J.D., Castrillon, S.K.I., Muniz, C.C.: Climate change reflected in one of the largest wetlands in the world: an overview of the Northern Pantanal water regime. Acta Limnologica Brasiliensia **32** (2020)
4. Greb, S.F., DiMichele, W.A., Gastaldo, R.A.: Evolution and importance of wetlands in earth history. Special Papers-Geol. Soc. Am. **399**, 1 (2006)
5. Bhowmik, S.: Ecological and economic importance of wetlands and their vulnerability: a review. Res. Anthology Ecosyst. Conserv. Preserving Biodiversity 11–27 (2022)
6. Milton, G.R., Prentice, R.C., Finlayson, C.: Wetlands of the world. In The Wetland Book II: Distribution, Description and Conservation, pp. 3–16. Springer (2018)
7. Declerck, S.A., Coronel, J.S., Legendre, P., Brendonck, L.: Scale dependency of processes structuring metacommunities of cladocerans in temporary pools of High-Andes wetlands. Ecography **34**(2), 296–305 (2011)
8. Segnini, A., Posadas, A., Quiroz, R., Milori, D.M.B.P., Saab, S.C., Neto, L.M., Vaz, C.M.P.: Spectroscopic assessment of soil organic matter in wetlands from the high Andes. Soil Sci. Soc. Am. J. **74**(6), 2246–2253 (2010)
9. Verzijl, A., Quispe, S.G.: The system nobody sees: Irrigated wetland management and alpaca herding in the Peruvian Andes. Mt. Res. Dev. **33**(3), 280–293 (2013)
10. Toulkeridis, T., Zach, I.: Wind directions of volcanic ash-charged clouds in Ecuador–implications for the public and flight safety. Geomat. Nat. Haz. Risk **8**(2), 242–256 (2017)
11. Camacho, M.: Los Páramos ecuatorianos: Caracterización y consideraciones para su conservación y aprovechamiento sostenible. ANALES. Universidad Central de Ecuador, p. 16. Quito, Ecuador (2013)
12. Molano, J.: Diversidad Ecosistémica y Territorial en Los Páramos: Territorios y Saberes en Las Altas Montañas Ecuatoriales. Universidad Central de Bogotá, Bogotá (2009)
13. Brassel, F., Ruiz, P., Zapata, A.: La Estructura Agraria en el Ecuador: Una Aproximación a su Problemática y Tendencias, 2nd edn. SIPAE, Quito (2010)
14. Hofstede, R.: Health state of Páramos: an effort to correlate science and practice. Lyonia a J Ecol Appl **6**, 61–73 (2004)
15. Mena, P., Hofstede, R.: Los Páramos Ecuatorianos. Universidad Mayor de San Andrés, La Paz (2006)
16. Borchart, C.: Llamas y Ovejas. El Desarrollo Lanar en la Audiencia de Quito. Abya Yala, Quito (1995)
17. WWF: Los humedales altoandinos, ecosistemas estratégicos que ofrecen servicios ambientales para el binestar de millones de personas. WWF, Colombia (2010)
18. Constituyente, A.: Constitución del Ecuador. Registro Oficial 449 (20 octubre), Quito (2008)
19. Cárdenas, F.M., Barrera, V.: Diagnóstico rural participativo desde la equidad de género, social y ambiental, en la subcuenca hidrográfica del rio. INIAP, Virginia Tech, USAID, SANREM, CRSP, Guaranda (2007)
20. Caranqui-Aldaz, J.M., Muelas-Domingo, R., Hernández, F., Martínez, R.: Chemical composition and polyphenol compounds of Vaccinium floribundum Kunth (Ericaceae) from the volcano Chimborazo Páramo (Ecuador). Horticulturae **8**(10), 956 (2022)
21. Vörösmarty, C.J., Syvitski, J., Day, J., De Sherbinin, A., Giosan, L., Paola, C.: Battling to save the world's river deltas. Bull. Atomic Scientists **65**(2), 31–43 (2009)
22. Zeballos, H., Pino, K., Medina, C.E., Pari, A., Chavez, D., Tinoco, N., Ceballos, G.: A new species of small-eared shrew of the genus cryptotis (Mammalia, Eulipotyphla, Soricidae) from the northernmost Peruvian Andes. Zootaxa **4377**(1), 51–73 (2018)

23. GAD: Plan estratégico de desarrollo provincial 2004–2024. Guaranda (2004)
24. Brinson, M.: Niveles extremos de variación de patrones y procesos en humedales. Bases ecológicas para la clasificación e inventario de humedales en Argentina, pp 19–24 (2004)
25. Galeano, J.P.: El uso del suelo en el caso de los humedales. Verba Iuris, pp 119–147 (2011)
26. Gavilanes, E., Martinez, M., Merino, O., Vizcarra, G., García, A., Paucar, A.: Protección de los humedales del páramo del cantón Guaranda, conservando la biodiversidad para la producción hídrica con niveles de pureza, calidad óptima, en el desarrollo del buen vivir: Primera parte perfil ambiental de los humedales, Guaranda: Universidad Estatal de Bolívar, Guaranda, Ecuador (2015)
27. Hernández, R., Fernández, C., Baptista, P.: Metodología de la investigación, 3rd edn. McGraw-Hill, Santiago de Chile (2004)
28. Aranda, J.: Introducción a la Contaminación. Universidad Técnica Particular de Loja, Ecuador, Loja (1997)
29. Bisbal, A., Picón, J., Casaverde, M., Jáuregui, F., Anchayhua, R., Sánchez, R., Musana, M.D.: Manual básico para la estimación del riesgo. INDECI, Lima (2016)
30. Torres, B., Cayambe, J., Paz, S., Ayerve, K., Heredia-R, M., Torres, E., Luna, M., Toulkeridis, T., García, A.: Livelihood capitals, income inequality, and the perception of climate change: a case study of small-scale cattle farmers in the Ecuadorian Andes. Sustainability **14**(9), 5028 (2022)
31. Manfredo, L.: Matemática, estadística y computación (2017). http://reyesestadistica.blogspot.com/
32. Echeverría-Puertas, J., Echeverría, M., Cargua, F., Toulkeridis, T.: Spatial dynamics of the shore coverage within the zone of influence of the Chambo river. Central Ecuador. Land **12**(1), 180 (2023)
33. Orbe, J., Herrera-Robalino, J.L., Ureña-Callay, G., Telenchano-Ilbay, J., Samaniego-León, S., Fienco-Bacusoy, A., Cando-Veintimilla, A., Toulkeridis, T.: An evaluation of radon in drinking water supplies in major cities of the province of Chimborazo, Central Andes of Ecuador. Water **15**(12), 2255 (2023)
34. Mora, M.J.M., González, C.A.L., Hidalgo, D.A.E., Toulkeridis, T.: Determination of altitudes of the three main Ecuadorian summits through GNSS positioning. Geodesy Geodyn. **13**(4), 343–351 (2022)
35. Macías, L., Quiñonez-Macías, M., Toulkeridis, T., Pastor, J. L.: Characterization and geophysical evaluation of the recent 2023 Alausí landslide in the northern Andes of Ecuador. Landslides, pp 1–12 (2023)

Climate Variability and Its Effect on Flood Risk Through Dynamic Modeling in the Quigushuaycu Stream

David Carrera-Villacrés, Ximena Chamorro, Allison Chaquinga, María José Correa, Jose Daniel Lara, and Diego Vinueza

Abstract Every year in the rainy season, Quito's ravines generate intense floods with considerable human and economic losses. Water is crucial for human development, the environment, and the economy. The water infiltrating the soil depends on the degree of salinity in it. This can generate adverse effects such as flooding. In the last decade, the Conocoto parish in the south of Quito city, Ecuador's capital, has gained importance. It has become a highly populated area where agricultural, industrial, and residential activities occur. This study evaluated the impact of the Sodium Adsorption Ratio (SAR) index on the risks of infiltration in the Quigushuaycu stream in the Conocoto parish. With the help of HEC-RAS modeling, it is determined that 4.7% of the stream has a potential risk of heavy flooding. A representative sample of water from the creek was taken, and the concentration of calcium, magnesium, and sodium and electrical conductivity were measured. It is worth mentioning that the best time to take the sample is in the rainy season because there is a more significant amount of water in the stream there. The sodium adsorption ratio index is calculated, with a value of 0.74 meq/L and an electrical conductivity of 410 μS/cm. When applying the Relative Reduction Diagram of infiltration caused by salinity and RAS, water with medium salinity and low sodium from the creek fell into the severe and light-moderate reduction categories. Regarding the anomalies of the accumulated annual rainfall, they presented a significant interannual variability associated with the occurrence of the phenomenon "El Niño Southern Oscillation."

Keywords Flood · SAR · Electrical conductivity · Climatic anomalies

D. Carrera-Villacrés (✉)
Universidad de las Fuerzas Armadas ESPE. Departamento de Ciencia de la Tierra y La Construcción, Av. Gral. Rumiñahui. Grupo de Investigación en Contaminación Ambiental (GICA), Sangolquí, Ecuador
e-mail: dvcarrera@espe.edu.ec

D. Carrera-Villacrés · X. Chamorro · A. Chaquinga · M. J. Correa · J. D. Lara · D. Vinueza
Facultad de Ingeniería en Geología, Minas, Petróleos y Ambiental, Ciudadela Universitaria, Universidad Central del Ecuador. Carrera Ingeniería Ambiental, Quito, Ecuador

G. F. Olmedo Cifuentes et al. (eds.), *Emerging Research in Intelligent Systems*, Lecture Notes in Networks and Systems 902, https://doi.org/10.1007/978-3-031-52255-0_13

1 Introduction

The Intergovernmental Panel on Climate Change (IPCC), made up of scientists from around the world 2004, stated that excessive industrial pollution directly increases the "greenhouse effect," which triggers an increase in temperature, better known as global warming, as well as sea level rise and climate change, with droughts, floods, violent storms, and heat waves [1].

Hence, climate change for humanity has become one of the most critical challenges in recent years due to the impacts it will have on all areas of society and the challenge represented by the development model of humanity since industrialization. Currently, in Ecuador, few studies consider the variability of the atmosphere and the behavior of specific meteorological variables in the country [2].

The variation in the rainy season from one year to the next is extreme. These irregularities result in droughts and floods, especially in flat and low-lying areas, causing severe human and economic deaths [3].

The city of Quito, the capital of Ecuador, faces several challenges during the rainy season, especially when the arrival of an El Niño phenomenon is anticipated, a weather pattern that can result in a significant increase in rainfall in the region. Quito's most common problems during this time included flooding in different parts of the city, especially in vulnerable areas such as low-lying areas and river banks. This can result in damage to homes, roads, and drainage systems. In addition, heavy rainfall may negatively affect the increased risk of landslides in mountainous and steep areas around the city and the collapse of infrastructure such as roads, bridges, and other infrastructure, making access to certain areas difficult for the town [4].

The El Niño climate phenomenon has proven to have a significant impact on the country's meteorological and climatic conditions. In particular, El Niño events have been observed to be closely related to extreme rainfall and flooding events. This climatic interaction is eloquently illustrated in Fig. 1, in the case of the flood in Quito on January 31, 2022 [5].

Fig. 1 Flood in Ecuador: at least 24 people die in Quito after heavy rains

Several protection and control works are implemented, such as the construction of dikes to protect cities and populated areas and drainage systems that help to evacuate excess water to reduce these problems.

Infiltration is when water, rain, or irrigation enters the soil. Since this water can complete a hydrological cycle, it can be stopped, causing an excess that accumulates on the surface only if the conditions slope allow it and evaporates, circulates on the earth's surface, or enters the ground [6]. The relative reduction graph of Infiltration due to Salinity and SAR, extracted from the study "Evaluation of water quality for irrigation and the use of water resources" in the Toglla-huayco stream, is used to classify the waters of the stream according to its sodium salinity level.

Regarding the Sodium Adsorption Ratio (SAR), the most relevant characteristics for its determination are the total concentration of soluble salts, the relative concentration of sodium concerning other cations, the concentration of bicarbonates about that of calcium and magnesium, in addition to the concentration of boron and other elements that may be harmful such as sodium and chlorine.

The objectives of this work are to model possible flooding of the Quigushuaycu stream based on the relationship between the Sodium Adsorption Ratio (RAS) index and the dangers of infiltration with the help of the HEC RAS software, a one-dimensional hydraulic modeling program, in addition to completing the data of the time series and analyzing the climate variability of the area and identifying the anomalies of the rainfall and temperature variables between the years 1990 and 2020 in the Excel program. The study is developed in 3 phases: the first is collecting meteorological data on the study area provided by REMMAQ. Additionally, data from the TerraClimate and ERA5 satellites are used to complete missing data and the time series from 1990 to 2020. The second phase consisted of relating climate variability and its influence on flood hazards in the Quigushuaycu stream to finally, in the third phase, model a possible flood. This research seeks to understand in depth how the climatic variability associated with the El Niño phenomenon has influenced the occurrence of avalanches and floods in this specific region of Ecuador, intending to improve preparation and future risk mitigation materials and methods of the study area.

Figure 2 shows the study area of the Quigushuaycu stream in the Conocoto parish of Quito. Which limits to the north with the city of Quito and the Cumbayá Parish, to the south to the Amaguaña Parish and the Rumiñahui Canton, to the east to the Guangopolo and Alangasí Parishes, and the Rumiñahui Canton and to the west with the Quito city [7], has an area of 51.46 km2 with a population of 82,072 inhabitants [8]. The climate of the study area is characteristic of the equatorial inter-Andean region: humid mesothermal, with an average annual temperature of 17 °C, and solar radiation has a value of 2.2 kWh/m2 /day [9], and the relative humidity is 66.90% as an annual average of the last 20 years [10] (Table 1).

The water sample was taken from the Quigushuaycu stream on August 19, 2022. It is analyzed in the Service and Product Offerings (OSP) laboratories of the Faculty of Chemical Sciences and the Chemistry laboratory of the Career of Environmental

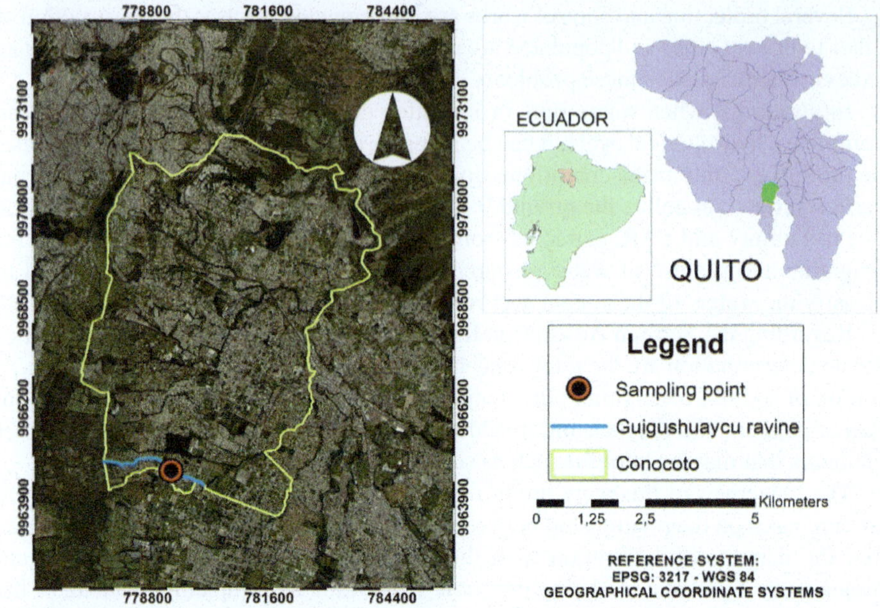

Fig. 2 Location of the sampling point in Conocoto

Table 1 Summary of the study area

Sample	Points	Coordinates UTM WGS84	Coordinates in degrees		
		X	Y	Latitude	Longitude
M1	Stream Quigushuaycu	779,263.5	9,964,658.9	−0.329441	−78.491102

Engineering from the Universidad Central del Ecuador. The physicochemical parameters measured during the sampling are Ca^{2+}, Mg^{2+}, Na^{+}, and electrical conductivity (EC). Table 2 presents a summary of the methods and references used.

1.1 Sodium Adsorption Ratio (SAR)

The concentration values in meq/L are used to determine the water quality for irrigation of the Quigushuaycu stream, obtained from the Ca^{2+}, Mg^{2+}, and Na^{+} cations. The sodium adsorption ratio is determined using (1) (where *SAR* is the sodium adsorption ratio [meq/L], Ca^{2+} is the calcium cation concentration [meq/L], Mg^{2+} is the magnesium cation concentration [meq/L], and Na^{+} is the sodium cation concentration [meq/L]) because it allows estimating the tendency of water to promote compaction and

Table 2 Physicochemical analysis methods and references

Determination	Method	References
Calcium (Ca)	M-GO-AM-05 apha 3111-B modified Absorption spectrophotometry atomic	Standard Methods 3500-Mg-B [11]
Magnesium (Mg)	M-GO-AM-19 apha 3111-B modified Absorption spectrophotometry atomic	Standard Methods 3500-Ca-B [12]
Sodium (Na)	M-GO-AM-27 apha 3111-B modified Absorption spectrophotometry atomic	Standard Methods 3500-Na-B [13]
Electrical Conductivity (EC)	Thermo scientific Orion STAR A212	Standard Methods 2510 B [14]

caking of the soil, where the higher the SAR value, the less suitable water will be for use in agricultural activities [15].

$$SAR = \frac{Na^{+}}{\sqrt{(Ca^{2+} + Mg^{2+})/2}} \tag{1}$$

1.2 Climate Variability Analysis

Daily rainfall data is recorded to analyze climate variability from 2004 to 2020. These data have been collected from the Los Chil-los-M5041 weather station, corresponding to the REMMAQ weather station network of the Quito Environment Secretariat. Similarly, data from 1990 to 2020 from remote sensors from the Google Earth Engine (GEE) platform, from the ERA-5 land data collection, and TerraClimate are used, with a monthly and spatial–temporal resolution of 27.8 km and 4.6 km, respectively, which are used to complete some missing data from the station and, in turn, meet the time series using linear regression and the correlation coefficient [16].

The daily rainfall data are processed and ordered chronologically in the EXCEL program. Then, the accumulated monthly totals are obtained. So, when comparing the data from the meteorological station and the data from the sensors, a better correlation with the TerraClimate sensor is obtained. An equation is obtained through linear regression that allowed for filling in the missing data and completing the time series from 1990 to 2020, given for an analysis of climate variability. It is necessary to have a series of consecutive data of at least 30 years, as recommended by the World Meteorological Organization [17].

Table 3 Intensity range of rainfall anomalies

Percentage of anomaly	Classification
I > 60%	Extremely excessive
60% > I > 40%	Intensely excessive
40% > I > 20%	Moderately excessive
20% > I > −20%	Normal
−20% > I > −40%	Moderately dry
−40% > I > −60%	Intensely dry
I < −60%	Extremely dry

1.3 Calculation of Rainfall Anomalies

This indicator is the percentage relationship between the annual rainfall value and the multi-annual average rainfall [18], which allows us to determine the occurrence of abnormalities below normal.

Rainfall anomalies are calculated from (2) (where *AP* is the annual rainfall anomaly [%], *P* is the total yearly rainfall [mm], and *Pn* is the normal rainfall [mm]).

$$AP = \left[\frac{P}{Pn} - 1\right] \cdot 100 \tag{2}$$

The scale of rainfall anomalies established [17], presented in Table 3, determines the climatic variability according to the anomalies' intensity range, where *I* represent the intensity of the anomalies.

1.4 Dynamic Flood Modeling

For modeling in HEC-RAS, various variables are used to model floods, including topographic, hydrological, and geomorphology data. Topographic data are obtained from ground surveys or satellite imagery to create a digital elevation model [19]. Hydrological data include flows, rainfall, and runoff rates, while geomorphology data includes information on channel shape and characteristics, such as cross-section and bottom roughness. These data are used in the hydraulic model to simulate flow conditions and water levels during a flood [20]. A study published in the "Journal of Hydrology" (2016) highlighted that dynamic modeling in HEC-RAS allows for a more accurate representation of flood events. By considering variations in river flow and conditions over time, more accurate flood forecasts can be made, which is essential for emergency management and protecting life and property.

2 Results and Discussion

2.1 *Sodium Adsorption Ratio in the Water Sample for Irrigation*

The results obtained from the laboratories are presented in Table 4, where the value of the cations from mg/L to meq/L is transformed, which is the number of units necessary to calculate the SAR.

With the calculated data, it is observed that at the stream Quigushuaycu sampling point, it has a SAR value of 0.74, a meager value, entering the "light to moderate reduction" category, that is, low Na^+ waters have a value less than <10 meq/L, unlikely to be percent exchangeable sodium (PES) [21].

A study published in the "Journal of Agricultural Science and Technology" (2018) found a strong correlation between sodium adsorption ratio and soil structure. High SAR can cause the dispersion of soil particles, resulting in reduced water infiltration, crusting of the soil surface, and decreased soil porosity, all of which negatively affect soil quality and the water retention capacity.

Infiltration hazards. The SAR index and electrical conductivity (EC) are related to determine infiltration risks. These values obtained from the laboratory are the following:

- SAR (meq/L) = 0.74
- EC (μS/cm) = 410.0

According to the classification proposed in [22] regarding the quality of water for irrigation and use of the water resource of the Togllahuayco stream (see Fig. 3), the SAR values concerning the EC values are represented in the relative reduction graph of the infiltration caused by salinity and the SAR of the water sampling, which is built into three categories according to the risk of salinization or concentration of salts and the risk of alkalinization or SAR [22].

The water from the Quigushuaycu stream falls into the categories of severe reduction and moderate light reduction (medium salinity water, low in sodium), which causes increased runoff, a situation that can affect the flow of surface bodies in the area and, in cases where there is no flow control, cause flooding downstream of the stormwater discharge.

Table 4 Sodium adsorption ratio

Parameter	Unity	Sample
		M1
SAR	meq/L	0.74

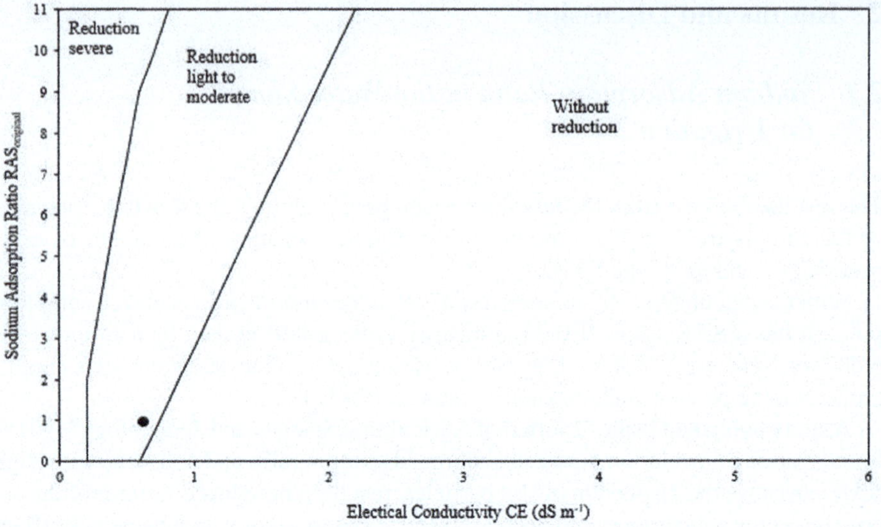

Fig. 3 Diagram of relative reduction of infiltration caused by salinity and the SAR on the water sampling

Table 5 Results of the correlation of the weather station and the sensor

	Equation	Determination coefficient (r^2)	Correlation coefficient (r)
Rainfall (mm)	$y = 0.8344x + 13.216$	0.79	0.89

2.2 Climate Data Correlation

The correlation between the data from the meteorological station and the data from the sensors is made, obtaining acceptable values of relationship with the Terra Climate sensor with a spatial resolution of 0.04°. As seen in Table 5, the value of r for the local station is more significant than 0.5. That is, the data have a strong to perfect positive correlation [23].

2.3 Calculation of Rainfall Anomalies

The correlation between the data from the meteorological station and the data from the sensors is made, obtaining acceptable values of relationship with the Terra Climate sensor with a spatial resolution of 0.04°. As shown in Fig. 4, the value of r for the local station is more significant than 0.5. The data have a strong positive correlation. Anomalies that are well below normal can cause impacts on crops and reduce

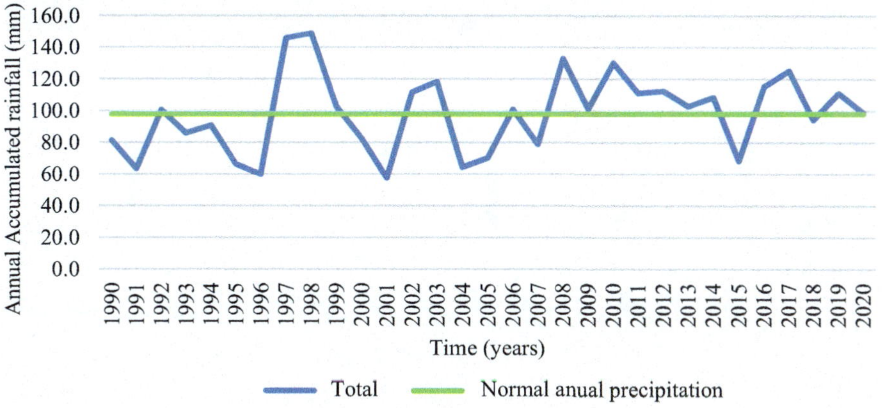

Fig. 4 Time series of accumulated and normal annual rainfall in the station

river levels due to rainfall deficiency [24]. These events are related to the ENSO phenomenon, which can cause fluctuations in rainfall patterns in the Andean regions [25].

Figure 5 shows the time series of the accumulated annual rainfall anomalies in the analysis period from 1990 to 2020, which indicates intensely dry anomalies in 2001 and modernly dry anomalies in 1991, 1995, 1996, 2004, 2005, and 2015—slightly cold in 1999, 2000, 2007, and 2008. Moderately excessive abnormalities were also noted in 2003, 2008, 2010, and 2017. In addition, intensely extreme anomalies were seen in 1997 and 1998. The latter is related to the El Niño phenomenon, which caused flooding, human deaths, and economic losses [26] in the coastal zone of Ecuador. On the other hand, in the inter-Andean part of the study area, the impact is different because the climatic event caused groundwater accumulation at the water table and threatened infrastructure collapse [27].

2.4 *Flood Modeling in HEC-RAS.*

Figure 6 represents the Quigushuaycu stream's flood risk and the area most prone to flooding.

According to the dynamic modeling in HEC-RAS, it is observed that areas most susceptible to flooding are at elevations 2,650.00–2,700.00 m, 2,580.00–2,600.00 m, and 2,610.00–2,620.00 m.

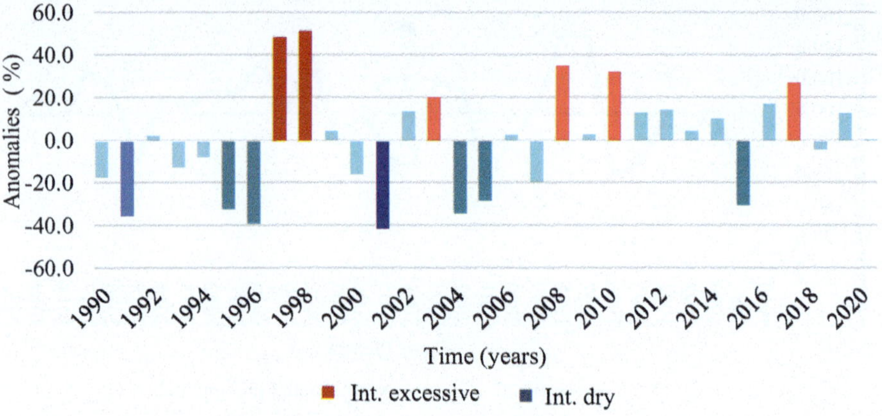

Fig. 5 Anomalies of the accumulated annual rainfall at the Los Chillos-M5041 station

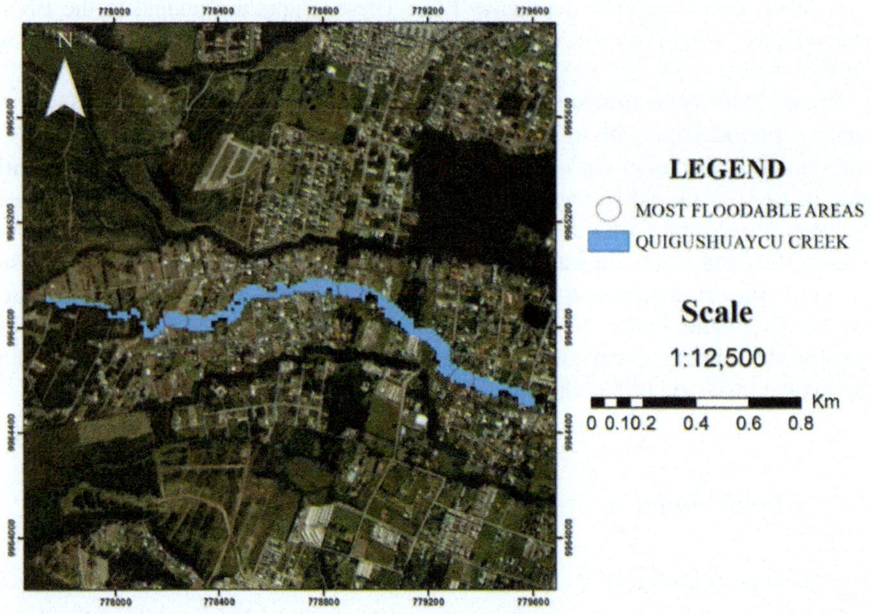

Fig. 6 Flood model in HEC-RAS

2.5 *Influence of Climatic Variability on Infiltration Hazards*

Climate variability and the potential effects of climate change are the main threats to flood risk. Decreased plant cover, increased evapotranspiration, and lower infiltration and recharge rates can cause salinity problems [28].

Climate variability and the potential effects of climate change are the main threats to flood risk. Decreased plant cover, increased evapotranspiration, and lower infiltration and recharge rates can cause salinity problems [29].

The opposite case occurred in Sonora, Mexico, where it is evidenced that the lack of rain in that place increased salinity problems between 15.8 and 24.1%, thus reducing the volume of floods by 18.9% in that area [30].

3 Conclusions

It has been determined that the water from the Quigushuaycu stream has a low sodium adsorption ratio (SAR of 0.74 meq/L), so there will be no problems of sodicity in the soils since said water falls into the classification "severe reduction and light to moderate," which represents waters of medium salinity with low sodium content. Therefore, there is a possible risk of flooding in the ravine.

The HEC-RAS tool has been used to determine the most floodable regions. This tool integrates topographic, hydrological, and geomorphology data to model the flooding, resulting in three vulnerable areas to possible flooding, whose elevations are 2,650.00–2,700.00, 2,580.00–2,600.00, and 2,610.00–2,620.00 m.

When analyzing the graphs and values of the anomalies of the annual rainfall averages in the Quigushuaycu stream, they have presented a tremendous interannual variability associated with the phenomenon "El Niño Southern Oscillation" since the severe rainfall anomalies occurred during this climatic event. In such a way that it can significantly affect the rainfall regimes in the analysis area, there is also the presence of El Niño events above normal and La Niña events below normal.

References

1. Espinosa-Contreras, A.E.: El caos y la caracterización de series de tiempo a través de técnicas de dinámica no-lineal. Mexico (2004). https://repositorio.unam.mx/contenidos/el-caos-y-la-caracterizacion-de-series-de-tiempo-a-traves-de-tecnicas-de-la-dinamica-no-lineal-3508942?c=LJplZd&d=false&q=*:*&i=2&v=1&t=search_0&as=0
2. Ayala, M.F., Tierra, A., Carrera-Villacrés, D.: Análisis de la dinámica caótica para las series temporales de variables meteorológicas en la estación climatológica de Chone. Ecuador-Chone, Geoacta (2017). https://revistas.unlp.edu.ar/geoacta/article/view/13319
3. Díaz-Borrego, L., Rodríguez-Infante, A.: Evaluación del riesgo por inundación en la comunidad Pradera Alta, municipio Maracaibo, Venezuela. Minería & Geología **32**(3), 1–15 (2016). https://www.redalyc.org/journal/2235/223547677001/html/
4. Paredes-Gómez, M., Molina-Estrella, M., Ceron-Carrera, M.: Aluvión de Quito: una mirada comunicacional del desastre. Tsafiqui **12**(8) (2022). https://revistas.ute.edu.ec/index.php/tsafiqui/article/view/1042
5. British Broadcasting Corporation (BBC) News Mundo: Aluvión en Ecuador: mueren al menos 24 personas en Quito tras intensas lluvias. (2022). https://www.bbc.com/mundo/noticias-america-latina-60212132

6. Mencías-Ibadango, D.: Determinación de la infiltración en el suelo, estudio de la calidad del agua para regadío y manejo ambiental del sistema de riego El Pisque-Pichincha. Quito (2015). https://repositorio.uisek.edu.ec/handle/123456789/1456
7. Gobierno Autonomo Descentralizado (GAD), Parroquial de Conocoto (2022). https://www.gadconocoto.gob.ec/parroquia/
8. Instituto Nacional de Estadística y Censo INEC (2022). https://www.ecuadorencifras.gob.ec/informacion-censal-cantonal/
9. Consejo de Nacional de Electricidad CONELEC Ecuador (2008). https://www.ariae.org/servicio-documental/atlas-solar-del-ecuador-con-fines-de-generacion-electrica. [Último acceso: 12-01-2024]
10. Guerrero-Bayas, W., Mora Villacís, M.: Instalación de sensores climáticos, generador atmosférico IoT y análisis lineal y no lineal de las variables generadas. Sangolqui: Universidad de las Fuerzas Armadas ESPE (2021). https://repositorio.espe.edu.ec/xmlui/handle/21000/26079
11. Standard Methods for the Examination of Water and Wastewater: APHA Method 3500-Mg, https://www.standardmethods.org/
12. Standard Methods for the Examination of Water and Wastewater: APHA Method 3500-Ca, https://www.standardmethods.org/
13. Standard Methods for the Examination of Water and Wastewater: APHA Method 3500-Na, https://www.standardmethods.org/
14. Standard Methods for the Examination of Water and Wastewater: APHA Method 2510, https://www.standardmethods.org/
15. Dunán-Avila, P., Fernández-Rodríguez, M., Riverón-Zaldívar, A., Bassas-Noa, P.: Evaluación de la calidad de las aguas del río Yamanigüey para el riego agrícola. Minería y Geología **38**(1) (2022). https://www.redalyc.org/journal/2235/223574495011/html/
16. Amat-Rodrigo, J.: Correlación lineal y regresión lineal simple (2016). https://www.cienciadedatos.net/documentos/24_correlacion_y_regresion_lineal,
17. Organización Mundial Metereológica (OMM 1203): Directrices de la Organización Meteorológica Mundial sobre el cálculo de normales climáticas (2017). https://library.wmo.int/viewer/60338?medianame=1203_es_#page=1&viewer=picture&o=bookmark&n=0&q=
18. Instituto de Hidrología, Metereologíay Estudios Ambientales (IDEAM): Variabilidad climática y el cambio climático en Colombia (2018). https://www.andi.com.co/Uploads/variabilidad.pdf
19. Alcantara-Boza, A., Castro-Tenorio, A.: Determinación de zonas inundables con HEC-RAS en zonas de la Cordillera Central de los Andes. Microcuenca del río Saraus-río La Llanga. Celendín. Peru. Revista de investigación UNMSM (2021). https://revistasinvestigacion.unmsm.edu.pe/index.php/iigeo/article/view/19852
20. GEASIG: Especialistad en SIG y Medio Ambiente: ¿Qué es HEC-RAS y para qué sirve? (2017). https://www.geasig.com/que-es-hec-ras-y-para-que-sirve/
21. Guerrón-Varela, E., Carrera-Villacrés, D., Torres, C., Barba, M., Monge, A., Hernández, J.: Multivariable relationship with the change of sulfate concentration and dynamic model for the water demand of the Huambi parish, Ecuador, South America. LACCEI (2020). https://laccei.org/LACCEI2020-VirtualEdition/meta/FP328.html
22. Carrera-Villacrés, D., Pazmiño-Montero, M., Duran, R., Mayorga, E.: Soil characterization from the Rio Grande high basin for preparing a technosol that retains phosphorus to prevent eutrophication in the dam water of the Presa Propósito Multiple Chone. IOP Conference Series: Earth and Environmental Science (2019). https://iopscience.iop.org/article/10.1088/1755-1315/344/1/012038/pdf
23. Anderson, D., Sweeney, D., Williams, T.: Estadística para Administración y Economía. Cengage Learning. 10 a edición (2003). https://www.upg.mx/wp-content/uploads/2015/10/LIBRO-13-Estadistica-para-administracion-y-economia.pdf
24. Instituto de Hidrología, Metereología y Estudios Ambientales (IDEAM): Cálculo de la anomalía de precipitación en Colombia: una propuesta de ajuste a índice actual. Bogota (2007). http://www.ideam.gov.co/documents/24155/123679/18-1.08_HM_Anomalia_precipitacion_3_FI.pdf/c60552b2-0808-4a1c-9c29-d34b3bb7662d

25. Hofstede, R., Calles, J., López, V., Polanco, R., Torres, F., Ulloa, J., Vásquez, A., Cerra, M.: Los Páramos Andinos ¿Qué sabemos? Estado de conocimiento sobre el impacto del cambio climático en el ecosistema páramo. (2014). https://portals.iucn.org/library/sites/library/files/documents/2014-025.pdf
26. Demoraes, F., D´Ercole, R.: Cartografía de las amenazas de origen natural por canton en el Ecuador (2001). https://horizon.documentation.ird.fr/exl-doc/pleins_textes/divers16-03/010065702.pdf
27. Corporación Andina de Fomento (CAF): El fenómeno El Niño 1997–1998: Memoria, retos y soluciones. https://scioteca.caf.com/bitstream/handle/123456789/675/Las%20lecciones%20de%20El%20Ni%C3%B1o.%20Ecuador.pdf?sequence=1&isAllowed=y
28. García-Hernández, M., Pérez-Magaña, A., Martínez-Corona, B., Gutiérrez-Villalpando, V.: Cambio de uso de suelo y variabilidad climática en Chiautzingo. Mexico. Revista Iberoamericana de Bioeconomía y Cambio Climático (6):11 (2021). https://dialnet.unirioja.es/servlet/articulo?codigo=8038318
29. Carrera-Villacrés, D., Guevara-García, P., Tamayo-Bacacela, L., Balarezo-Aguilar, A., Narváez-Rivera, C., Morocho-López, D.: Relleno de series anuales de datos meteorológicos mediante métodos estadísticos en la zona costera e interandina del Ecuador, y cálculo de la precipitación media. IDESIA, **34**(3), 81–90 (2016). https://www.scielo.cl/pdf/idesia/v34n3/art10.pdf
30. Pulido-Madrigal, L.: Cambio climático, ensalitramiento de suelos y producción agrícola en áreas de riego. Terra Latinoamericana **34**(2), 207–218 (2016). https://www.scielo.org.mx/scielo.php?script=sci_arttext&pid=S0187-57792016000200207

Network Neutrality Policies in the 5G South American Market

Roberto Triviño, Antonio Franco-Crespo, and Leonardo Ochoa-Urrego

Abstract The implementation of 5G is in process in South America and promises significant innovations, enabling a wide range of services and new business models. This new generation of mobile networks is designed to offer low latency, high speed, and reliability, as well as the ability to create logical network slices to provide differentiated packet treatment according to service requirements. However, these differentiation practices could violate the strict principle of network neutrality, which is present in some South American laws, and thus conflict with the deployment of new 5G services. This article analyzes the potential effect of network neutrality policies on the new 5G services in four countries. The results reveal that a flexible interpretation of network neutrality should accommodate virtualization, network slicing, and other new innovative services in 5G. At the same time, a strict application could restrict innovation in services and investment in South America.

Keywords 5G · Internet · Network neutrality · South America · Innovation

1 Introduction

Although the fifth generation of broadband mobile networks (5G) is still under development in South America, it promises to be more disruptive, with ultra-reliable communications, low latency, and high availability for massive communications, allowing for the development of new and innovative services [1, 2]. South America's interest in 5G is growing, and its deployment and adoption present a genuine chance to promote economic recovery and social inclusion across the region [3]. By 2030, 5G

R. Triviño (✉)
Universidad de las Fuerzas Armadas, Sangolquí 17110, Ecuador
e-mail: rdtrivino@espe.edu.ec

R. Triviño · A. Franco-Crespo
Escuela Politécnica Nacional, Quito, Ecuador

L. Ochoa-Urrego
Universidad de Santiago, Santiago, Chile

G. F. Olmedo Cifuentes et al. (eds.), *Emerging Research in Intelligent Systems*, Lecture Notes in Networks and Systems 902, https://doi.org/10.1007/978-3-031-52255-0_14

technology will represent nearly 60% of total mobile connections in Latin America [4], generating over USD 28.4 billion in increased mobile broadband revenues [3].

5G is not just an incremental advancement from the fourth generation. Still, it extends to multiple environments that allow companies to integrate new and improved innovative components [5] to achieve high levels of flexibility in the network, creating a new platform that meets the technical requirements of the so-called vertical industries: transportation, health, energy, etc. According to [6], platform innovations significantly change technological infrastructures that enable further innovations in processes or services. On the other hand, innovation in the mobile telecommunications market generally requires a high level of coordination between network infrastructure, devices, and applications, where exclusivity agreements prevail [5]. However, other authors argue for the benefits of open platforms [7, 8]. To achieve this goal, the International Telecommunication Union developed the IMT-2020 standard, which established a series of specifications and parameters required for fifth-generation mobile networks [9] based on the use of innovative technologies such as network virtualization (NFV), software-defined networks (SDN), cloud computing, and artificial intelligence, among others. This allows the creation of a network-slicing architecture, which enables the execution of multiple logical networks as virtually independent business operations on a single shared physical infrastructure.

Each mobile generation has emerged new business niches, market opportunities, and competition. The fifth generation is no exception, allowing the possibility of creating multiple virtual networks on demand, where "anything" can be offered as a service [1]. From a commercial perspective, this ability to provide on-demand services enables innovative business models that will generate more sources of income for mobile network operators (MNOs) than traditional broadband subscriptions. From a technical perspective, increasing the sharing of physical network resources can change how communications are provided, increasing competition and innovation in the mobile market.

On the other hand, most of the research on 5G has focused on its potential uses, scenarios, or economic benefits [2, 10, 11]. However, from a regulatory standpoint, little has been studied about how 5G could fit into the current regulations in the region. Sharing physical resources among networks, services, or applications has given rise to numerous debates around paid prioritization or data discrimination in a neutral network environment. For several authors [1, 12], one of the most concerning aspects that require further analysis regarding 5G is the principle of network neutrality (NN), which states that all Internet traffic should be treated equally without discrimination or prioritization [13]. Although this debate is not new, the development of new technologies and services has led several countries in South America to analyze and implement policies to guarantee this principle. Furthermore, they require national regulatory agencies (NRAs) to monitor the actions of network operators seeking to increase their revenues by charging for prioritized access or guaranteeing a certain level of quality of service (QoS) to application and content providers (CAPs), but at the expense of competition and innovation in the mobile telecommunications market.

Although NN laws aim to protect and defend the end-users' choice and maintain innovation in Internet services and applications [14], it is crucial to understand the regulatory effect on the challenging management of the mobile telecommunications market [15]. Considering that mobile networks are in a state of regulatory maturity, the innovative technical features of new mobile generations could challenge current regulations, and these regulations could hinder the development of new products or services. This is why this research aims to explore the potential impact of developing innovative 5G products and services in four mobile markets in South America (Brazil, Chile, Colombia, and Ecuador) due to current NN regulations. Thus, the following research question is explored: How does network neutrality regulation in South America affect services in 5G networks? Although the region is adopting various policies to boost the digital market, considering 5G as the driver of this digital transformation, it is unclear how these policies align with the current regulatory frameworks. Therefore, analyzing how the current NN regulations in the region can influence 5G services is of particular relevance, as the region needs to position itself better in numerous industries to remain competitive—taking into account that since the industrial revolution, countries have required to provide high-added value to economic sectors to stay competitive in a globalized world.

The remainder of the article is organized as follows: Sect. 2 describes 5G and the technologies associated with its implementation. Section 3 analyzes the development of 5G in four South American countries. Section 4 explores the most relevant aspects of NN laws in the region. Section 5 provides a brief discussion, and finally, the conclusions are presented.

2 Methodology

The objective of this work is to identify the development process of 5G networks in four South American countries and explore the potential impact of current NN regulations on future innovative 5G services.

The study uses an empirical methodology because its primary area of interest is the context of 5G in South America. The intricacy of NN and its impact on 5G, where the context affects the case in a situation with many uncontrollable and perceptible events occurring over time, makes a multiple case study with a single unit of analysis the chosen technique for this study. The authors in [16] state that numerous case studies are more thorough and enable the production of reliable outcomes.

The data sources for the article come from secondary sources through the review and analysis of official records and academic material on the subject, as well as regulatory and industry reports in the context of 5G and NN for each of the four cases studied. This approach is used due to the need to describe the process of 5G deployment and explore whether the development of new products or services aligns with compliance with NN policies based on combining theoretical understanding and empirical data to identify regulatory ambiguities or uncertainties.

3 The Concept of 5G

The fifth generation of broadband mobile networks will feature technological innovations and new features that will enable a wide range of new types of applications and services. Although the technical and commercial details are still evolving [17], a fundamental principle of its architecture is that 5G is not a one-size-fits-all network but that network functions can be adapted to a specific use case to facilitate innovations [14].

Thus, 5G offers lower latency and higher speed for a reliable network, capable of managing massive connectivity and accommodating millions of devices (see Fig. 1). To achieve this, extreme network densification is performed at the air interface using smaller cells, millimeter wave spectrum (mmW), massive MIMO, beamforming, and coordinated multipoint processing (CoMP). On the other hand, to make it scalable and flexible, using Software-Defined Networking (SDN) and Network Function Virtualization (NFV) allows replacing specific hardware functions with software that runs on general-purpose machines. This enables network capabilities through an application programming interface (API), where the application layer can manipulate services grouped in a portion of the 5G network (slices) for a specific use case [18].

From a technical perspective, most of these new features aim to increase resource sharing to improve network performance. But they also aim to reduce the cost of implementation, reconfiguration, and transitions that follow technological cycles. NFV seeks to use ready-made equipment to perform most of the computing, switching, and storage processing in 5G using cost-effective software in a data center,

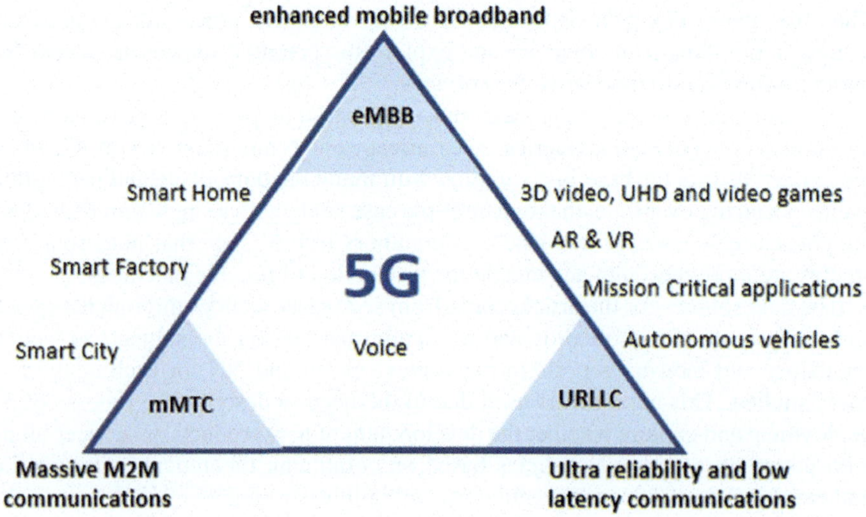

Fig. 1. 5G usage scenarios

network node, or near the user. Thus, cloud computing and edge computing can optimize traffic flows, applications, and systems near the network's edges. It is worth noting that SDN is one of the technologies intertwined with NFV. In SDN, network functions manage data and control, allowing different levels of quality of service (QoS) to be provided to different applications through software configurations. The scalability and flexibility provided by SDN, NFV, and other innovative technologies allow various network functions to be provided as flexible and scalable services [19]. One of the new concepts in 5G networks is the ability to slice the network and create end-to-end virtual network slices with independent control and management, capable of meeting specific requirements on demand and based on the same physical equipment [20]. This has profound implications for the system's organization, where complementary and interdependent innovations will be more critical than on the internet.

From a business perspective, "network slicing" is at the core of new solutions. This means that 5G applications and services will be diverse, fully customizable, and variable in prioritizing requirements, low latency, high speed, or availability. Therefore, mobile network operators (MNOs) could expand and create new business models [21] by offering innovative services and solutions for different market scenarios with diverse QoS or performance requirements, such as virtual mobile network operators (MVNOs), content and application providers (CAPs), or different vertical industry segments seeking differentiation and customization in their operations [22]. For [1], this on-demand service customization allows for innovative business models in the mobile telecommunications market, with differentiated pricing based on the resources used, generating more sources of income for MNOs, new investment opportunities in the network, and possible new challenges and regulatory needs.

Regarding regulatory aspects, ignoring the potential of 5G for South America means missing out on a considerable opportunity or leaving it in the hands of competitors. Allowing the implementation of 5G in the region is a prerequisite for attracting investments, as it is promoted as a means to reach the mass market, as a substitute for fixed broadband, and in the business market to adopt the digital transformation that keeps the region competitive in the global economy [23]. For this reason, the different National Regulatory Agencies (NRAs) in South America aim to provide legal certainty for the constantly evolving telecommunications sector. They recognize the growing role of the internet, applications, and content and the need for neutral technology regulation to balance the competition between traditional operators and CAPs. At the same time, they establish frameworks for spectrum planning and coordination for 5G without neglecting infrastructure policy, competition regulation, consumer protection, and collaboration with regulators and sector-specific legislation to make 5G a reality in South America.

4 5G Networks in South America

This new generation of mobile networks is more disruptive than previous ones, as it introduces a significant increase in data transmission speed and highly transformative specifications in mobile wireless communications. Generally, it can be observed worldwide, especially in Latin America, that policies in different countries favor the completion of 4G deployments, with varying degrees of support for 5G, as shown in Table 1 for the cases studied. To have a broad view of 5G network deployment in Latin America, [24] shows that the percentage of 5G networks compared to LTE is low, around 17%. Therefore, the deployment level is lower than North America, Europe, and the global average (31%).

However, 5G continues to make inroads in South America. As of March 2023, commercial 5G services were available in Argentina, Brazil, Chile, and Perú, with more markets expected to follow soon [4]. On the other hand, to promote investment and innovation, several countries are reviewing and adapting their regulatory approaches for 5G [17]. One of the main concerns of South American National Regulatory Authorities (NRAs) is ensuring that the entire population has broadband access and how to close the digital divide. In this regard, Mobile Network Operators (MNOs) are using 5G for new opportunities. For example, they initially employed fixed wireless access (FWA) 5G to drive the first broadband adoption in markets like Brazil and Colombia [25]. Also, the study identified that users are willing to pay for 5G access. However, it is vital to understand the market context in which the network is deployed, as there are constant pressures on MNOs to generate additional revenue and invest in infrastructure. Therefore, mobile operators aim to diversify their income with new business models allowing network slicing, traffic optimization, private networks, and other special services. Additionally, discussions on how current regulations may impact 5G and vice versa have only begun in the four selected countries (Brazil, Chile, Colombia, and Ecuador) at different stages of 5G deployment.

Currently, there are disparate situations in the region where the technology has been implemented, and growth has been accelerated. Where the implementation has not been orderly, growth has been more gradual. In this regard, there are two major groups of countries. On the one hand, Chile and Brazil have followed a similar roadmap in structure: conducting spectrum auctions, having commercial launches, and generating deployments. Another group of countries, Colombia and Ecuador, had plans to create auctions and spectrum licensing and have activated and illuminated

Table 1 Deployment of LTE and 5G networks per country [24]

Country	LTE	5G	5G bands
Brazil	6	4	0.7, 2.3, 3.5 y 26 GHz
Chile	4	4	0.7, 2.1, 3.5 y 26 GHz
Colombia	5	0	3.5 GHz
Ecuador	3	0	N/D

specific 5G sites, but with primarily private initiatives. It is worth noting that each country has its regulatory framework that establishes the conditions under which the processes of concession or allocation of radio spectrum should be carried out, as well as the steps that must be followed for their completion. Additionally, the existing circumstances in each market also influence these processes [4].

4.1 Brazil

The development of 5G is included in the state's growth planning for the telecommunications sector. With the approval of the 2021–2028 Radio Spectrum Usage Plan, the 5G spectrum auction for various bands took place (see Table 1) [24]. The frequency allocation to operators included service provision and coverage commitments across all bands for municipalities and highways. This achieved a balanced solution that serves citizens, operators, and government development goals. Thus, Brazil has the highest number of operating 5G networks in the region [2], with four companies having active networks. However, there are still many challenges to the complete deployment of 5G in Brazil, such as the lack of telecommunication infrastructure in some areas, the limited availability of 5G devices, and the need to allocate more spectrum for 5G services in the future [25]. However, according to [18], implementing the 5G network introduces new regulatory challenges, particularly in applying NN principles. The technical features of this new generation of networks raise questions about balancing network management, network slicing, and Quality of Service (QoS) with the preservation of NN. Therefore, a balanced approach is needed to promote innovation and technological development without compromising the fundamental principles of NN and Internet user rights.

4.2 Chile

In 2020, a public tender for 5G was opened for the installation, operation, and exploitation of high-speed wireless networks in various bands [24]. Like Brazil, it included 5G coverage obligations in remote, rural, and non-profitable urban areas. By the end of 2021, deployment processes were authorized, and in the first quarter of 2022, three operators began offering commercial services. In Chile, more than 2 million lines have been enabled by the end of 2022 [24]. According to [12], 5G also brings various regulatory challenges, such as the potential conflict between a 5G-driven Internet and the vision of a global, open, and non-discriminatory Internet. This is because 5G allows for creating multiple end-to-end virtual networks on a shared physical infrastructure that is managed independently. While differentiated traffic management is fundamental to 5G technology and drives significant technological innovations and new business models, this same arbitrary differentiation of traffic is also what NN aims to prevent, as both content and application providers and end users of the internet should receive equal treatment from network providers.

4.3 Colombia

With the "5G Colombia Plan," policy guidelines and strategies for 5G deployment in the country have been established since 2019 [24]. This includes updating public policies and regulatory and legal frameworks to enable 5G deployment and mass adoption. In 2020, the regulatory agency published a document called "Study on the Regulatory Conditions to Foster the Adoption of 5G Technology in Colombia," which includes monitoring the commitments established for mobile network modernization, regulating operator participation in the tender, spectrum allocation, and the relevance of monitoring the regulations associated with NN considering the technical characteristics of 5G technology and the possibility of network slicing. Although the spectrum auction scheduled for the end of 2021 has been delayed, operators continue with pilot tests [25]. Colombia is also estimated to have its initial deployments in 2023 in the frequency bands of 700 MHz, 3500 MHz, and 26 GHz.

4.4 Ecuador

This nation lags in the implementation of 5G technology [26]. Before the pandemic, the plan was to expand 4G coverage by 2021 and have 5G services in 2022 in the 700 MHz and 3.5 GHz bands. Although 5G trials began in 2021 by the state operator, band concession processes have not yet started [25]. However, measures have been taken to promote investment in telecommunications network deployment, including reducing taxes, reducing equipment import tariffs, simplifying regulatory simplification, and updating the National Frequency Plan [2]. Thus, to establish the rules for new spectrum assignments, the NRA decided to conduct a valuation based on technical–economic criteria and with the advice of international organizations, which would favor industry interest in the currently available spectrum for 5G allocation. However, other rules or regulations that could interfere with deployments have not been reviewed yet. Therefore, the competitive spectrum allocation processes will start in the last quarter of 2023 through public bidding or auction [4].

The regulation of 5G influences the market, strategies, and decisions of MNOs to explore innovation opportunities in new services and applications [12]. Most of the early 5G deployments in the region began with a non-standalone architecture, using a 4G core network for control functions and 5G radio access network equipment for connectivity [25]. This allowed MNOs to launch 5G services quickly and at a lower cost. Standalone (SA) 5G technology is only available in Brazil. Therefore, it's also important to highlight that deploying SA 5G networks will generate interest in fully customizable applications and services network slicing provides. However, the fact that MNOs are still conducting network slicing trials indicates much work to be done before commercialization [25].

On the other hand, each country has unique characteristics. However, South America's demand for 5G among enterprises aligns with more advanced 5G markets [4]. According to [23], the retail sector will be the most significant vertical to benefit from the deployment of 5G in the mass market, followed by manufacturing and ICT in terms of impact, and then agriculture and government (smart cities). Nevertheless, user protection is an important point in a service that aims to be essential for economic activities. Many of the existing regulations will be maintained, but in some cases, regulatory simplification or modernization is necessary. For example, the long-standing debate on NN restricts arbitrary traffic discrimination and allows service optimization based on special technical requirements or the prioritization of emergency services. Since several of the new 5G services offered would apply network optimization and segmentation, regulators must be vigilant to guarantee that there are no anticompetitive behaviors with arbitrary acts that could affect horizontal and vertical business relationships among ecosystem actors while preserving the principle of an open Internet.

5 Network Neutrality in South America

The sharing of physical resources, discrimination, and prioritization among different applications and services initiated the debate on Network Neutrality (NN), which refers to the principle that all data packets transmitted over the internet should be treated equally without discrimination or prioritization [13]. The basic rules are no blocking, no throttling, and no paid prioritization [27], and these guidelines govern the relationships between network operators and actors in other layers of the ICT ecosystem. The effects of these rules on innovation and the development of new businesses are influenced by their implementation. A strict enactment of NN requires that every data packet should be treated equally, without discrimination, regardless of its origin, destination, or content. In contrast, a flexible implementation of NN allows for the differentiation of services and pricing as long as they are applied non-discriminately to all network actors [17].

Depending on the nation, NN regulations vary from prohibited practices to permissive or reactive approaches [28]. Therefore, any form of discrimination regarding QoS and pricing should not be allowed in strict implementation [29]. In recent years, as a commercial practice to increase revenue, some operators have attempted to charge CAPs to ensure specific QoS. However, prioritization involves creating a fast lane on the network where privileged traffic flows with higher quality while other types of data experience congestion (e.g., delays, variability, or speed reduction). With this commercial practice, providers could directly influence competition among and discriminate against CAPs, violating the principle of NN.

NN is a current topic of discussion regarding discriminatory traffic management policies and how they could threaten competition, consumer choice, and innovation

Table 2 NN laws in the case of countries [31]

Country	Law/Resolution	Type
Brazil	Law N° 12.965/2014	Strict
Chile	Law 20.453/2010 (first in the world)	Strict
Colombia	Law No.1450 and resolution CRC 3502	Flexible
Ecuador	Art 3 Telecommunications Law/2015	Flexible

[30]. In response to concerns about the potential abuse of providers to discriminate against or prioritize content and services, several governments worldwide have implemented NN laws that differ between countries where mobile networks have less restrictive NN rules than their fixed counterparts. In 2015, the European Union (EU) regulatory framework and the Open Internet Rules proposed by the Federal Communications Commission (FCC) in the United States approved laws to ensure an open state of the Internet. However, even when the FCC revoked the 2015 NN rules in December 2017 [31], in Europe, the Regulation on the Single Market for Telecommunications allows for NN and specialized services that are different from Internet access services but optimized for specific content, applications, or services with a precise level of quality.

In South America, the NN laws and regulations enacted in the four countries studied aim to maintain the principle that "all traffic should be treated equally" without discrimination (see Table 2). The implementation of NN in Brazil and Chile strictly adheres to the NN principle. Colombia has a flexible enactment, while Ecuador does not have an NN law, but compliance with the principle is in its Organic Telecommunications Law.

Although the implementation of NN regulation has different forms, South America seeks to maintain the basic rules in its laws, which state that Internet access providers shall not block, interfere, discriminate, or restrict the right of any Internet user to use, send, receive, or offer any legal content or service through the internet, regardless of the content, origin, destination, service, device, or application. Additionally, traffic management should be reasonable and based on technical criteria [31]. Among the differences, Chile has the strictest regulation. Brazilian regulation is the only one that establishes exceptions, where traffic discrimination can occur only for temporary essential technical requirements for the proper provision of services and applications or during the prioritization of emergency services. On the other hand, Colombia states that providers can make offers according to the market segments' needs based on usage and consumption profiles without being considered discriminatory.

In contrast, Brazil prohibits prioritizing data packets based on commercial arrangements or privileging applications offered by an infrastructure network operator [32]. Meanwhile, Ecuador includes the principle of NN in its Organic Telecommunications Law to protect users and competition without limiting arbitrary discrimination. Table 3 presents the list of case countries in the region and the differences between their regulations and laws [31].

Table 3 Policies comparison among countries

Criteria	Country			
	CH	CO	BR	EC
Regulation	Y	Y	Y	N
Define NN and other concepts	•	o	•	–
Guarantees to access or use of services or content	•	•	•	•
Prohibits blocking or discriminating access or use	•	•	•	•
Transparency of the Information	•	•	•	•
Use of HW to access	•	•	•	•
Set price by content, service, or market	–	•	▲	o
Implement ITM techniques	•	•	▲	•
Exceptions (technical requirements, emergency)	–	–	•	–
Zero Rating	▲	o	▲	–
Control of the obligations	•	o	•	–

• Specified, o Partially Specified, Not Specified, ▲ Prohibits.

While in South America, each country has different contexts and specific technological access and market needs to deal with, current regulatory policies regarding NN seem to affect the development of 5G and new business models. For instance, network slicing allows for providing different network requirements according to commercial service needs, but this is a form of paid prioritization prohibited by most current laws.

6 Discussion

Much of the multidisciplinary research on 5G has focused on its potential uses, scenarios, and economic benefits without considering how 5G services could fit within existing regulatory frameworks. However, in recent years, the implications of 5G for some industries have become more relevant from the perspective of competition, investment, and innovation. For [1], one of the most concerning aspects that require a more in-depth analysis regarding 5G is the principle of NN. The debate on NN has gone through different stages. Still, its intention has always been to prevent the possibility of a network operator discriminating against services or content, creating differentiated Internet access in favor of large companies, thereby affecting the open access, openness, and innovation that characterize the Internet ecosystem.

On the other hand, considering that several current laws in the region are technologically neutral, from an NN view, 5G networks could face some restrictions in most countries, especially regarding:

1. Internet traffic management practices oriented towards prioritization. Given that NN laws establish that network providers must treat any data packet equally,

regardless of its content, origin, destination, service, or application. This provision is relevant for 5G since it must comply with "reasonable" prioritization while ensuring transparency, temporality, and non-discrimination.

2. Prioritization must have a technical basis in quality of service and not a commercial one to avoid anticompetitive practices. However, in regulations that include exceptions, it is determined that they can be implemented only due to indispensable technical requirements (network congestion) and emergency services. This could be a challenge for 5G networks since, from a technical perspective, their design offers low latency and high speed for a reliable network. Additionally, they can create logical networks with different requirements in the various layers that constitute them. Under a strict NN regime, these characteristics are the principal concern for implementing network slices, as well as potential over-optimizations, costs, and ensuring fair competition among new services within the same vertical industry.
3. Differentiation of services for individual users or use cases is not permitted. NN laws regulate the commercial relationships between the network operator and actors in the logical layer, so it is not possible to "compromise the public character of Internet access," "prioritize data packets due to commercial agreements," or privilege applications offered by MNOs [32]. Since network slices in 5G permit the provision of different network capabilities according to service requirements, it can be considered a class of paid prioritization prohibited by strict NN laws. However, for flexible implementations of NN with price differentiation based on consumption or market profiles, 5G and network slices could be permitted for certain services as long as the network has sufficient capacity.

Moreover, it is necessary to analyze that 5G requires investment in cloud service technologies and virtualization to improve network efficiency, perceive it as having infinite capabilities, and enable more disruptive business models. While the current NN regulations in the region do not address these technologies, they limit the actions of MNOs to leverage these investments and offer differentiated services. When virtual machines or cloud computing run on MNO-owned equipment, they become part of the network, and their use is regulated under NN. Therefore, equal treatment is required regardless of their content, service, or application, potentially limiting the benefits of these technologies to a single use. On the other hand, when a cloud service provider owns the equipment, it can run any software without regulatory interference for different benefits, similar to an end-user-owned device, under which traffic management is allowed under NN. However, considering economies of scale and the goal of maximizing the return on investment made by MNOs, there could be incentives for major cloud service providers (AWS, Google, Meta, and Baidu) [14] to take over some core network functions and thus some business models to the detriment of third-party innovation or the MNOs them-selves.

Undoubtedly, 5G represents an opportunity for MNOs to create new business models with differentiated services to serve many use cases [17]. However, as observed, the current NN regulations in the region create uncertainty about how and if MNOs can use 5G to generate new revenue streams with network segmentation and service differentiation based on traffic type. This aspect discourages innovation and restricts investment in these networks, hindering their deployment and potential success in different vertical industries. Therefore, several authors agree that various elements of NN regulation should be reviewed and analyzed in light of new technological trends and take a more open stance on data prioritization. Thus, there are two scenarios identified for the advancement of 5G networks concerning network neutrality:

- Maintain partial or flexible NN, allowing for the coexistence of NN and 5G. Nowadays, the internet is not neutral in South America. Various forms of discrimination are applied even in countries with strict NN implementations, such as Brazil and Chile. For example, the "zero rating" is incompatible with most established NN regulations [33]. In 5G, the application of NN becomes complex since prioritization or differentiation based on network segments is intrinsic to its architecture. As a result, it is advised that a partial or flexible implementation of NN could be considered, ensuring that data is treated equally inside the same network segment regardless of its content, access device, application, hardware, operating system, origin, or destination. It can be achieved by (a) analyzing different types of traffic, their restrictions, and requirements; (b) determining network segments based on traffic type in terms of providing a level of quality of experience (QoE); and c) if congestion occurs and quality degrades, differentiating services to provide at least sufficient QoE for each slice while treating all packets equally.
- Abandon NN, claiming technically and economically that it is not in line with current technological developments since they are designed to offer low latency and high speed. Moreover, implementing 5G networks within this framework permits MNOs to intelligently segment network traffic to secure differentiation in QoS in their networks between (a) different network segments and (b) among services within the same network segment. This scenario allows MNOs to create differentiated plans for distinct classes of services and special plans for the same service for subscribers with the financial capacity to pay for differentiated treatment, thus departing from the ideas of arbitrary discrimination and paid prioritization.

7 Conclusions

Although the fifth generation of broadband mobile networks is in the process of implementation in South America, they promise significant innovations and improved characteristics that will enable a wide range of new business models and unknown services. Their architecture aims to increase resource sharing to enhance network performance and management by using technologies such as NFV, SDN, and cloud and creating network slices for different use cases. The successful implementation of 5G requires a solid regulatory framework and collaboration among stakeholders and regulators to guarantee legal certainty, promote competition and innovation, and coordinate infrastructure and spectrum planning.

To maintain an open Internet as an instrument for competition and innovation in South America, the objective is for all Internet traffic to be treated equally, without discrimination or paid prioritization. The implementation of NN differs among countries, varying from strict approaches that do not allow discrimination or differentiation of data (Brazil and Chile) to flexible procedures that permit certain forms of service differentiation and offers based on usage and consumption profiles (Colombia), or only expressing the principle without a specific law (Ecuador). In this context, the capacity of 5G networks to create new business models and offer differentiated services in network slices based on QoS characteristics could not be fully realized in South America under a regulatory vision restricted to "best effort" and equal treatment, thus limiting their innovative potential despite commercial deployments already available in the region, particularly in countries with strict NN laws.

Thus, a conflict between 5G services and the current NN regulatory frameworks is identified in South America, creating uncertainty. The standpoint toward data prioritization permits two possible scenarios. The first, which is more feasible, considers maintaining flexible or partial NN, along with a guarantee of equal treatment for all data within the same logical network slice that could allow for positive effects by providing some security to invest in the network. The second scenario is to abandon NN because network operators have evolved from having "dumb networks" to having "smart networks" capable of managing traffic better, making discriminatory or anticompetitive practices unnecessary and letting the market regulate them.

References

1. Frias, Z., Pérez, J.: The challenge of net neutrality policies for 5G networks. IEEE Internet Initiative eNewsletter (2017)
2. De León, O.: Redes 5G en América Latina: desarrollo y potencialidades, Documentos de Proyectos (LC/TS.2022/181/Rev.1), 1st edn. Comisión Económica para América Latina y el Caribe, Santiago (2023)
3. Ericsson: The state of 5G in Latin America: a closer look, https://www.ericsson.com/en/reports-and-papers/mobility-report/closer-look/latin-america. Accessed 20 May 2023
4. GSMA Intelligence: 5G in Latin America—Unleashing the potential, https://www.gsma.com/latinamerica/wp-content/uploads/2023/07/290623-5G-in-Latam-ENG.pdf. Accessed 18 June 2023
5. Bauer, J.,Woohyun, S.: Regulation and digital innovation: theory and evidence. 23rd European Regional Conference of the International Telecommunication Society, Austria (2012)
6. Krancke, J., Muller, C.: Innovation and regulated industries, Schmalenbachs Zeitschrift für betriebswirtschaftliche Forschung, vol. 63, pp. 32–51. Springer (2011)
7. Lemley, M., Lessig, L.: The end of end-to-end: preserving the architecture of the internet in the broadband Era. UCLA Law Rev. **48**(4), 925–972 (2001)
8. Van Schewick, B.: Internet Architecture and Innovation. Cambridge. 1st edn. MA: MIT Press, Cambridge (2010)
9. Craven, C.: What Is IMT-2020. SDX Central. https://www.sdxcentral.com/5g/definitions/what-is-5g/5g-standards/imt-2020/ (2018)
10. 5G NORMA.: Deliverable D2.1. Use cases, scenarios and requirements. 5G Novel Ratio Multi-service adaptive network Architecture (2015). https://5gnorma.5g-ppp.eu/wp-content/uploads/2016/11/5g_norma_d2-1.pdf. Accessed 15 April 2023

11. Yoo, C., Lambert, J.: 5G and net neutrality. In: The future of the internet—innovation, integration and sustainability 1st edn, TPRC47: The 47th Research Conference on Communication, Information and Internet Policy 2019, U of Penn, Inst for Law & Econ (2019)
12. Alexiadis, P., Shortall, T.: The advent of 5G: Should technological evolution lead to regulatory revolution?, CPI Antitrust Chronicle NOVEMBER, Vol. 3, (2016)
13. Wu, T.: Network neutrality, broadband discrimination. SSRN Scholarly Paper Rochester, NY: Social Science Research Network (2013)
14. Kantola, R.: Net Neutrality Under EU Law—a Hindrance to 5G Success, 30th European Regional ITS Conference, International Telecommunications Society (ITS), Helsinki (2019)
15. Yoo, C.: Wireless Networks: Technological Challenges and Policy Implications, Inst for law & Econ research, U of Penn (2013)
16. Yin, R.: Case Study Research: Design and Methods, 4th edn. SAGE, Los Angeles USA (2009)
17. Bauer, J. ,Bohlin E.: Regulation and innovation in 5G markets. Telecommun. Policy **46**(4) (2022)
18. Moraes, A.: 5G Network neutrality in Brazil: novel technological challenges to the Brazilian civil rights framework for the internet. The 48th Research Conference on Communication, Information and Internet Policy (2019)
19. Ordonez-Lucena, J., Ameigeiras, P., Lopez, D., Ramos-Munoz, J., Lorca, J., Folgueira, J.: Network slicing for 5G with SDN/NFV: concepts, architectures, and challenges. IEEE Commun. Mag. **55**(5), 80–87 (2017)
20. NGMN: A delivery by the NGMN 5G White paper2015, https://www.ngmn.org/work-programme/5g-white-paper.html. Accessed 25 February 2023
21. Iwamura, M.: NGMN view on 5G architecture. In: 2015 IEEE 81st Vehicular Technology Conference (2015)
22. Jiang, M., Condoluci, M., Mahmoodi, T.: Network slicing management & prioritization in 5G mobile systems. In: European Wireless 2016; 22th European Wireless Conference (2016)
23. Swain, W., Lopes, A., Agnese, S.: Why 5G in Latin America? A call to action for Latin American operators and policymakers, An Omdia–Nokia research paper (2020)
24. 5G Americas.: 5G & LTE Deployments, https://www.5gamericas.org/resources/deployments/, (2022). Accessed 27 December 2022
25. GSMA Intelligence.: La Economía móvil en américa latina, https://www.gsma.com/mobileeconomy/wp-content/uploads/2022/11/SPANISH_GSMA_LATAM_ME2022_R_Web.pdf. (2022), Accessed 14 December 202
26. Pazmiño, L., Flores, F., Becerra, F., Cevallos, G., Rivera, R.: Challenges and opportunities of 5G deployment in Ecuador, Incodtrin 2022. Kalpa Publications in Computing **15**, 58–70 (2022)
27. Soares-Ramos, P.: Toward a developmental framework for net neutrality: The rise of sponsored data plans in developing countries. TPRC Conference2014, http://papers.ssrn.com/sol3/papers.cfm?abstract_id=2418307. Accessed 15 February 2022
28. Garrett, T., Setenareski, L., Peres, L., Bona, L.,Duarte E.: A survey of network neutrality regulations worldwide. Comput. Law Secur. Rev. **44** (2022)
29. Krämer, J., Wiewiorra, L., Weinhardt, C.: Net neutrality: a progress report. Telecommun. Policy **37**(9), 794–813 (2013)
30. Garrett, T., Dustdar, S., Bona L., Duarte E.: Ensuring network neutrality for future distributed systems. 2017 IEEE 37th International Conference on Distributed Computing Systems (ICDCS), pp. 1780–1786 (2017)
31. Triviño, R., Franco, A., Ochoa L.: Network neutrality: the case of five South American countries. Proceedings of CIT 2020, Advances in Intelligent Systems and Computing (AISC) series, Springer (2020)
32. Derechos Digitales.: Neutralidad de red en América Latina: Reglamentación, aplicación de la ley y perspectivas, 1era. edn, Sao Paulo, Brasil (2017)
33. Triviño, R., Franco, A., Ochoa L.: Zero rating effects in South American Countries, Ninth International Conference on eDemocracy & eGovernment (ICEDEG), pp. 1–6. IEEE, Quito, pp. 1–6 (2023)

Evaluation of the Internal Temperature and Hours of Thermal Discomfort Produced by Sunshine: Study of a Hot Dry Climate

Javier G. Ascanio-Villabona, Brayan E. Tarazona-Romero, Nilson Y. Castillo-Leon, and Omar Lengerke-Pérez

Abstract This paper presents the advances in forecasting mechanical properties by applying machine learning regressors in advanced manufacturing processes. Computational intelligence is advancing by leaps and bounds in artificial intelligence and data science, which is how data is currently considered the new gold. In addition to advances in information technology, manufacturing processes have also evolved. In that sense, additive manufacturing has several advantages over conventional manufacturing processes. On the one hand, it offers the possibility of processing any material: metals, ceramics, polymers, and composite materials. On the other hand, these technologies make it possible to optimize the use of the material, reducing its impact on the environment. Due to its versatility, it is possible to improve design to save material and reduce assembly complexity. This work presents a series of cases of applying machine learning regressors in forecasting mechanical properties in advanced manufacturing processes, from friction welding, laser welding, selective laser melting, wire arc additive manufacturing, and fused filament fabrication. The results show that the gradient-boosting regressors have greater precision in predicting several mechanical properties than conventional artificial neural networks. These algorithms are more effective than using the traditional design of experiments, reducing costs and allowing information about the physical phenomenon involved in each manufacturing process.

Keywords Heat index · Thermal sensation · Monitoring · Microclimate · Thermal discomfort

J. G. Ascanio-Villabona (✉) · B. E. Tarazona-Romero · N. Y. Castillo-Leon · O. Lengerke-Pérez
College of Natural Sciences and Engineering, Unidades Tecnológicas de Santander, Bucaramanga, Colombia
e-mail: btarazona@correo.uts.edu.co

G. F. Olmedo Cifuentes et al. (eds.), *Emerging Research in Intelligent Systems*, Lecture Notes in Networks and Systems 902, https://doi.org/10.1007/978-3-031-52255-0_15

1 Introduction

Energy is an essential resource in almost every aspect of life. In developed countries, energy demand and consumption now have critical environmental, social, and economic impacts due to population growth, technological development, and economic progress [1, 2]. As the building sector is a significant contributor to energy consumption, various innovative strategies have been proposed to reduce the energy consumption of buildings. One of these strategies involves studying the effects of shading on structures and the use of solar energy [3].

In general terms, shadow analysis describes the process by which the shadow profiles of an object and its environment are studied at a particular time of day or during certain days of the year to determine the incidence of solar radiation correlation to its temperature generated to take advantage of it or block it [4]. Fundamentally, there are three types of tools for shadow analysis: solar path diagrams, heliodon devices, and software specialized in the energy simulation of buildings [5]. Solar trajectory diagrams are graphically formatted tools that determine apparent changes in the sun's position at any time of day throughout the year [6].

These tools are durable but expensive to manufacture. Software packages have been introduced to address a wide range of energy simulations and analyses of buildings. These tools have significantly improved designer productivity [7].

Energy simulation software is a comprehensive environmental and energy design tool that covers a wide range of simulation and analysis functions needed to truly understand how a building's design will work and perform. Climate design principles, such as solar, thermal, shading, lighting, and airflow energy, are most effectively addressed early in the design process [8]. With these types of tools, architects and engineers can measure fundamental criteria that will affect the performance of the building in the conceptual and detailed phases of the design. Thanks to these tools, techniques are worked more efficiently in 3D CAD to interpret results [9]. From the above, this project analyzes the incidence of sunlight in a case study house in a hot, dry climate. This is done through the experimental research methodology with a quantitative approach, which begins with collecting the data required to enter the analysis software. Once the data is entered into the program, the house is modeled. Then, the incidence analyses are carried out for one year, obtaining the figures of the trajectories of incidence of the shadows on the house and the internal temperatures in it, starting from the external temperatures in the environment. Likewise, the internal temperature is when there is a higher outside air temperature. The more direct effect of the sun's rays generates more hours of thermal discomfort for people in living spaces. This increase in temperature and thermal discomfort indirectly increases energy consumption for cooling. It reduces air quality in cities, so it is essential to calculate the hours of thermal discomfort generated with direct sunlight during the study year. Different authors and cases have used methodologies to perform the hours of thermal discomfort [10–14]. For this case, the method of cooling power to determine the hours of occupational thermal discontent proposed by Colombia's Institute of Hydrology, meteorology, and Environmental Studies [13] is used.

2 Methodology

This research is based on an experimental methodology due to the possibility of choosing and modifying variables within the design tool. They are generating simulations in the Desing Builder software to obtain the behavior of the house case study and using a quantitative approach.

According to Hernández et al. [15], the quantitative approach consists of a sequence of processes that are tested, so each stage precedes the next, where the hypothesis is established, and the relevant variables are determined for its subsequent design and from there, analyze using statistical methods, the values obtained in the measurements. Energy is essential in almost every aspect of life [2]. In developed countries, energy demand and consumption now have critical environmental, social, and economic impacts due to population growth, technological development, and economic progress. As the building sector is a significant contributor to energy consumption [16], various innovative strategies have been proposed to reduce the energy consumption of buildings. One of these strategies involves studying the effects of shading on structures and the use of solar energy [3]. In general terms, shading temperature analysis describes how the shadow profiles of an object and its surroundings are studied at a specific time of day or during certain days of the year to take advantage of solar radiation or block it [4]. Fundamentally, there are three types of tools for shadow analysis: solar path diagrams, heliodon devices, and software packages. Solar path diagrams are graphically formatted mechanisms that determine apparent changes in the sun's position at any time of day throughout the year [6]. A heliodon is a mechanical device simulating the sun's seasonal and hourly position changes through an artificial light source and positioning mechanism. These tools are durable but expensive to manufacture. Software packages have been introduced to address various building energy simulations and analyses. These tools have significantly improved designer productivity [17]. With the above, the procedural methodology for developing the research is established and structured in stages table. As initiation stage (A) Preparation of model, where the climatic parameters of the area and information for the case modeling are established. Following (B) Construction of the model: The model's technical, structural, and solar orientation characteristics are verified. Once the model is built, the third stage of (C) Simulation is carried out, where the replica of the dynamic behavior of the house is carried out to finally obtain the microclimatic data of the temperature and internal humidity of each of the areas of the house to perform (D) Analysis of the behavior of the temperature as a function of sunlight, also calculate the hours of thermal discomfort through the methodology used by the Institute of Hydrology, Meteorology and Environmental Studies of Colombia (Fig. 1).

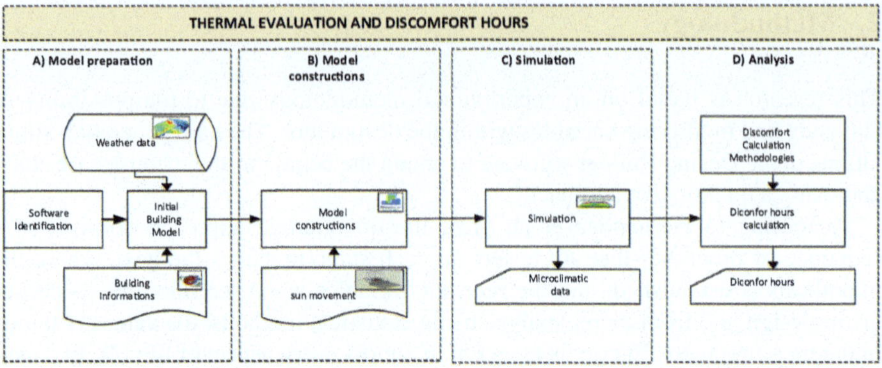

Fig. 1 Methodology for determining hours of discomfort due to the effect of sunlight

2.1 *Identification del Software*

The methodology to analyze the incidence of sunlight using software on a building begins with selecting a computational tool specialized in energy studies of the building, taking into account the geographical position with the latitude and longitude of the site of the land and observing the shadow projected on the model. The analysis of shading or sunlight represents a strategy to reduce electricity consumption by cooling and ensuring the living room's thermal comfort [7, 18–21]. This sunlight in the house is directly dependent on orientation, the envelope of the building, and the relationship between the solid wall and the transparent wall or opening to identify the energy balance by Direct Solar Gain, as well as the energy load requirements for air conditioning [8, 22–24]. In addition, shading analysis identifies the possible error of overestimating cooling or heating loads, which would result in unnecessary energy consumption for the air conditioning of buildings. On the other hand, it is also possible to estimate the impact of light or the use of natural light inside living spaces without increasing the thermal load by radiation [25]. Shading analysis tools make it possible to assess the effect of direct solar incidence on buildings and verify, during the architectural design stage, the potential impact on the cost of air conditioning, thermal comfort, and even aspects of natural lighting [26]. They also allow the proper design of the shading elements themselves, such as eaves, parasols, and openings or windows of a building, so that there is a balance between Direct Solar Gain (SGD) and light comfort to avoid glare or shadows, especially in desert or high radiation areas [27]. Design Builder is a BIM environmental simulation tool that can analyze thermal loads, lighting design, shadows and reflections, sun shading devices, and solar radiation [28, 29].

Table 1 General description of the house

Basic parameters	Specification
Location	7°08′N 73°08′TO
Total area	126 m^2
Height of the house	2.5 m on each story
Structure	Concrete/steel/stucco
Surround	Brick/cement/stucco

2.2 *Case Study Area Selection*

The house has a total area of 126 m2 distributed over two stories. The kitchen, dining room, main bedroom with private bathroom, and internal stairs leading to the second story are allocated on the first story. On the second story, there are two bedrooms and a general bathroom. General information about this building is shown in Table 1.

2.3 *Climate Data*

Bucaramanga has a very unpredictable, dry, warm climate, so its climatic conditions vary according to the year's season. These local conditions interact with humans and their homes, so there is a need to improve the comfort of buildings and the quality of life by increasing efficiency in energy consumption in the residential sector. From there, the mitigation measures to be implemented are related to the design, construction, equipment, and operation that privilege the lighting and ventilation of buildings.

During the year, it presents rains distributed in dry and rainy seasons. December, January, and February are the driest months, unlike June, July, and August, which occur in smaller proportions. The rainy season starts from March to May and September to November [30].

Its average temperature is 22.6 °C, and the presented sunshine is 4 h a day or less during the rainy months. For the dry months, it is between 5 and 6 h a day [30]. Figure 2 shows the location of the weather station.

2.4 *Modeling in the Software*

Building orientation refers to how the dwelling is situated on a site and the positioning of windows, roof lines, and other features. The software analyzes the model for the best orientation using the sun's path, allowing the façade to be projected to provide openings in a building. Figure 3 shows the model created in the program with the specifications of the house case study mentioned above [30].

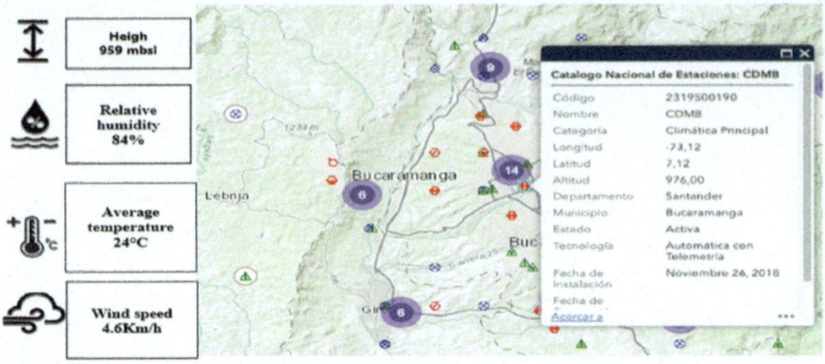

Fig. 2 Location of weather station

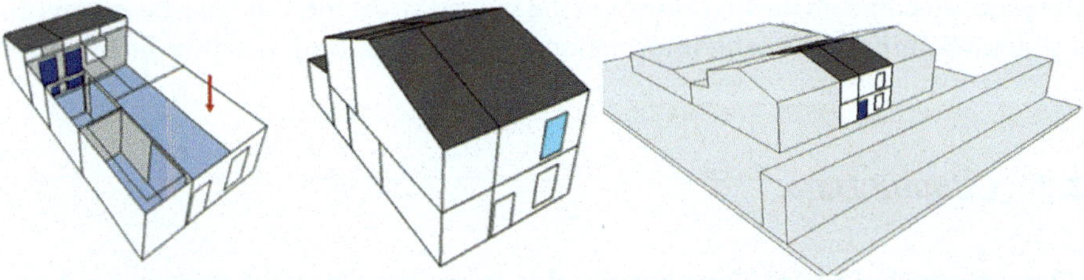

Fig. 3 3D drawing of the studio house

2.5 *Structural Characterization*

Once the studio home is modeled, building materials are assigned to improve calibration and bring them closer to the conditions of the natural environment. The design-builder has, by default, a library with various materials classified according to the surface, enclosure, and roof. The materials are assigned to the program based on Table 2, which is compiled from the actual model's construction materials.

Table 2 Building materials of the house

Element	Refinement
Outer wall	2cm stucco + 4cm concrete + 2cm steel + 4cm concrete + 2cm stucco
Interior wall	1cm stucco + 2cm concrete + 10cm red brick + 2cm concrete + 1cm stucco
Soil	7cm cast concrete + 2cm cement plastic + 2cm ceramic earthenware
Cover	1.91cm wood panels + 1cm PVC waterproofing + 2cm lightweight concrete + 3cm clay tiles
Glazing	3mm clear glass (external windows)

2.6 Defining the Location

Subsequently, the geographical location where the study model will be is configured, which in this case will be Colombia exactly located at 7°03′32″ from the North and 7°04′54″ W and 930m at sea level. This item will be of great importance for the analysis to be carried out. This software will place the house in a solar reality depending on its geographical position so that the solar rays can be projected in the period to be studied. Climatological data contain variables such as temperature, relative humidity, heat index, wind speed, etc. For this shadow analysis, this data is necessary, so it is loaded into the model made in the software.

3 Results and Discussion

A single-family house model is created and simulated with time intervals per hour during the 12 months of 2022. Table 3 shows a 3D view and trajectory diagram of solar incidence on housing incidence at the initial and final peak hour for each month of the year. Project information, such as site location, orientation, altitude, and terrain, is entered into the model—the weather file is imported from the software.

The above illustrations allow the visual interpretation of where the sun is and its impact on the different building parts during other hours of the day. As can be seen, the shaded area of the story is more significant in the final rush hour, contrary to the initial rush hour. The software can calculate the operating and external temperatures in the house during the simulated year, as shown in Table 4. These graphs show the evolution of the housing model's temperatures during one month, from May 1 to 30 of these months.

The observed behavior of the previous graphs is characterized by the variable trend of the operating temperature on the external temperature that covers the house. From these graphs, the operating temperature inside the house is determined and previewed, where a pattern of 23 °C is presented as a minimum temperature and does not exceed 35 °C. The trend in the May graph indicates an increase in outdoor temperature exceeding 35 °C throughout the month. On the other hand, in January, February, March, April, July, September, October, November, and December, it is observed that the environments of the house reach the maximum temperature inside, exceeding the maximum outside temperature, where it is presented that the environments in these periods in the home remain warmer during this time. The analysis of the thermal sensation establishes formulas adapted to climatic seasons (spring, summer, autumn, and winter). However, according to the above, the methodology used is based on the fact that it is not strictly for areas with a diversity of climatic seasons due to the geographical location of the study area with proximity to the equator. Also include the study variables generated by the simulation, such as temperature and relative humidity, which influence the thermal sensation of a given area. Finally, the thermal sense described the results qualitatively through those established by a quantitative

Table 3 Shade from sunlight

Month	Initial rush hour	End rush hour	Month	Initial rush hour	End rush hour
1	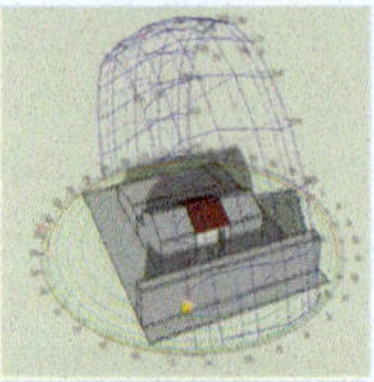	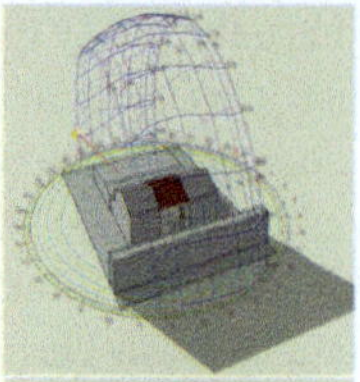	6	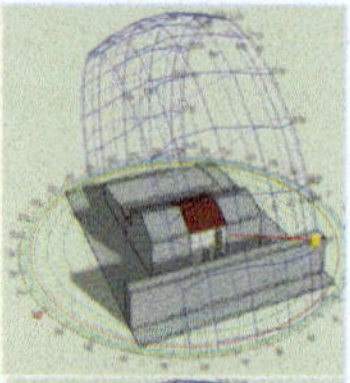	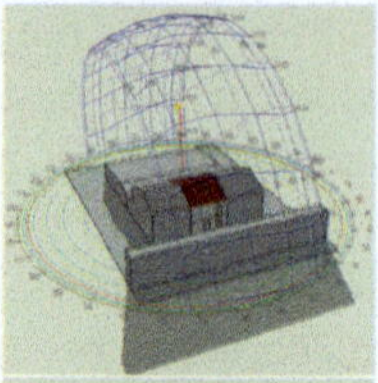
2	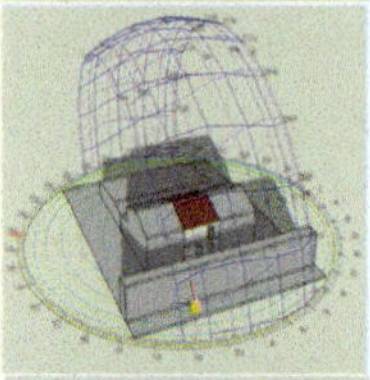	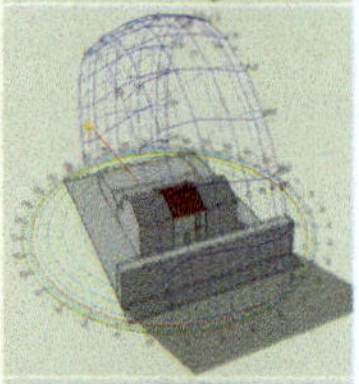	7	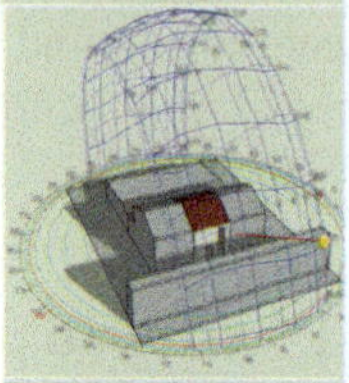	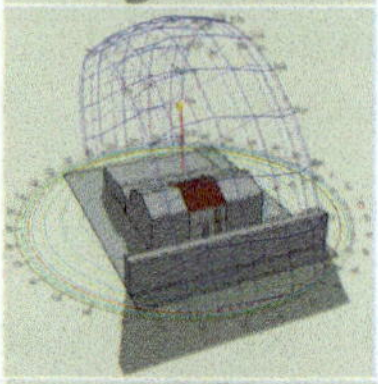
3	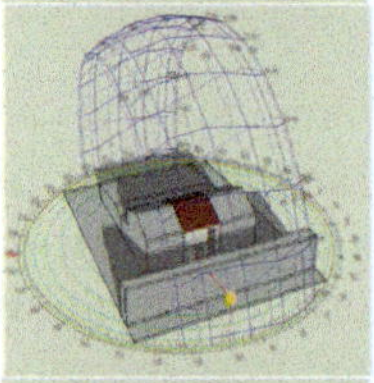	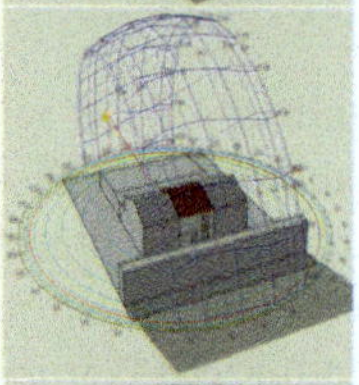	8	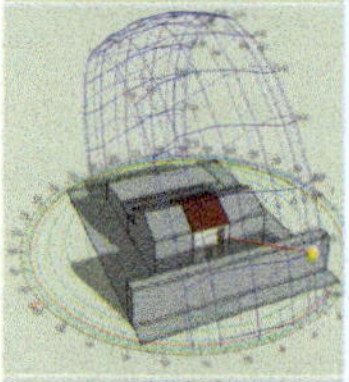	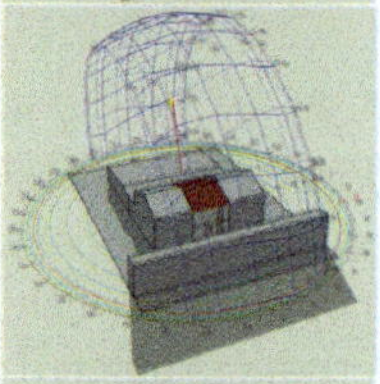
4	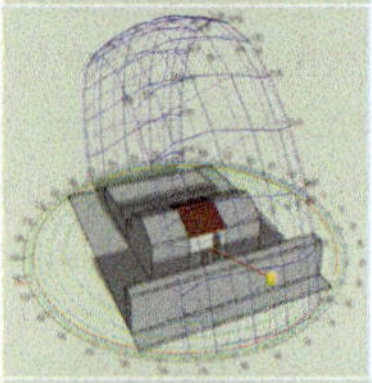	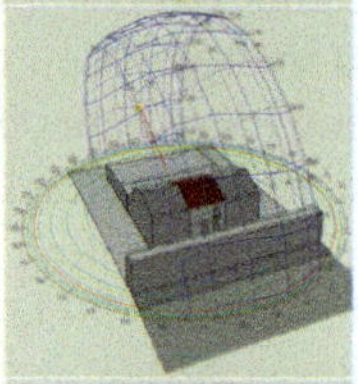	9	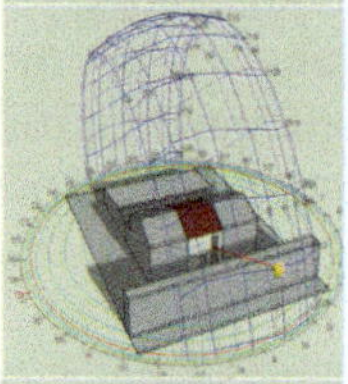	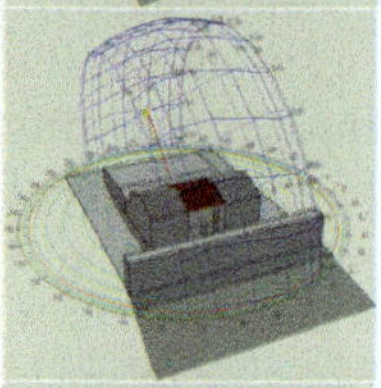
5	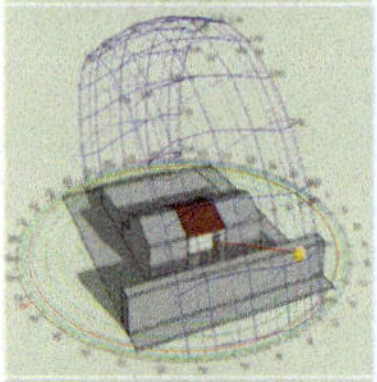	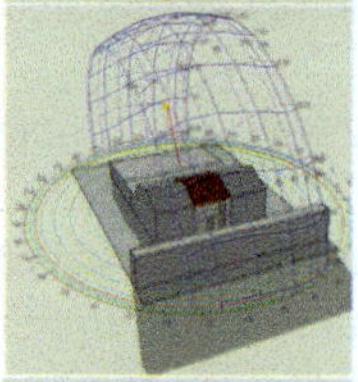	10	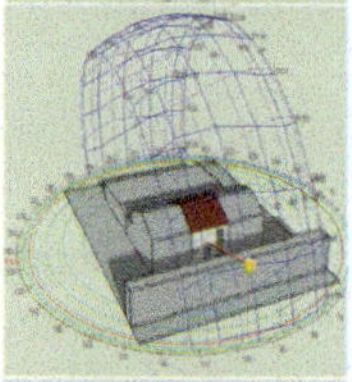	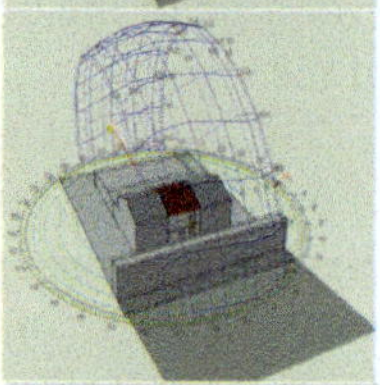
11	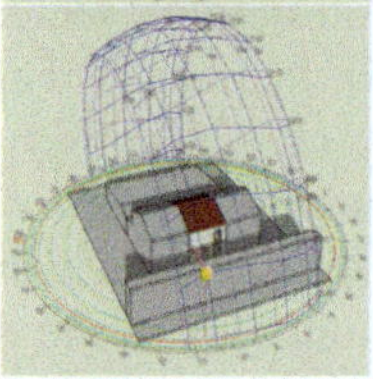	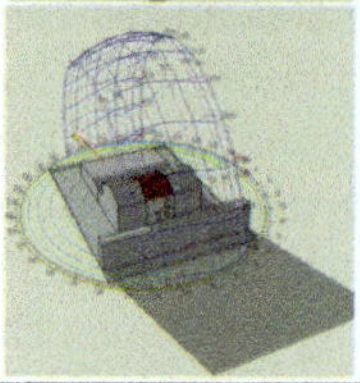	12	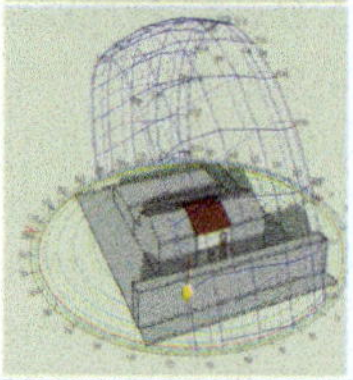	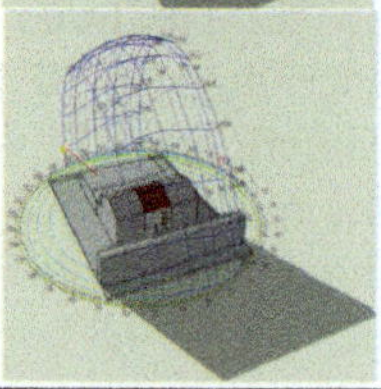

Table 4 Operating temperature behavior as a function of outside temperature

Month	External temperature as a function of operating temperature
1	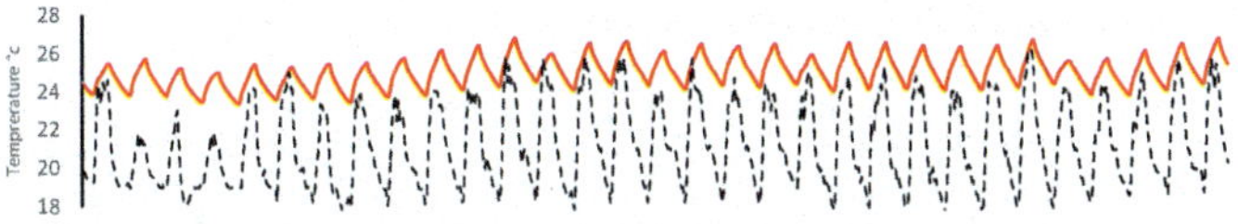
2	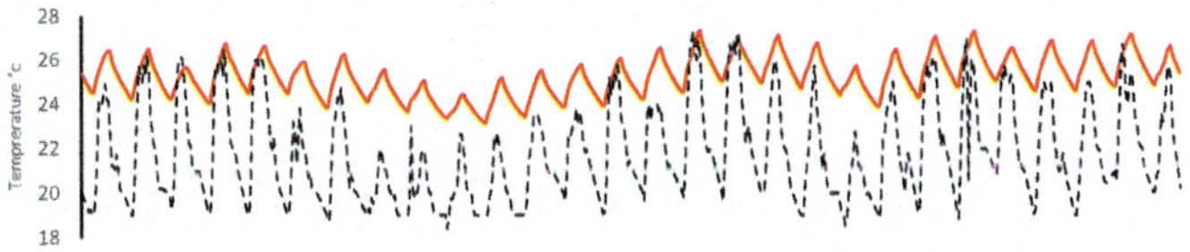
3	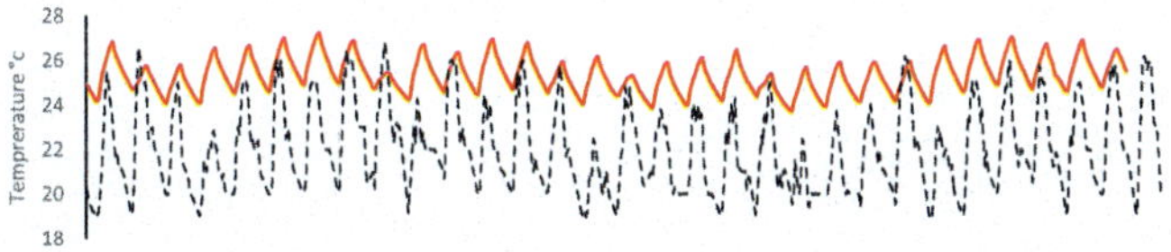
4	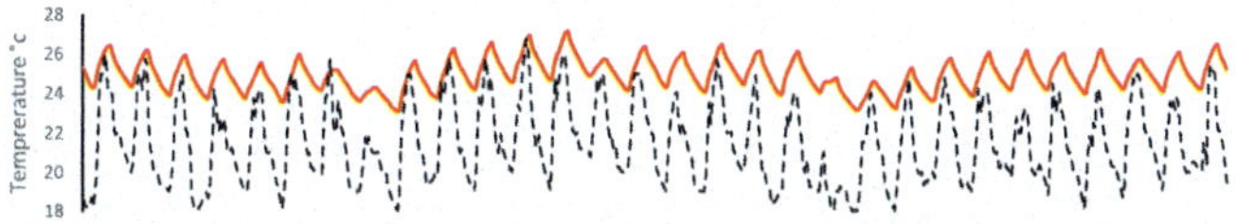
5	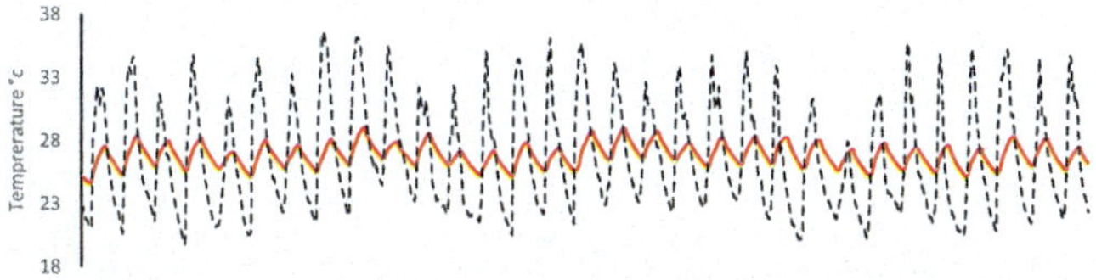
6	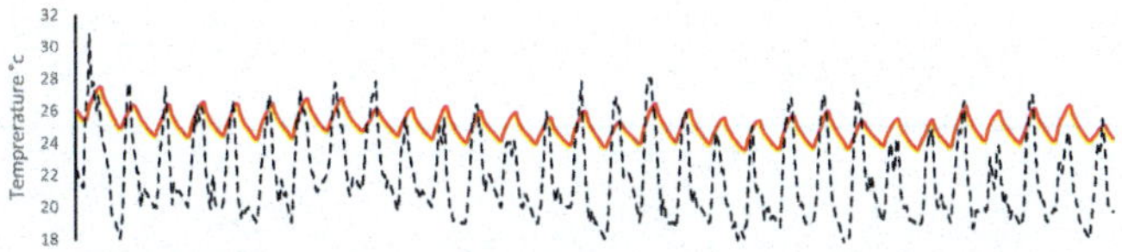
7	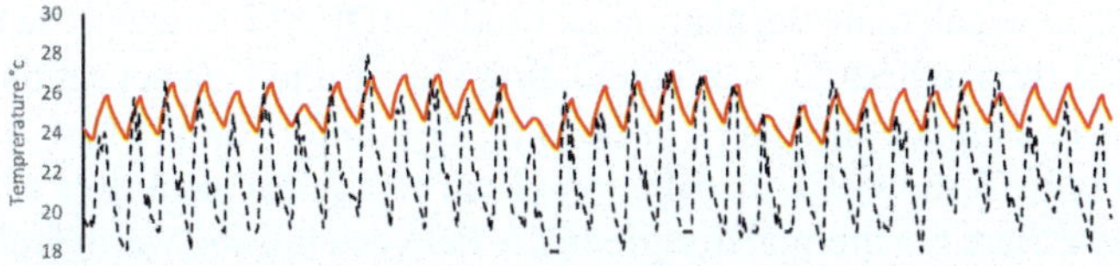

(continued)

Table 4 (continued)

Month	External temperature as a function of operating temperature
8	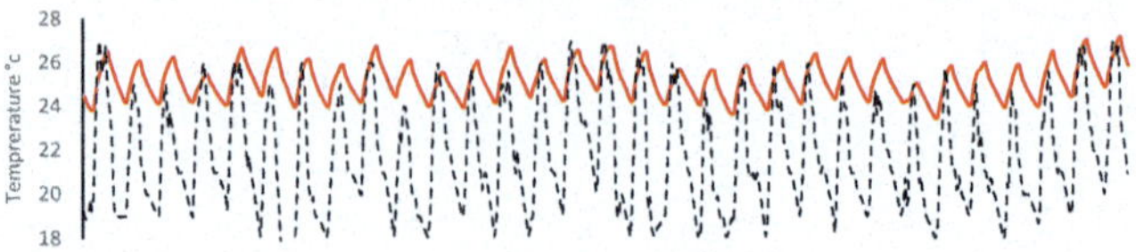
9	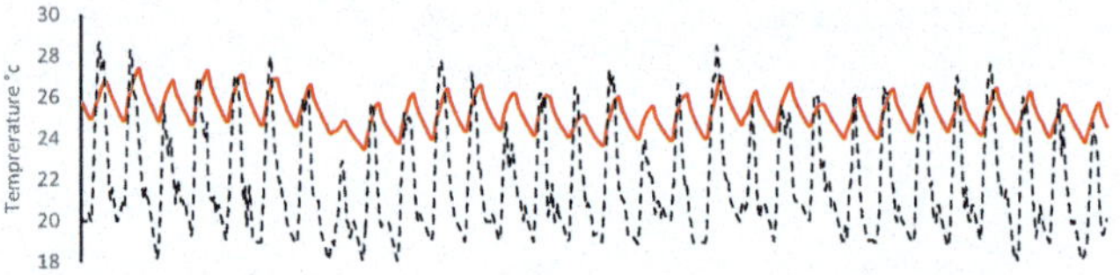
10	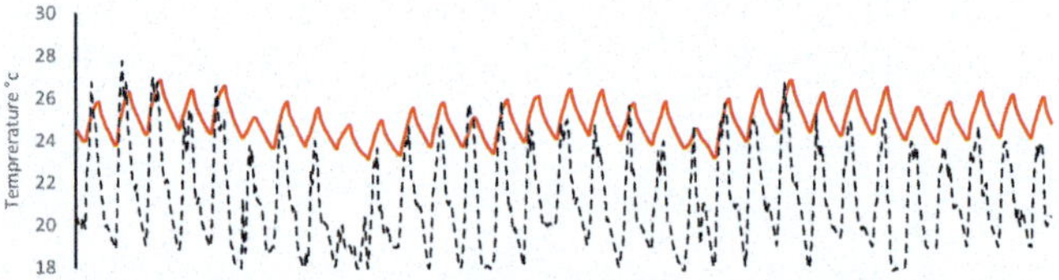
11	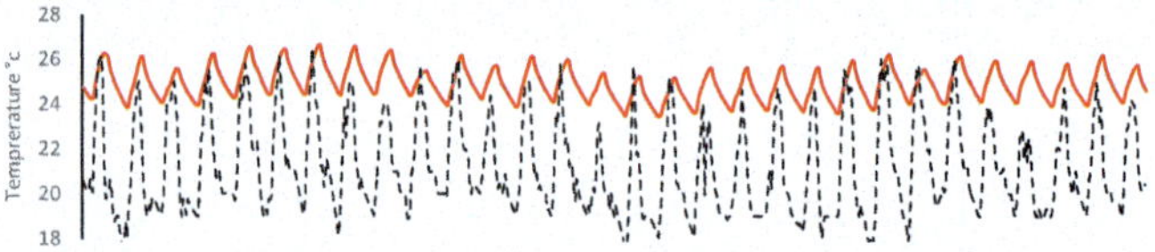
12	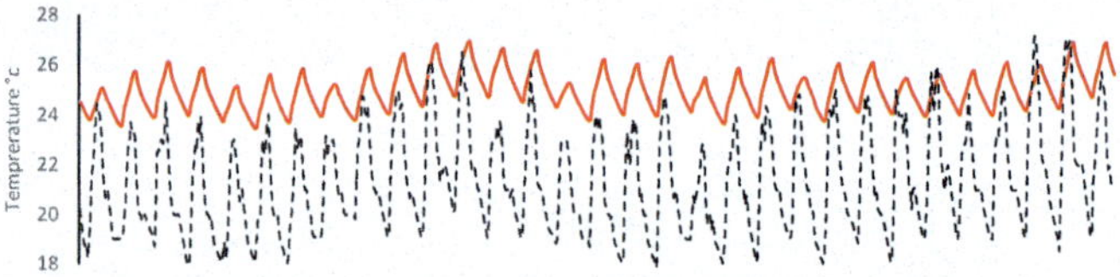

method that supports the implemented methodology [31–34]. Figure 4 shows the method used to calculate the heat index used by IDEAM to select the heat sensation.

Applied the methodology of calculating the CI, the heat sensation is established each year to evaluate the hours of housing discomfort on the first story one of the houses. Figure 5 shows the distribution of the heat index according to its heat sensation. Where there is a thermal disagreement between hot and warm feelings, likewise, it is calculated for plant two, as presented in Fig. 5, which shows a higher indication of nonconformity between her and uncomfortably hot (Figs. 6 and 7).

When analyzing the thermal sensation graphs, there is more dispersion in plant one and warm and hot thermal sensations. At the same time, in story two, there

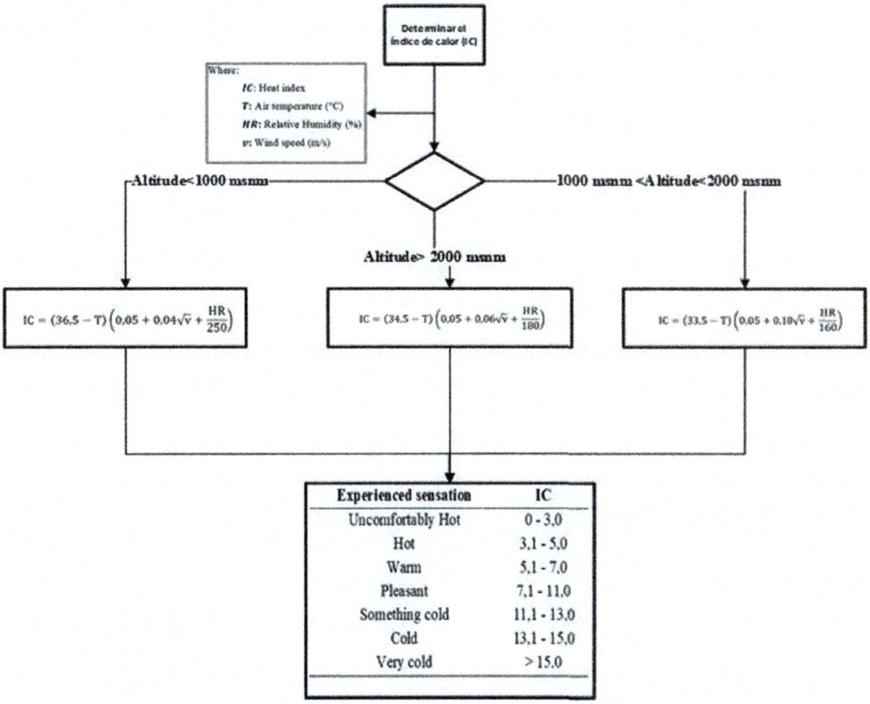

Fig. 4 Methodology for the determination of heat sensation [35]

Fig. 5 Heat index of hours calculated outside the first story

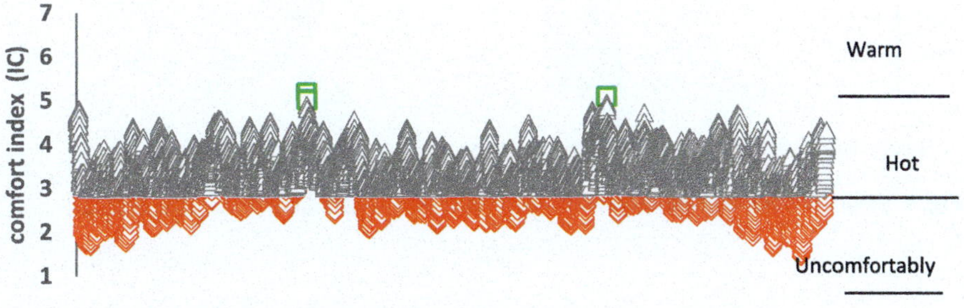

Fig. 6 Heat index of hours calculated outside the second story

Fig. 7 Heat index of hours calculated outside and inside the house

is a greater hot and uncomfortably hot thermal sensation since a greater surface is exposed to direct sunlight in the second story. The hours of thermal discomfort are presented in Figs. 8 and 9.

It can be established with the hours of thermal sensation that 100% of the hours are dissatisfied by the internal temperatures and relative humidity, demonstrating that sunlight provides significant loads inside the house.

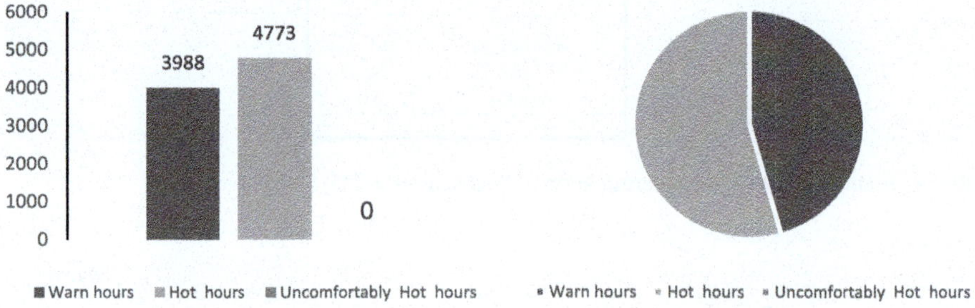

Fig. 8 Hours of wind chill story 1

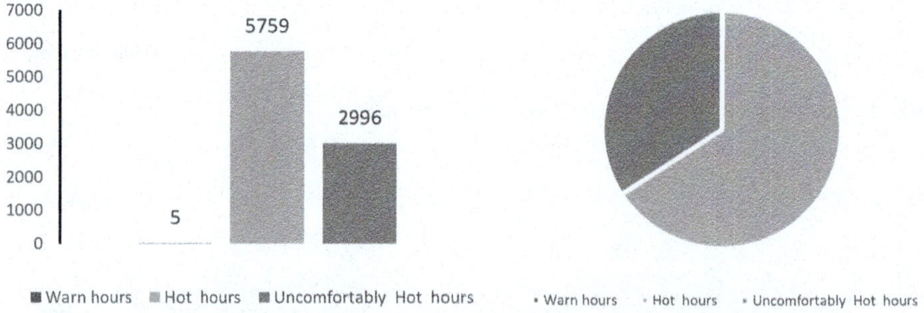

Fig. 9 Hours of wind chill story 9

4 Conclusions

In this work, a case study house is modeled, which is located at coordinates 7°08′ N 73°08′W presenting a dry, warm climate, in the computational tool Desing Builder. The software was fed with data regarding the geometry and dimensions of the house, construction materials, and the climatological database corresponding to the location of the study. From the data entered in the software, the house case study was modeled to later perform the analysis through the simulations allowed by the tool. This tool allowed for determining that the envelope of the house case study is not homogeneous. This is due to the losses and gains of heat detected during the year analyzed.

It is determined that sunlight presents heat gains and is vital in overheating buildings, particularly in regions with direct incidence. In areas with no natural radiation, such as the internal made limits the solar transmission and improves its internal temperatures. On the other hand, heat gains through glazed surfaces increase the internal temperature due to its high heat transfer capacity.

A case study model was analyzed using simulated sun exposure to analyze thermal comfort and detect occupational thermal inconvenience factors. For the analysis of thermal housing comfort, the heat index model proposed by the Hydrological and Environmental Institute of Colombia is implemented, which allows analyzing the data if considering the changes of climatic season, an important criterion because the case location is located near the equator. When the hours of thermal sensation were examined, it was established that 99% of discomfort was due to the high temperatures and relative humidity generated inside the house. It was eventually discovered that sunscreen devices would help reduce the hours of thermal nonconformity due to no direct sun exposure.

References

1. Harrison, S., Jiang, L.: An investigation into the energy performance gap between the predicted and measured output of photovoltaic systems using dynamic simulation modelling software—a case study. Int. J. Low-Carbon Technol. **13**(1), 23–29 (2018). https://doi.org/10.1093/ijlct/ctx016
2. Nematchoua, M.K. et al.: A review on energy consumption in the residential and commercial buildings located in tropical regions of Indian Ocean: a case of Madagascar island. J. Energy Storage **24**, 100748 (2019). https://doi.org/10.1016/j.est.2019.04.022
3. Bellia, L., Marino, C., Minichiello, F., Pedace, A.: An overview on solar shading systems for buildings. Energy Procedia **62**, 309–317 (2018). https://doi.org/10.1016/j.egypro.2014.12.392
4. Babota, F., Manea, D.L., Aciu, C., Munteanu, C., Cobîrzan, N., Tămaș-Gavrea, D. R.: Determination of optimal dimensions of fixed shadowing systems (pergolas) to reduce energy consumption in buildings in Romania. Procedia Manuf. **22**, 358–363 (2018). https://doi.org/10.1016/j.promfg.2018.03.054
5. Palmero-Marrero, A.I., Oliveira, A.C.: Effect of louver shading devices on building energy requirements. Appl. Energy **87**(6), 2040–2049 (2010). https://doi.org/10.1016/j.apenergy.2009.11.020

6. Ralegaonkar, R.V., Gupta, R.: Review of intelligent building construction: a passive solar architecture approach. Renew. Sustain. Energy Rev. **14**(8), 2238–2242 (2019). https://doi.org/10.1016/j.rser.2010.04.016
7. Pawar, B.S., Kanade, G.N.: Energy optimization of building using design builder software **4**(1), 5 (2018)
8. Sokkar, R., Alibaba, H.Z.: Thermal comfort improvement for atrium building with double-skin skylight in the mediterranean climate. Sustainability **12**(6), Art. n.o 6 (2020). https://doi.org/10.3390/su12062253
9. Khosla, S., Sharma, U.: Modelling and analysis of sustainable green building design using Autodesk Ecotect analysis, en Sustainable Engineering, Agnihotri, A.K., Reddy, K., Bansal, A. (Eds.), en Lecture Notes in Civil Engineering. Singapore: Springer (2019). pp. 235–242. doi: https://doi.org/10.1007/978-981-13-6717-5_23
10. Ruivo, C.R., da Silva, M.G., Broday, E.E.: Methodology for calculating an atmospheric pressure-sensitive thermal comfort index PMVaps. Energy Build. **240**, 110887 (2021). https://doi.org/10.1016/J.ENBUILD.2021.110887
11. Azcarate, I., Acero, J.Á., Garmendia, L., Rojí, E.: Tree layout methodology for shading pedestrian zones: Thermal comfort study in Bilbao (Northern Iberian Peninsula). Sustain. Cities Soc. **72**, 102996 (2021). https://doi.org/10.1016/J.SCS.2021.102996
12. Lopes, H.S., Remoaldo, P.C., Ribeiro, V., Martín-Vide, J.: A comprehensive methodology for assessing outdoor thermal comfort in touristic city of Porto (Portugal). Urban Clim. **45**, 101264 (2022). https://doi.org/10.1016/J.UCLIM.2022.101264
13. González, O.C., Meteorología, S.D.: Metodología para el Calculo del Confort Climático en Colombia
14. Barile, C.J., Slotcavage, D.J., Hou, J., Strand, M.T., Hernandez, T.S., McGehee, M.D.: Dynamic windows with neutral color, high contrast, and excellent durability using reversible metal electrodeposition. Joule **1**(1), 133–145 (2017). https://doi.org/10.1016/j.joule.2017.06.001
15. Hernández, R., Fernández, C., Baptista, P.: "Capítulo 1. Definiciones de los enfoques cuantitativo y cualitativo, sus similitudes y diferencias" en Metodología de la investigación, Sexta. McGraw Hill Education (2014)
16. Ramesh, T., Prakash, R., Shukla, K.K.: Life cycle energy analysis of buildings: an overview. Energy Build. **42**(10), 1592–1600 (2019). https://doi.org/10.1016/j.enbuild.2010.05.007
17. Sheng, Y., Yapo, T.C., Young, C., Cutler, B.: Virtual heliodon: spatially augmented reality for architectural daylighting design. IEEE Virtual Reality Conference, pp. 63–70 (2019). https://doi.org/10.1109/VR.2009.4811000
18. Aram, R., Alibaba, H.Z.: Thermal comfort and energy performance of atrium in mediterranean climate. Sustainability **11**(4), Art. n.o 4 (2019). https://doi.org/10.3390/su11041213
19. Esabegloo, A., Haghshenas, M., Borzoui, A.: Comparing the results of thermal simulation of rasoulian house in Yazd by design builder software, with experimental data. Iran Univ. Sci. Technol. **26**(2), 121–130 (2016). https://doi.org/10.22068/ijaup.26.2.121
20. Moussa, R.R., Moawad, D.R.M.: Investigating the efficiency of building energy simulation software on architectural design process. In: Proceedings of the 2020 9th International Conference on Software and Information Engineering (ICSIE), en ICSIE 2020. New York, NY, USA: Association for Computing Machinery, pp. 37–40 (2020). https://doi.org/10.1145/3436829.3436860
21. Valladares-Rendón, L.G., Schmid, G., Lo, S.-L.: Review on energy savings by solar control techniques and optimal building orientation for the strategic placement of façade shading systems. Energy Build. **140**, 458–479 (2017). https://doi.org/10.1016/j.enbuild.2016.12.073
22. Al-Masrani, S.M., Al-Obaidi, K.M.: Dynamic shading systems: a review of design parameters, platforms and evaluation strategies. Autom. Constr. **102**, 195–216 (2019). https://doi.org/10.1016/j.autcon.2019.01.014
23. Bournas, I., Abugabbara, M., Balcerzak, A., Dubois, M.-C., Javed, S.: Energy renovation of an office building using a holistic design approach. J. Build. Eng. **7**, 194–206 (2016). https://doi.org/10.1016/j.jobe.2016.06.010

24. Peng, F., Xiong, Y., Zou, B.: Identifying the optimal travel path based on shading effect at pedestrian level in cool and hot climates. Urban Clim. **40**, 100988 (2021). https://doi.org/10.1016/j.uclim.2021.100988
25. Liu, H., Pan, Y., Yang, Y., Huang, Z.: Evaluating the impact of shading from surrounding buildings on heating/ cooling energy demands of different community forms. Build. Environ. **206**, 108322 (2021). https://doi.org/10.1016/j.buildenv.2021.108322
26. Aram, R., Alibaba, H.Z.: Analyzing atrium volume designs for hot and humid climates. Sustainability **11**(22), Art. n.o 22 (2019). https://doi.org/10.3390/su11226213
27. Kirimtat, A., Krejcar, O., Ekici, B., Fatih Tasgetiren, M.: Multi-objective energy and daylight optimization of amorphous shading devices in buildings. Sol. Energy **185**, 100–111 (2019). https://doi.org/10.1016/j.solener.2019.04.048
28. Autodesk: Autodesk empowers innovators everywhere to make the new possible, Architecture, Engineering & Construction, 2021. https://www.autodesk.com/. accedido 31 de mayo de 2022
29. Attia, S.G., Herde, A.D.: Early design simulation tools for net zero energy buildings: a comparison of ten tools, p. 8 (2017)
30. IDEAM: Características climatológicas de ciudades principales y municipios turísticos (2019)
31. Liu, W., Tian, X., Yang, D., Deng, Y.: Evaluation of individual thermal sensation at raised indoor temperatures based on skin temperature. Build. Environ. **188**, 107486 (2021). https://doi.org/10.1016/J.BUILDENV.2020.107486
32. Kuru, M., Calis, G.: Understanding the relationship between indoor environmental parameters and thermal sensation of users via statistical analysis. Procedia Eng. **196**, 808–815 (2017). https://doi.org/10.1016/J.PROENG.2017.08.011
33. Xie, Y., et al.: Outdoor thermal sensation and logistic regression analysis of comfort range of meteorological parameters in Hong Kong. Build. Environ. **155**, 175–186 (2019). https://doi.org/10.1016/J.BUILDENV.2019.03.035
34. Ocampo-Marulanda, C. et al.: Missing data estimation in extreme rainfall indices for the Metropolitan area of Cali—Colombia: an approach based on artificial neural networks. Data Brief **39**, 107592 (2021). https://doi.org/10.1016/J.DIB.2021.107592
35. Ascanio-Villabona, J.G., Terés-Zubiaga, J., Muñoz-Maldonado, Y.A., Lengerke-Pérez, O., Del Portillo-Valdés, L.A.: Assessing the thermal performance of a conventional architecture in dry warm climate. In: Botto-Tobar, M., Cruz, H., Diaz Cadena, A. (Eds.) Recent Advances in Electrical Engineering, Electronics and Energy, en Lecture Notes in Electrical Engineering. Cham: Springer International Publishing, pp. 47–58 (2022). https://doi.org/10.1007/978-3-031-08288-7_4.

Review of Biomass and Biopolymers Synthesized in Ecuador

Jessenia López Ortega, Romel Carrera Tapia, Liseth Chacón Molina, Jonathan Sayavedra Delgado, Freddy Chipugsi Calero, and Javier Culqui Tipán

Abstract This bibliographic review identified the biopolymers extracted from biomass sources in Ecuador, used as raw material to synthesize high-value petrochemical products, such as bioplastics. In addition, this study investigated the synthesis methodology of bioplastics studied in research centers within the country. Biomass, a complex mixture of organic and inorganic substances, represents a promising source for the extraction of biopolymers. The study focused on three generations of biomass: permanent and transitory crops (first generation), agricultural and forestry residues (second generation), and algae (third generation). In addition, biomass consumption in various industrial sectors is analyzed, identifying opportunities to produce biopolymers from the waste generated. Biomass collection and biopolymer synthesis in Ecuador present significant economic and environmental opportunities by diversifying the supply chain and reducing dependence on non-renewable resources. Despite research into biopolymers such as starch, cellulose, alginate, carrageenan, lactic acid, and polyols, which are fundamental to the synthesis of bioplastics, industrial-scale production has not yet been achieved. To make the most of these opportunities, it is crucial to establish incentives, regulations, and educational programs on recycling. In addition, it is essential to create biomass storage and sorting management systems, thus facilitating the extraction of biopolymers and the synthesis of bioplastics on an industrial scale. This could position Ecuador as a leader in sustainable innovation. In addition, close collaboration between industries and research centers is essential. Therefore, this paper represents an updated field view and suggests directions for future research in Ecuador.

Keywords Biomass · Biopolymers · Bioplastic synthesis

J. L. Ortega (✉)
Universidad de Sevilla, Camino de los Descubrimientos, 41092 Sevilla, Spain
e-mail: ojesseniaestefania@us.es

J. L. Ortega · R. C. Tapia · L. C. Molina · J. S. Delgado · F. C. Calero · J. C. Tipán
Universidad de las Fuerzas Armadas, Av. Gral. Rumiñahui s/n, Sangolquí 171103, Ecuador

G. F. Olmedo Cifuentes et al. (eds.), *Emerging Research in Intelligent Systems*, Lecture Notes in Networks and Systems 902, https://doi.org/10.1007/978-3-031-52255-0_16

1 Introduction

Biomass is a complex mixture of organic and inorganic compounds generated by natural processes. Historically, this resource has been used as a source of electrical energy production or to generate heat through incineration, and it contributes approximately 14% of the world's energy generation [1]. Agricultural residues such as wood, straw, dry leaves, soybean grains, corn, and wheat, among others, are considered cellulosic biomass with great potential for renewable energy generation. However, the availability of this resource depends on geography and biomass sources. This energy source does not contribute to the greenhouse effect as it produces carbon dioxide that will be used again in the energy cycle [2]. Biomass can be classified into first, second, third, and fourth generations. First-generation biomass is obtained from crops with a high content of starch, sugars, and oils. Low molecular weight biofuels such as ethanol, methanol, and n-butanol can be obtained from this type of biomass. In recent years, the use of vegetable oils to obtain biodiesel and the mixture of methane and carbon dioxide to produce biogas has been investigated. Still, its disadvantage is the diversion of food resources to produce energy [3].

Second-generation biomass comes from non-edible sources, such as forest residues or woody matter. As it is waste, it does not need to be cultivated, does not compete with food production, and does not require water supply or fertilizers for its cultivation. A convenient way to produce bioethanol is through cellulosic biomass such as wood from industrial waste, dead trees, and domestic waste, among other sources [4].

Third-generation biomass refers to micro and macroalgae, which have a faster growth rate than other plants and do not require land for cultivation. Algae reduce the greenhouse effect by reducing the amount of carbon dioxide. With the need to generate energy that does not compete with food production, one current option is using microalgae to develop biofuels [5]. The production of biodiesel from microalgae has advantages over conventional diesel due to its zero content of toxic components and sulfur. This option to produce petrochemical compounds from this type of biomass is advantageous [6]. Finally, fourth-generation biomass encompasses the advance of genetic engineering technology by improving the characteristics of growth, development, many compounds of interest, and resistance to unfavorable environments, including removing genes that do not contribute directly to the development of the plant to obtain the highest yield in the shortest possible time, not only at the research level but also at the industrial level [7]. However, the present research focuses on using first, second, and third-generation biomass.

Biopolymers are large chains of carbon molecules comprising monomers synthesized in metabolic processes. As indicated in their name, the prefix bio means they are polymers produced from living beings. An essential characteristic of conventional polymers that gives them an advantage over biopolymers is the usability and durability they provide to the structure [8]. This research aims to know the biomass resources available in Ecuador from different regions of the country that could be focused on synthesizing biopolymers from different matrices. In addition, a bibliographic compilation of the researched studies of biopolymer synthesis in Ecuador from 2017 to 2022 is conducted to establish the advances and challenges in this research area, giving originality and benefit to future research to produce biopolymers. This qualitative research is based on the description, analysis, and interpretation of agricultural data of the biomass available in Ecuador and biopolymer synthesis studies conducted in research centers within the country. This study is also documentary-bibliographic research that will help to make decisions based on research carried out by other authors. The information gathered in the studies is previously organized and valued to establish the perspective of classifying the types of biomasses.

2 Biomass Available in Ecuador

Ecuador, a tropical country with a great diversity of flora, has become a source of raw materials for bioplastics. Plant biomass has essential characteristics such as moisture content, calorific value, fixed carbon and volatile material amount, and ash content. Such features can be determined by proximal analysis. On the other hand, biomass characterization allows the establishment of the amount of cellulose, lignin, and hemicellulose existing.

2.1 First Generation Biomass

The largest agricultural land area is in the country's coastal region, specifically in the province of Manabí, with cultivated and natural pastures. Three sectors significantly impact biomass production in Ecuador: the agricultural industry, with highly diversified products such as rice, cocoa, and coffee, and the livestock sector, which includes poultry and cattle. And finally, the forestry sector, which exploits the country's timber sources [9]. With the growth of Ecuador's population, clean and viable energy sources have been sought to meet the growing demand. However, biomass production is insufficient to meet this objective since occupying part of the crops destined for food products to use as biomass would be necessary. Another disadvantage is the use of resources, mainly water.

According to the ESPAC (Survey of Surface Area and Continuous Agricultural Production), Ecuador's land use is classified as shown in Fig. 1, where the most extensive area is devoted to forests and woodlands, with 47.91% of the national total [10].

Among the areas planted with permanent crops, cocoa is in first place with 41.83% compared to African palm, banana, plantain, and sugar cane. In the area planted with transitory crops, dry hard corn is in first place with 37.12% compared to rice, dry soft corn, dry beans, corn, potatoes, and dry peas. In 2021, sugarcane, banana, and African palm are recorded as the crops with the highest production at the national

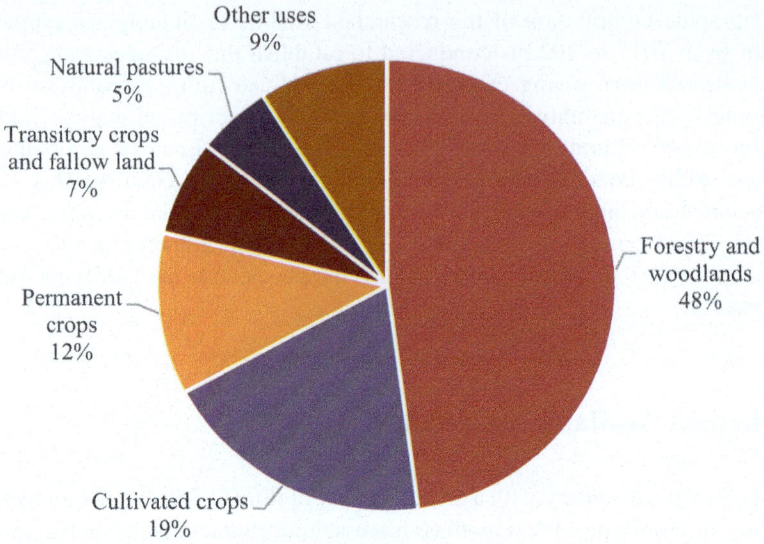

Fig. 1 Percentage of land use participation [10]

level. In addition, other crops are in different regions, such as potatoes, dried peas, avocado, broccoli, cereals, and seeds. Vegetable oils are also extracted from the seeds, which are used to produce classified into edible and non-edible according to the total exports of edible vegetable oil extracts show no growth in recent years [11]. In addition, there is no record of exports of non-edible oils in Ecuador that can be used to synthesize polyurethanes.

Among pasture crops, sobaya occupies the first place with 37.67% compared to mixed pasture, honey grass, brachiaria, and gramalote. In the floricultural area, permanent plantations are 88.94%, and transitory plantations are 11.06%. The pioneer is the cultivation of roses.

2.2 *Second Generation Biomass*

According to Ecuador's bioenergy atlas, second-generation biomass is produced by agricultural activities, forestry, and urban waste. The central provinces that generate this type of biomass are Azuay, Bolivar, Cañar, Carchi, and Chimborazo [9]. With its large extensions of forests and forest areas, Ecuador has many second-generation biomass. This type of biomass is also called forest biomass, a great energy source generated naturally over time by the action of climate or animals. The disadvantage of this type of biomass is the complexity of its correct conversion and preparation for energy extraction, considering that currently, there is no adequate technology for this purpose. Sugarcane is regarded as the most viable source of biomass to be used

Table 1 Agricultural waste in Ecuador

Crop	Agricultural waste
Sugar cane	Bagasse, stems, and leaves
Banana	Leaf, pseudostem
Cocoa	Pruning, discarded cob, cob husk
Oil palm	Stalks and flowers
Dried hard corn	Stems and leaves
Rice	Husk
Banana	Leaves and husk
Orange	Husk
Flowers	Stems, leaves

as raw material in the biorefining process. However, it has a lower calorific value than other species studied in this research.

Despite the technological limitations involved in transforming forest biomass, it is an attractive source due to its availability and zero cost, in addition to its diverse characteristics depending on the area in which it is found. It is for this reason that projects are currently being carried out in public and private educational institutions in Ecuador to use sustainably and responsibly all the residues of the agricultural industry in the country due to its accessibility in almost all provinces of the country, which can accommodate energy generation projects and biofuels, contributing not only to economic recovery but also to mitigate the effects caused by the combustion of fossil fuels. Table 1 shows the agricultural residues of the largest biomass produced in Ecuador.

2.3 *Third Generation Biomass*

Ecuador has a great diversity of algae in the coastal zone and microalgae in lagoons and lakes, from which polysaccharides can be obtained that can be extracted to produce biopolymers. Limited information about the country's marine ecosystem regarding its temporal composition exists. However, there is evidence of many species accumulated on the shores of the coastal region of the country that has been recorded in different investigations the presence of macroalgae in other climatic storms [12], mainly in the province of Manta in the coastal area of Manabí evidencing a total of 17 species of algae, eight belonging to the Rhodophyta division (red algae), five to the Chlorophyta (green algae) and four to the Phaeophyta (brown algae). The presence of microalgae has been recorded in different lagoons in Ecuador, such as Lomincocha [13], Chinchillas [14], Cajas National Park [15], Yambo [16], Limpiopungo [17], Cuicocha [18], among others.

2.4 Biomass as an Alternative to Produce Clean and Efficient Technologies

The creation of clean and efficient technologies has been an attraction of study in the last decade, evidenced by the collection and consumption of different types of biomasses in other countries. As shown in Table 2, using natural resources and waste is pioneering Europe. The price varies from country to country depending on the collection type and treatment technique [19]. Among the countries with the highest production of biofuels, Brazil and the United States are the leaders [20].

In Ecuador, the use of biofuels became evident in 2010 with the transformation of sugarcane into bioethanol, reducing imports of high-octane naphtha [21]. Currently, the project "Proyecto Piñón en Ecuador" is being implemented to obtain biodiesel from pine nut oil on an industrial scale to generate electricity on Floreana Island in the Galapagos [22]. However, no company is dedicated to using biomass to produce biopolymers. There is only laboratory-scale research on transforming the material into chemical products such as biofuels, low molecular weight chemical compounds, and biopolymers. Globally, the biopolymers market grew 12.6% between 2021 and 2026. In addition, companies in the market have focused on increasing biopolymer production capacities. One of the main objectives is to create antibacterial and antiviral materials to contribute to the COVID-19 era. Figure 2 shows the production of biopolymers on a global scale. India is the pioneer in research, followed by the European Union, which estimates the use of 100% recyclable packaging. As seen in the image, Latin American countries have a low production of these materials [23].

Table 2 Countries using biomass [20]

Countries	Type of Biomass	Use
Ironbridge (UK)	Wood pellets	Energy generation
Alholmenskraft (Finland)	Cellulose	Paper company and human heating
Toppila (Finland)	Turba	Fuel
Polaniec (Poland)	Agricultural by-products and wood residues	Power generation
Kymijärvi (Finland)	Plastic, paper, cardboard, and wood	Fuel gas
Brazil	Sugar cane	Bioethanol Biodiesel
Argentina	–	Biodiesel
Colombia	–	Biodiesel
USA	Sugar cane	Bioethanol Biodiesel
China	–	Bioethanol
Canada	–	Bioethanol

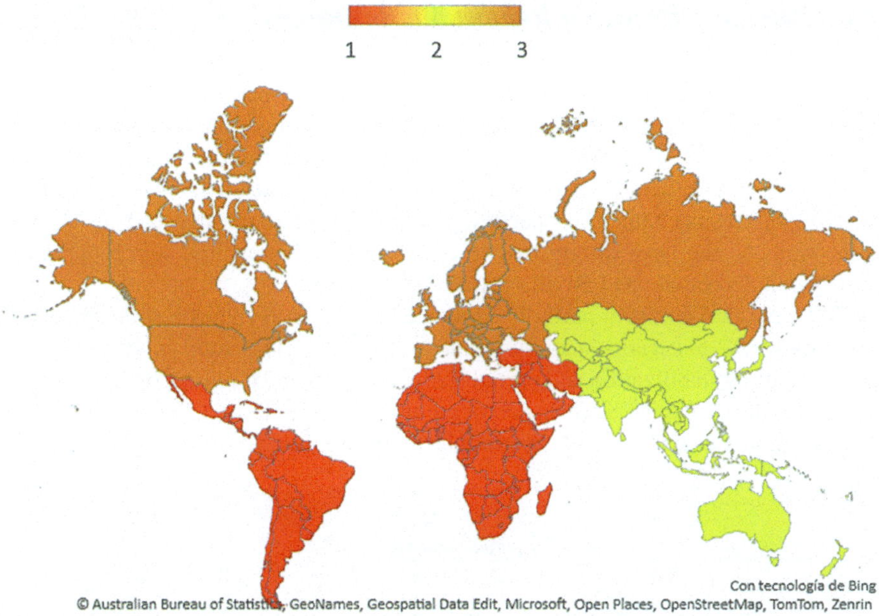

Fig. 2 Biopolymer production—growth rate by region (2020- 2025), 1-low, 2-medium, and 3-high [23]

The following are the leading producers of biopolymers worldwide:

- United States: Some companies, such as Nature Works LLC [24].
- Germany: BASF company [25].
- Japan: Mitsubishi Chemical, Showa Denko, and Teijin [26].
- China: Companies like Kingfa Sci. & Tech. Co. Ltd. and Tianan Biologic Materials Co. Ltd. [27].
- Brazil: Braskem company [28].
- Italy: Novamont company [29].
- United Kingdom: Biome Bioplastics company [30].

3 Biopolymers Synthesized in Ecuador

In this bibliographic research, biopolymers have been divided into three groups based on renewable resources, biodegradable based on biodegradable monomers (vegetable oils and lactic acid), and algae.

Before obtaining any biopolymers from the biomass, the chemical characterization of the biomass (moisture, extractable compounds, ash, lignin, cellulose, hemicellulose) is essential.

3.1 Polymers Based on Renewable Resources

Starch is a mixture of amylose and amylopectin. The percentage of each varies according to the raw material to be extracted. Some biopolymers are obtained from agricultural residues (second generation), and others directly from starch (first generation) extracted from the raw material. To get this biopolymer, reagents must be added to improve the mechanical properties. Three types of starch that can be obtained from biomass can be distinguished (Fig. 3): partially fermented starch derivative, pure starch derivative, and modified starch derivative.

Polymers derived from partially fermented starch are synthesized using second-generation raw materials, such as potato waste fluid. This fluid is stored for two weeks to allow stabilization and partial fermentation of the biomass. In fermentation, bacteria in the medium transform part of the starch into lactic acid. Finally, the product is dried and extruded to obtain thermoplastic properties. To improve the properties

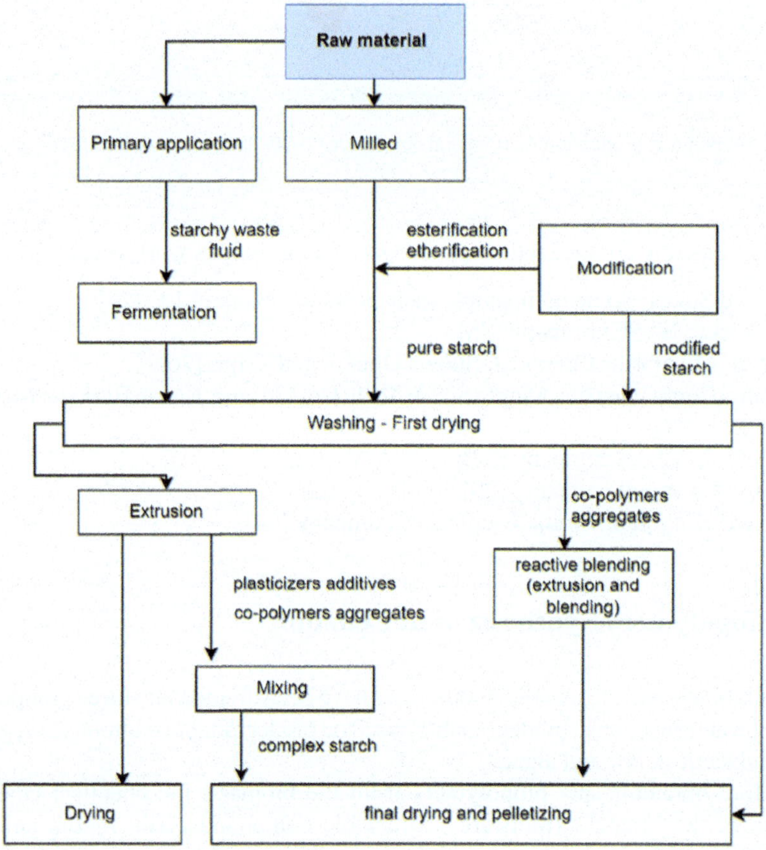

Fig. 3 Synthesis of biopolymers from starch [31]

of the biopolymer, additives such as titanium dioxide (TiO_2) and calcium carbonate ($CaCO_3$) can be added to the extrusion process. In the case of polymers derived from pure starch, the polysaccharide must be extracted directly from the first-generation biomass. For example, corn starch is removed through a wet mill previously softened with an acid medium and centrifuged, separating the denser protein and the less dense starch. The starchy liquid is washed in a centrifuge, dehydrated, and dried before extrusion. Additives can also be added before or after drying to improve the properties of the bioplastic. Modified starch has been treated with chemicals so that other groups, such as esters, replace some hydroxyl groups. Modified starch can be used directly and chemically dried to obtain biopolymer. Table 3 shows the biopolymers synthesized in Ecuador from different types of starch. Adding plasticizers such as water, alcohols, ketones, amines, and polyols to the structure improves the mechanical properties. In addition, the use of oils provides microbial activities.

Table 3 Starch biopolymers

Biomass	Starch	Additive	References
White malanga	Pure	Glycerin, Urea	[1]
Purple malanga	Pure	Glycerin, Urea	[1]
Banana waste	Pure and modified	Polycaprolactone	[2]
Banana peel	Partially fermented	Vinegar and glycerin	[3]
Mango husk	Partially fermented	Water, glycerin	[4]
Mango peel, rice husk, and cassava starch	Pure	Glycerin	[5]
Avocado seeds	Pure	Water, vinegar, glycerin, ascorbic acid, citric acid, and potassium sorbate	[6]
Corn	Pure	Water, glycerin	[7]
Cassava	Pure	Water, glycerin	[4]
Oryza sativa husk	Pure	Water, glycerin	
Potato	Pure	Tea tree extract and acacia gum	[8]
Cassava starch and chitosan	Pure	Water, glycerin, celery oil, thyme, bay laurel	[9]
Nopal cactus mucilage (Opuntia spp.)	Pure	Water, glycerin, mucilage oil and rosemary oil	[10]

Table 4 Cellulose biopolymers

Biomass	Type of cellulose	Additives	References
Common grass (Festuca arundinacea)	Cellulose acetate	Acetic acid	[11]
Banana peel rejects	Cellulose acetate	Acetic acid	[12]
Banana waste	Cellulose acetate	Vegetable oils	[2]
Cocoa husk	Cellulose	Acetic acid	[13]
Pineapple and banana	Bacterial cellulose	–	[14]
Orange peel	Nitrodeculose	–	[15]
Coconut tow fibers	Microcrystalline cellulose	Nitric acid	[16]
Calamagrostis species	Cellulose	–	[17]
Cigarette butts	Cellulose acetate	Acetic Acid	[18]

3.2 Biopolymers from Cellulose

Cellulose has a linear structure composed of β-glucose molecules through β-1,4-O-glycosidic bonds. Biopolymers obtained from cellulose are produced by chemical modification of natural cellulose (cellophane, cellulose acetate, cellulose ester, regenerated cellulose for fiber, among others). From different types of biomass, cellulose can be obtained practically pure, as in the case of cotton, or with impurities, as in the case of cellulose present in wood, which is found together with lignin and other polysaccharides. Cellulose can be obtained by acid or basic hydrolysis and then by one or several stages of bleaching for the synthesis of ethylene acetate or nitrocellulose precursors such as acetic acid and nitric acid are needed along with cellulose to verify the synthesis of these biopolymers it is necessary to analyze their structure by quantification by Uv–Vis and IR analysis of their properties such as moisture, density, solubility, intrinsic viscosity and molecular weight (Table 4).

Starch and cellulose bioplastics are more expensive than traditional plastics due to more complex extraction and production processes that require more energy, water, and reagents. In addition, bioplastics have limitations in mechanical properties such as hardness and durability. To overcome these challenges in the industry, it is necessary to invest in research and development to improve bioplastic properties, create infrastructure, and raise awareness of environmental benefits in society. It is essential to create policies that regulate the responsible and sustainable use of bioplastics.

3.3 Biodegradable Polymers Based on Bio-Based Monomers

Few studies have been reported in Ecuador on synthesizing polymers obtained from monomers. Among the research is the production of polyurethanes from palm oil, a research opportunity because of the diversity of non-edible and used oils. Other

Table 5 Biodegradable polymers based on bioderived monomers synthesized in Ecuador

Biomass	Primary product	Secondary product	References
Brown seaweed	Alginate	Lactic acid	[33]
Palm oil	Polyols	Polyurethanes	[34]

Table 6 Algae-based biopolymers

Biomass	Biopolymer	References
Brown seaweed	Sodium alginate	[36]
Brown seaweed	Calcium alginate	[37]
Freshwater algae	Sodium alginate	[38]
Red algae	Carrageenan	[38]

research is the production of lactic acid from alginate obtained from algae. Although this production is low, it opens a gap in the research to modify the extraction conditions and increase the production of this biopolymer. The use of vegetable oils to produce polyols has attracted the attention of several researchers [32] (Table 5).

3.4 Algae-Based Biopolymers

Fish farming has recently increased, with seaweed being the aquatic life scenario. Seaweeds contain polysaccharides, proteins, and lipids as fish food [35]. Research on algae and the products that can be extracted from them may not have been a priority in Ecuador. This leads to a lack of detailed knowledge about the algae species available and their properties in algae production. The lack of infrastructure for collecting, classifying, and studying algae hinders technological progress in extracting biopolymers.

For this reason, there is little research on third-generation bioplastics. The current techniques observed in Table 6 to produce bioplastics based on microalgae and macroalgae are at laboratory scale and focus on the feasibility of mixture design. However, there is little evidence of research on these biopolymers from third-generation biomass.

4 Conclusions

Biomass harvesting and biopolymer synthesis represent a significant opportunity for several economic sectors in Ecuador, promoting environmental, economic, and technological sustainability in the country. Bioplastics are synthesized from biopolymers obtained from different types of biomasses. First-generation biomass is grown mainly in the coastal zone, especially Manabí, with permanent and transitory crops. Second-generation biomass comes from agricultural activities, forestry, and urban waste in provinces such as Azuay, Bolívar, Cañar, Carchi, and Chimborazo. Third-generation biomass includes algae, although more information is still needed on the species in Ecuadorian areas. So far, Ecuador has not fully exploited these resources, representing an opportunity to diversify the supply chain and reduce dependence on non-renewable resources. This initiative could stimulate economic growth in the country's rural, agricultural, and forestry areas. To achieve this, it is essential to reduce dependence on petroleum as a source of raw material by promoting the production of bioplastics from renewable sources. In addition, cross-sector collaboration is critical to this process. Cooperation between the agricultural, chemical, and manufacturing industries creates an environment conducive to developing new business opportunities, fostering innovation, and stimulating the local economy.

From a technological innovation perspective, there has been a constant search to improve the properties of bioplastics using biopolymers derived from various types of biomasses. These advances have allowed bioplastics to be applied in multiple applications. Research centers have investigated and synthesized several biopolymers in Ecuador, including starch, cellulose, alginate, carrageenan, lactic acid, and polyols. These biopolymers have served as the basis for applications in coatings, tableware, sorbets, and the production of other compounds, such as lactic acid used to synthesize polylactic acid and polyols used in the production of polyurethanes.

However, it is essential to note that this production has been limited to the laboratory scale and has not yet been implemented at the industrial level. The government and other entities must create economic incentives, establish effective policy regulations, and educate citizens about recycling systems to overcome these challenges. In addition, close collaboration between the agricultural industry and research and development centers is essential to improve the technical properties of bioplastics, thus ensuring their viability in various sectors. Addressing these challenges will allow Ecuador to take full advantage of bioplastics' economic and environmental benefits, placing the country at the forefront of sustainable innovation.

5 Future Research

Research the chemical characterization of the country's most productive biomass and agricultural waste to determine the amount of cellulose, lignin, and hemicellulose it contains, thus choosing its application in producing biopolymers. It is essential to have a source of information on the algae in Ecuador's coastal zone, lakes, and lagoons since there is no classification of the species. However, the massive existence of this resource in shrimp and fish farming has been evidenced.

The production of biopolymers for industrial use must be competitive with traditional polymers. A market study of the costs and benefits of polymers made from renewable raw materials is necessary. Finally, it is required to establish agreements between research centers and the country's productive sector since manufacturing polymers on an industrial scale requires research, infrastructure, and initial investments.

References

1. Kumar, A., Kumar, N., Baredar, P., Shukla, A.:A review on biomass energy resources, potential, conversion and policy in India. Renew. Sustain. Energy Rev. **1**(45), 530–539 (2015)
2. Nhuchhen, D., Basu, P., Acharya, B. A comprehensive review on biomass torrefaction. Int. J. Renew. Energy Biofuels, 1–56 (2014)
3. Maciel, C.: Biocombustibles: desarrollo histórico-tecnológico, mercados actuales y comercio internacional. Economía **1**(59), 63–89 (2009)
4. Osorio, N., Yánez, M.: Obtención de un recubrimiento comestible a base de Mucílago de Nopal (opuntia spp.) y Aceite Esencial de Romero (rosmarinus officinalis) [Tesis de ingeniería, Universidad Técnica de Cotopaxi]. Repositorio institucional (2018). http://repositorio.utc.edu.ec/handle/27000/5445
5. Faba, L., Díaz, E., Ordóñez, S.: Transformación de biomasa en biocombustibles de segunda generación. Madera y bosques **20**(3), 11–24 (2014)
6. Almenares-Verdecía, L., Ngoma-Presline, D., Serrat-Díaz, M.: Aspectos tecnológicos generales para la conversión a etanol de la biomasa lignocelulósica II. Tecnología Química **31**(3), 392–407 (2011)
7. Dutta, K., Daverey, A., Lin, J.: Evolution retrospective for alternative fuels: first to fourth generation. Renewable Energy **69**, 114–122 (2014)
8. Smith, A., Moxon, S., Morris, G.: Biopolymers as wound healing materials. Wound Healing Biomaterials **2**(1), 261–287 (2016)
9. Guzmán, R., Albornoz, E., Palacios, S.: Atlas Bioenergetico en Ecuador" 2nd edn. ARIAE, Ecuador (2014)
10. INEC: Encuesta de Superficie y Producción Agropecuaria Continua 2021 (2022)
11. Mayorga, F., Freire, C.: El arroz en Ecuador (2020)
12. Franco, A.: Caracterización ecológica de las comunidades intermareales rocosas a lo largo de una gradiente de productividad en las costas del Ecuador [Tesis de licenciatura, Pontificia Universidad Católica Del Ecuador]. Repositorio institucional (2012). http://repositorio.puce.edu.ec/handle/22000/10760
13. Castillejo, P., Bravo, A.: Caracterización de microalgas de la laguna de Limoncocha y optimización de las condiciones de cultivo in vitro para la obtención de biomasa [Tesis de ingeniería, Universidad Internacional SEK Ecuador]. Repositorio institucional (2017). https://repositorio.uisek.edu.ec/handle/123456789/2213
14. Córdova, L., Vera, V.: Diseño de una laguna de microalgas para la producción de biodiesel como estrategia generadora de fuentes alternas de energía en la porvincia de santa Elena [Tesis de ingenieria, Universidad Agraria del Ecuador] Repositorio institucional (2021). http://repositorio.uae.edu.ec/handle/230890/5320
15. Cabrera, M., Pulla, M.: Linea base para el aprovechamiento de microalgas sistema de tratamientos de agua [Tesis de ingeniería, Universidad de Cuenca]. Repositorio institucional (2014). https://hdl.handle.net/20.500.12692/112322
16. Taopanta, M.: Determinación del estado trófico de laguna de Yambo a través de cuantificación de la clorofila A [Tesis de ingeniería, Universidad Central de Ecuador] Repositorio institucional (2017). http://dspace.ups.edu.ec/handle/123456789/19563
17. Brito, G.: Caracterización fisicoquímica y bacteriológica de las aguas de la laguna de Limpiopungo de la zona central del Ecuador [Tesis de licenciatura, Universidad Técnica de Ambato] Repositorio institucional (2019). https://repositorio.uta.edu.ec/jspui/handle/123456789/30427

18. Alarcón, D.: Microbiota bacteriana de la laguna Cuicocha y sus aplicaciones biotecnológicas [Tesis de licenciatura, Universidad Central del Ecuador] Repositorio institucional (2019). http://repositorio.espam.edu.ec/handle/42000/1431
19. Cerdá, E.: Energía obtenida a partir de la biomasa. Cuad. Económicos ICE, 1–24 (2011)
20. Gonzáles, A., Castañeda, Y.: Biocombustibles, biotecnología y alimentos: Impactos sociales para México. Cris. Aliment. abundancia y Aliment. **21**(57) (2008)
21. Martínez, H.: Análisis del uso de biocombustibles en Ecuador periodo 2010–2017 (2018)
22. Nasimba, L., Soraya, D.: MERNNR-Ministerio de Energía y Recursos Naturales no Renovables datos generales. Ficha Inf. del Gob. por Result. **1**(1), 1–19 (2019)
23. Mordor Intelligence, Mercado de envases de biopolímeros" https://www.mordorintelligence.com/es/industry-reports/biopolymer-packaging-market. Accessed 25 May 2022
24. "NatureWorks | Home", https://www.natureworksllc.com. Accessed 5 June 2023
25. "BASF – We Create Chemistry.", https://www.basf.com/global/en.html. Accessed 5 June 2023
26. "Materials | News | TEIJIN." https://www.teijin.com/news/material/. Accessed 5 June 2023
27. "Kingfa", https://www.kingfa.com.cn/. Accessed 5 June 2023
28. "Braskem", https://www.braskem.com/produtos. Accessed 5 June 2023
29. Novamont - Company Profile, https://www.novamont.com/eng/company-profile. Accessed 5 June 2023
30. Biome Bioplastics, https://biomebioplastics.com/. Accessed 5 June 2023
31. Textos Científicos, "Polímeros derivados del almidón", https://www.textoscientificos.com/polimeros/almidon. Accessed 21 June 2022
32. López, J.: Estudio de la producción de alginato de sodio derivado de macroalgas pardas y su conversión hidrotérmica catalizada a ácido láctico utilizando digestión por microondas., [Tesis de ingeniería, Universidad de las Fuerzas Armadas ESPE]. Repositorio institucional (2018). http://repositorio.espe.edu.ec/xmlui/handle/21000/15199
33. Peso-Echarri, P., Frontela-Saseta, C., González-Bermúdez, G., Ros-Berruezo, F., Martínez-Graciá, C.: Polisacáridos de algas como ingredientes funcionales en acuicultura marina: alginato, carragenato y ulvano. Rev. Biol. Mar. Oceanogr. **47**(3), 373–381 (2012)
34. Cuéllar, M., Rivas, A.: El uso del aceite de palma en la producción de poliuretanos. Palmas **25**(1), 422–427 (2004)
35. Robalino, M.: Extracción y caracterización de alginatos derivados de algas pardas, para la producción de ácido láctico. [Tesis de ingeniería, Universidad de las Fuerzas Armadas ESPE]. Repositorio institucional (2018). http://repositorio.espe.edu.ec/handle/21000/13799
36. Loja, C.: Obtención de un polímero biodegradable a partir del alginato de calcio extraído de la biomasa del alga parda (Sargassum Ecuadoreanum), [Tesis de ingenieria, Universidad Politécnica Salesiana Sede Cuenca]. Repositorio institucional (2021). http://dspace.ups.edu.ec/handle/123456789/18680
37. Coral, K., Naranjo, D.: Obtención de alginato de sodio a partir de algas de agua dulce de la familia Zygnemataceae, mediante reacciones de intercambio iónico para ser utilizado como gelificante., [Tesis de ingeniería, Universidad Internacional SEK]. Repositorio institucional (2019). http://repositorio.uisek.edu.ec/handle/123456789/3527
38. Lasluisa, G.: Estudio de la producción de la carragenina y su demanda a nivel mundial y nacional, [Tesis de ingeniería, Universidad de las Fuerzas Armadas ESPE]. Repositorio institucional (2021). http://repositorio.espe.edu.ec/handle/21000/25022

Solar Energy Conversion Systems Thermal Management Applying Constructal Theory: A Comparative Analysis

Francisco J. Montero, Wilson Navas Pinto, and Alejandro Aljure

Abstract This paper analyses the impact on the performance of a Solar Thermoelectric Generator coupled with a Latent Heat Storage and Cooling System (STEG + LHSCS) by applying the constructal theory in designing a copper heat sink embedded in the LHSCS. The performance of the STEG + LHSCS system follows the capacity of the copper heat sink to efficiently transfer the waste heat of the STEG system to the LHSCS, to store this remanent heat and maintain a temperature difference between the thermoelectric generator (TEG) included in the STEG. It is observed that the application of the constructal theory in the design of the copper heat sink does not have a significant impact on the performance of the STEG + LHSCS system, showing a reduction in the TEG electricity generation, if it is compared with the design of the copper heat sink which includes radial fins. The model presented in this paper will be adapted to study the performance of the STEG + LHSCS in various geographical locations and to study other variations of the application of constructal theory in latent heat storage systems.

Keywords Solar thermoelectric generator · Latent heat storage · Constructal theory

F. J. Montero (✉) · W. N. Pinto · A. Aljure
Departamento de Ciencias de la Energía y Mecánica, Universidad de las Fuerzas Armadas ESPE, Sangolquí 171103, Ecuador
e-mail: fjmontero1@espe.edu.ec

F. J. Montero
Department of Mechanical and Metallurgical Engineering, School of Engineering, Pontificia Universidad Católica de Chile, Av. Vicuña Mackenna, 4860 Santiago, Macul, Chile

G. F. Olmedo Cifuentes et al. (eds.), *Emerging Research in Intelligent Systems*, Lecture Notes in Networks and Systems 902, https://doi.org/10.1007/978-3-031-52255-0_17

1 Introduction

Constructal theory describes how a finite-size flow system develops in time to improve its outcome and persist in time [1]. This theory has been successfully observed in nature, where configurations that provide a more straightforward outcome have replaced previous configurations for different systems [2]. For instance, the anatomy of lungs with their primary, secondary, and tertiary bronchus follows a dendritic architecture, improving airflow in the human body [3]. In addition, trees and leaves' structures have been improving in time in a point-to-volume architecture to be more efficient when receiving solar light [1] or how water flow divides into smaller dendritic architectures in a river basin to reach its tributary [4, 5].

A fundamental result obtained from the constructal theory states that the geometry of a system does not develop randomly. Still, they are the result of the system's struggle within a period to become more efficient [6]. For this reason, natural systems, both animate and inanimate, show a great variety of geometries that are far from perfect from an aesthetic point of view [7]. Also, it has been observed that fractal structures allow a significant improvement in the efficiency of different systems [8]. This asymmetry shows that systems tend to minimize the opposition they face to overcome obstacles and keep improving their efficiency [4].

Engineering is not apart from using the constructal theory to study different phenomena, considering that efficiency and optimization are fundamental principles in other engineering fields. Specifically, within the various engineering fields, heat transfer has become a field in which the mentioned theory has been used to improve different systems outcomes. For instance, Nakayama [9] investigated different heat flow paths and geometries and concluded that different lattices can cause a difference of approximately 11% in the equivalent thermal conductivity of the system. In addition, Rocha et al. [10] studied the optimization of the inclusion of tree-shaped inserts manufactured with a high thermal conductivity material to improve the cooling process in electronics. They concluded that different tree architectures enhance the performance of conduction cooling of bodies.

Furthermore, Ziaei et al. [11] simulated the outcome of various degrees of T-shaped bifurcations in the heat transfer process between a higher temperature (353 K) invading line with a different number of bifurcations and an isothermal and solid material with a melting temperature of 303 K. This study concluded that an increasing number of T-branching bifurcations and adding more freedom for the design produces a more efficient heat transfer and a more rapid melting process [11]. Chen and Cheng analyzed the heat transfer process in channel nets with fractal branches by varying the dimensions of the mentioned branches and the number of branching levels. They concluded that a branched network produces a higher convective heat transfer when compared to a parallel channel without a fractal architecture [12].

Numerous efforts have been dedicated to applying a constructal optimization for improving heat transfer [13, 14] and developing constructal trees of different geometries' fins to minimize thermal resistance [15, 16]. For instance, Almongbel and Bejan [13] applied a constructal optimization process to develop a nonuniform lattice to reduce thermal resistance. They concluded that a nonuniform tree-shaped structure caused a slight decrease in the global thermal resistance of the system. Moreover, Chen et al. [14] applied a respective constructal optimization process for a disc-point heat conduction system. They observed that a tree-shaped disc structure with variable cross-sections produced a higher maximum temperature difference.

On the other hand, Alebrahim and Bejan [15] investigated the thermal performance of different arrays of circular fins to minimize the thermal resistance between a volume and one point. The researchers concluded that the system's thermal conduction improves while the lattice's complexity increases. In addition, Ledezma and Bejan [16] investigated the performance of different arrays of three-dimensional trees. They concluded that by optimizing the other geometrical parameters, a tree-shaped array can outperform an array without bifurcations.

Considering the extensive use of constructal theory in the field of heat transfer optimization and the promising results obtained from a group of authors, this work focuses on analyzing the possible changes in the heat management of a thermoelectric generator operation temperature included in a Solar Thermoelectric Generator (STEG) coupled with a Latent Heat Storage and Cooling System (LHSCS), using two different heat sink geometries considering a tree-shaped fins structure instead of a radial fins heat sink. The comparison is developed under the desert location environment conditions and using a previously validated numerical model that permits quantifying the variation in the electricity generation of an STEG + LHSCS system. The latter will help choose the correct design of latent heat storage systems for particular applications and environmental conditions.

2 Methodology

2.1 System Configuration and Operation

The referenced model proposed previously by Montero et al. in [17] and [18] couples a solar thermoelectric generator (STEG) and a latent heat storage and cooling system (LHSCS). The main advantage of this hybrid STEG + LHSCS system is the possibility to generate electricity 24 h a day, using the solar radiation during the insolation period and the heat stored during the absence of solar radiation to generate electricity in the Thermoelectric Generator (TEG) included in the STEG system. The proposed model in this study contains modifications in the heat sink design according to the constructal theory. In addition, an axisymmetric 2D model of the STEG + LHSCS system was used in the numerical model analyzed in COMSOL Multiphysics, as shown in Fig. 1.

The analyzed STEG + LHSCS system includes a Bismuth Telluride (Bi_2Te_3) TEG and an optical system consisting of a Fresnel lens collector. The TEG cold side is coupled to the LHSCS. The LHSCS consists of an insulated acrylic plastic container filled with a Phase Change Material (PCM) and an embedded copper heat sink,

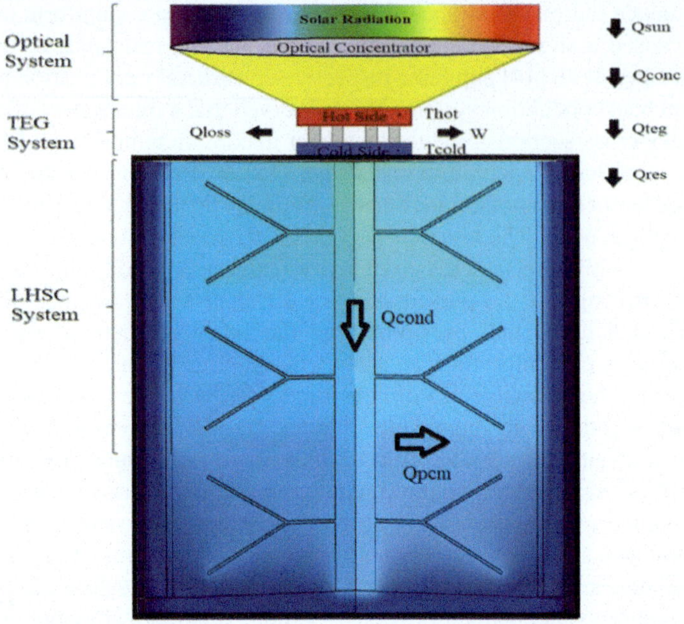

Fig. 1 Solar thermoelectric generator coupled latent heat storage and cooling system designed according to the constructal theory

designed following the constructal theory. Puretemp 48X organic paraffin is used as PCM (6 kg), with a melting temperature of around 48 °C. This study uses the ambient conditions of the Atacama Desert and the same border conditions presented in [17, 18] to compare the present study's results with those obtained in the reference analysis. In the proposed STEG + LHSCS system, the TEG hot side receives the insolation (Q_{sun}) as a concentrated heat flux (Q_{conc}) which causes a higher temperature on its surface (T_{hot}). The TEG cold side is coupled to the LHSCS to maintain its temperature lower than the TEG hot side temperature. A fraction of the Q_{conc} is rejected to the environment (convective and radiative heat losses), and the other fraction is transferred by conduction to the TEG's hot side (Q_{teg}). A Q_{teg} fraction generates electricity, and the LHSCS rejects the residual heat. This residual heat is conducted to the PCM during the charging process. The PCM stores the residual energy as a latent heat of fusion. The PCM temperature (T_{pcm}) increases in the charging period and starts to melt when the melting temperature is achieved. Consequently, the PCM maintains a lower temperature on the TEG's cold side (T_{cold}). A reversion of the heat flux occurs during the PCM discharging period, causing a reversion in the operation of LHSCS. Heat flows from the PCM into the TEG's cold side during this period. The charging process in the LHSCS is coupled with the hours of incident solar radiation in the selected geographic location. It determines the amount of the PCM and the LHSCS's container dimensions. The discharging process in the LHSCS is coupled with the absence of solar insolation. Accordingly, the stored energy is conducted to

TEG's cold side to produce more electricity. In the discharging period, as the ambient temperature is lower than the PCM temperature, the TEG hot side becomes the cold side. This direction of heat flow in the TEG induces the reversion of the electron flux.

2.2 Governing Equations

The STEG system receives a concentrated heat flux of an optical concentrator (Q_{conc}) in the TEG hot side due to the inlet solar radiation, and it is calculated as [19]:

$$Q_{conc} = C_{opt} \cdot \eta_{opt} \cdot Q_{sun} \tag{1}$$

where C_{opt} is the optical concentration factor, η_{opt} is the optical concentrator efficiency, and Q_{sun} is the inlet solar irradiation.

A part of the inlet concentrated heat flux is wasted by radiative and convective heat losses from the TEG's hot side to the ambient (Q_{loss}). The remaining fraction of the inlet heat flux is used in the TEG's module (Q_{teg}), calculated as [19]:

$$Q_{teg} = Q_{conc} - Q_{loss} \tag{2}$$

A fraction of Q_{teg} is converted into electricity [W], and the residual heat (Q_{res}) is rejected from the TEG cold side. The net energy balance is calculated as [19]:

$$Q_{teg} = W + Q_{res} \tag{3}$$

The TEG's efficiency (η_{TEG}) is evaluated by a function of the Seebeck coefficient (S) and the dimensionless figure of merit (ZT). The TEG's maximum conversion efficiency (η_{maxTEG}) is evaluated as the product of the Carnot efficiency and the TEG's efficiency, calculated using the following equation [20]:

$$\eta_{maxTEG} = \left(\frac{T_h - T_c}{T_h}\right) \cdot \frac{\sqrt{1 + ZT_m} - 1}{\sqrt{1 + ZT_m} + \frac{T_c}{T_h}} \tag{4}$$

where T_c and T_h are the TEG's cold and hot side temperatures, and T_m is the average temperature.

The STEG's conversion efficiency is a function of the temperature difference between the TEG's hot and cold sides (T_h–T_c), the concentrated heat flux (Q_{conc}), the Seebeck coefficient (S), the electrical current (I), the TEG's internal resistance (R_i), and the TEG's absorptance area (A_S), it is calculated from [19]:

$$\eta_{STEG} = \frac{I[S(T_h - T_c) - IR_i]}{Q_{conc}A_S} \tag{5}$$

The heat transfer in the STEG + LHSCS system is evaluated as conductive heat transfer in solids and conductive-convective heat transfer in the PCM, using Fourier's´ law and the energy conservation equation [21]:

$$\rho C_p \frac{\partial T}{\partial t} + \rho C_p \mu \nabla T + \nabla q = Q \tag{6}$$

$$q = -k \nabla T \tag{7}$$

where ρ is the fluid density, C_p is the constant pressure-specific heat capacity, T is the fluid temperature, and q is the heat flux. The laminar flow is dominated by the continuity and Navier Stokes equations, assuming a Newtonian and incompressible fluid flow [16]:

$$\rho \frac{\partial u}{\partial t} + \rho (u \nabla) u = \nabla\left[-p\mathbf{I} + \mu\left(\nabla u + (\nabla u)^T\right)\right] + F \tag{8}$$

$$\rho \nabla u = 0 \tag{9}$$

where μ is the dynamic viscosity, u is the fluid velocity, **I** is the identity matrix, and F represents the buoyancy forces. The natural convection inside the PCM is modeled using a buoyancy force in laminar flow, following a Boussinesq approach. Consequently, the variation in the PCM density induced by the temperature variation is considered for the force term only. The gravity force inside the PCM causes the liquid PCM to flow upward in the LHSCS compartments. The material properties are a function of the temperature (T), and the phase change phenomenon defined by Kim and Kaviany [22] is used:

$$\rho = \theta_{\rho_{phase1}} + (1-\theta)_{\rho_{phase1}} \tag{10}$$

$$C_p = \frac{1}{\rho}\left[\theta_{\rho_{phase1}} C_{p,phase1} + (1-\theta)_{\rho_{phase2}} C_{p,phase2}\right] + L\frac{\partial \alpha_m}{\partial T} \tag{11}$$

$$k = \theta k_{phase1} + (1-\theta) k_{phase2} \tag{12}$$

$$\alpha_m = \frac{1}{2}\frac{(1-\theta)_{\rho_{phase2}} - \theta_{\rho_{phase1}}}{(1-\theta)_{\rho_{phase2}} + \theta_{\rho_{phase1}}} \tag{13}$$

where α is the mass fraction, θ is the PCM solid fraction, k is the thermal conductivity, and the subscripts *phase*1 and *phase*2 refer to solid and liquid phases, respectively.

2.3 Numerical Model

The STEG + LHSCS system is simulated in the COMSOL Multiphysics software as a 2D and time-dependent axisymmetric model. The model uses non-isothermal flow multiphysics to simulate the fluid flow and heat transfer coupling. The TEG's hot side heat flux is simulated as the concentrated solar radiation for the selected geographical zone. The environmental data of the selected location (Calama, El Loa, 22.483S, 68.9W, 2291 MAMSL) includes the solar radiation, ambient temperature, and wind speed for each hour of the day for January (summer) as shown in Fig. 2 [23]. The TEG includes temperature-dependent alumina and Bi_2Te_3 material properties in the software database [24]. The PCM is modeled as a fixed grid.

Montero et al. previously validated the numerical model [17, 18] through experimental results. This numerical model was also used in an energy and exergy efficiency analysis of the STEG + LHSCS system [25]. The STEG + LHSCS computational model used in the current analysis consists of a time-dependent simulation for 86,400 s (24 h).

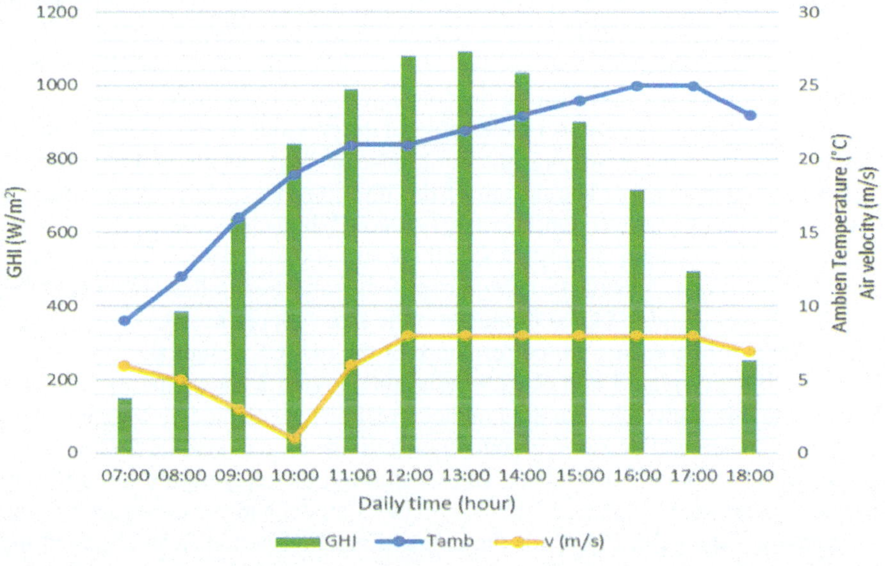

Fig. 2 Hourly variation of ambient temperature, wind speed, and global horizontal irradiation for January in the selected geographical location [23]

3 Results and Discussion

The previous model of STEG + LHSCS developed by Montero et al. [17, 18] has been modified following the constructal theory to verify the possible improvements in the heat flux between the STEG and LHSCS systems. The transient analysis is carried out using COMSOL Multiphysics software. Following the constructal theory, the modification in the copper heat sink fins wants to maintain and control the TEG's cold side temperature better than the original model with radial fins. Further, this study tried to verify if the constructal theory design modifications allow enhancing the capacity of the STEG + LHSCS system to generate electricity during the absence of solar radiation, or in other words, increase the temperature difference between the TEG's hot and cold sides during this period. In this way, the STEG + LHSCS system is numerically simulated considering the same environmental and solar irradiation conditions as the previous STEG + LHSCS with radial fins. In this way, all the border conditions of the STEG + LHSCS previous model [17, 18] are identical. The numerical model includes the same 6 kg of Puretemp 48X Organic Paraffin and its physical properties defined by the manufacturer. The Bi2Te3 TEG model is simulated considering the temperature-dependent thermophysical properties from the COMSOL Multiphysics software. The software also defined all other materials' physical properties. A 75,168 triangular element mesh has been used for all the computational simulations, showing a mesh quality of 0.95 and ensuring the convergence of the numerical results.

This study considers the exact dimensions of the previous STEG + LHSCS model (standard model), considering only the variation in the heat sink fins as the application of the constructal theory. The design of the fins considers a branching pattern in which the fins have one (version 1) or two (version 2) branch divisions. As explained previously, the fin length is constant for all the successive divisions. This study also considers the dimensions of the central copper bar attached to the TEG's cold side and the number of fins adapted to maintain the same copper heat sink weight. The latter allows us to compare the performance of the LHSCS designed by constructal theory with the original LHSCS with radial fins. Figure 3 shows the three models analyzed in this study: The standard model, previously presented by Montero et al.; version 1 (constructal design with one branch division); and version 2 (constructal design with two branch divisions). The variation in the temperature of TEG's hot and cold sides for the standard STEG + LHSCS model and the constructal design variations are shown in Fig. 4 for the entire simulation period. It is observed that the design's variation negatively impacts the variation of the temperature difference between the TEG's hot and cold sides, observing a decrease in both temperatures during the charging period. On the other hand, during the discharging period, this temperature reduction allows the LHSCS to maintain a higher temperature variation in the TEG's hot side (inversion of the heat flux), which can elevate energy production during this period. It is also observed that the temperature curves follow the same pattern due to the non-variation of the LHSCS geometry and border conditions.

It is necessary to analyze the variation in the temperature of PCM and its melting behavior to determine the impact on the TEG sides' temperature due to the change in

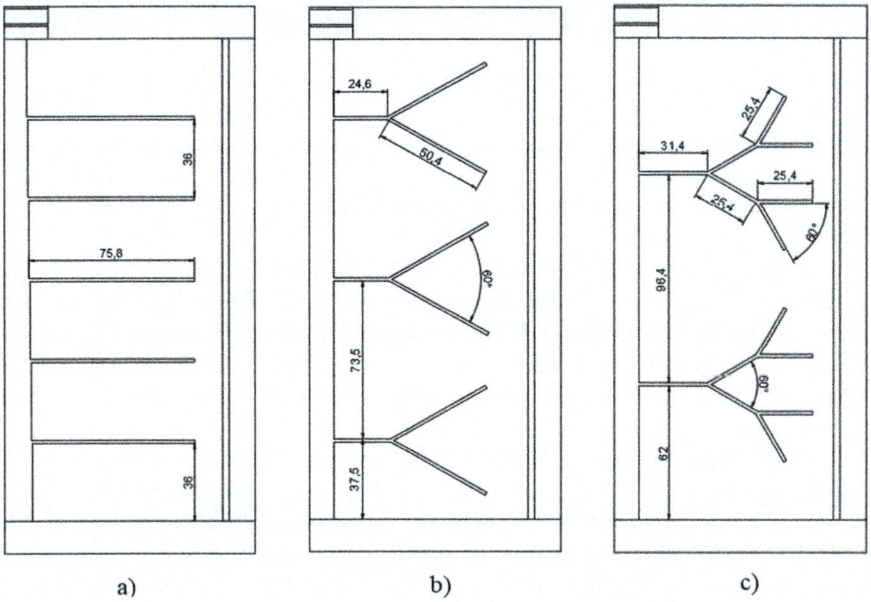

Fig. 3 STEG + LHSCS models considered in this analysis: **a** standard model, **b** version 1 constructal design, c) version 2 constructal design

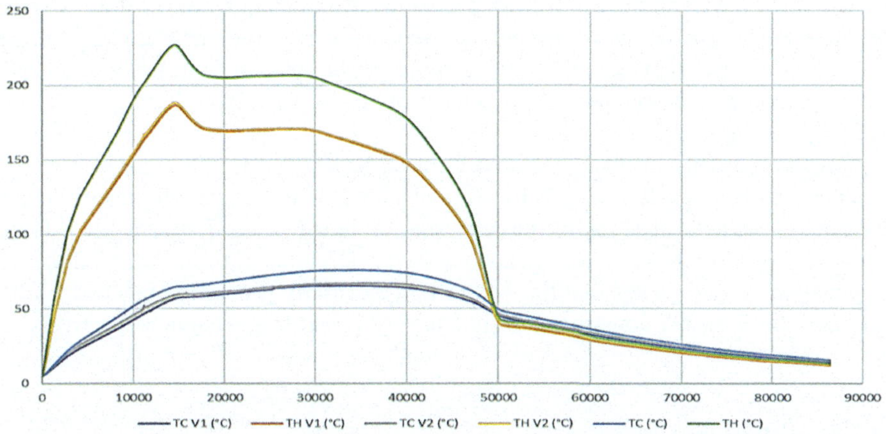

Fig. 4 STEG + LHSCS models hot side (TH) and cold side (TC) temperatures for **a** standard model, **b** version 1 constructal design, and **c** version 2 constructal design

the heat sink design following the constructal theory. Figure 5 shows the variation in the PCM melting for the three LHSCS designs during the charging and discharging periods. It is observed that the original design has better performance in melting the PCM during the charging period. It is a consequence of compartments that improve the heat transfer by convection and allow the melting process to occur more effectively along the container. As shown in Fig. 4, the more effectively the heat transference to the PCM occurs, the higher the temperature difference between the TEG's hot and cold sides.

Nevertheless, it is also observed that the V2 constructal design has a better thermal performance than the V1 constructal design since more PCM melted in the charging period (25,800 s and 36,600 s). In the discharging period, all three designs melted all the PCM during the observed simulation time (59,000 s). During this discharging time, the PCM delivers the heat stored to the environment through the TEG, which generates more electricity without solar radiation.

It is essential to analyze the impact of the constructal design in the convection heat transfer in the PCM. It is reflected as the heat flux inside the melted PCM and the velocity of the liquid PCM caused by the difference in density and the buoyancy force. It is observed in Fig. 6 that the different heat sink designs govern the heat transfer between the TEG's cold side and the PCM in the container. As observed previously, compartments inside the container create an independent region where the heat is transferred more properly through conduction by the radial fins and then dispersed in the PCM by convection. On the other hand, the constructal design fins do not allow the creation of those compartments. Consequently, the heat transferred by convection is lower, as confirmed by a low velocity in the melting PCM for the two constructal designs. During the charging period and for both times analyzed, the V1 constructal design performed poorly, observing the liquid PCM's lower flow velocity.

Finally, the daily energy generated by each STEG + LHSCS system model is evaluated and compared in Fig. 7. Because of a lower temperature difference reached by both constructal designs in the heat sink included in the LHSCS, there is no observed increase in the energy generated in the TEG due to the constructal theory application. It is observed that the decrease in electricity generation reached 24.0% and 24.7% for the V1 constructal design and V2 constructal design, respectively.

4 Conclusions

This study analyzed the impact on the performance and electricity generation of the design of the copper heat sink embedded in the LHSCS following the constructal theory for an STEG + LHSCS system that operates under the ambient conditions of a selected location. The results were compared with a previously analyzed model of STEG + LHSCS using the exact geographical location, considering the same geometry and border conditions (materials conductivity, heat losses, heat

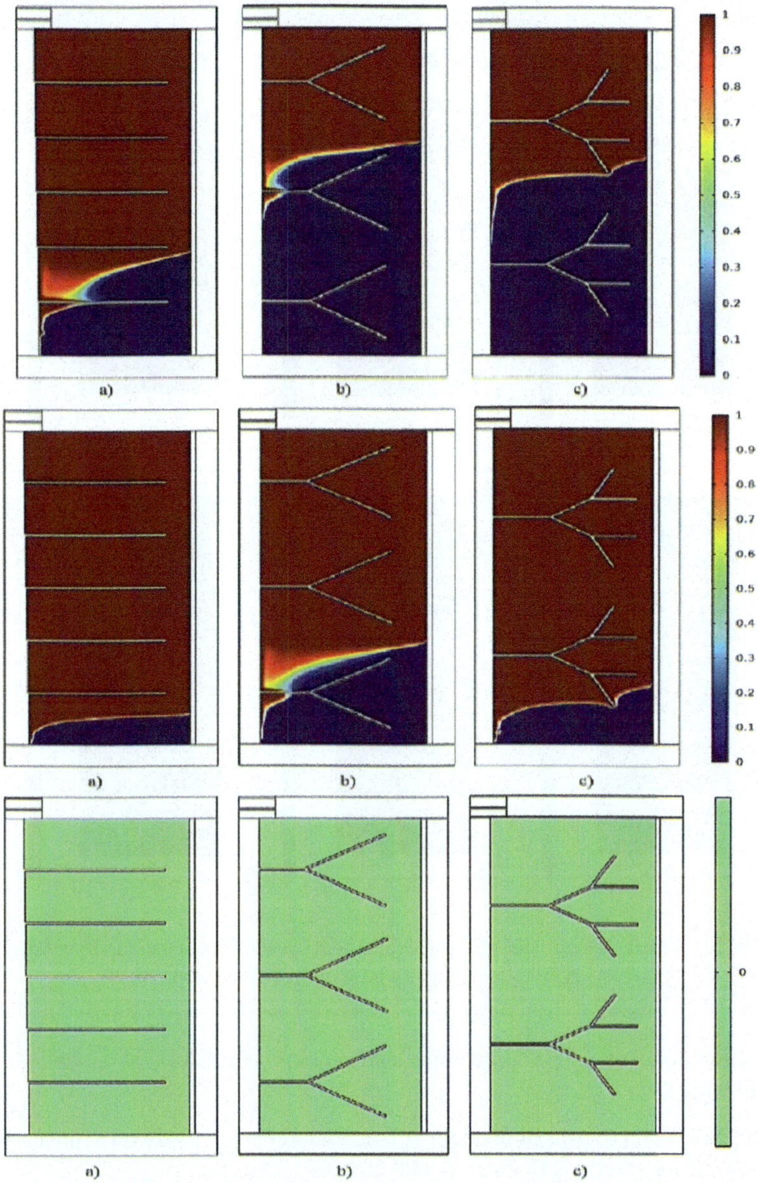

Fig. 5 PCM melting for the three analyzed models: **a** standard, **b** constructal V1, **c** constructal V2, and for 25,800 s (charging period), 36,600 s (charging period), and 59,000 s (discharging period), respectively

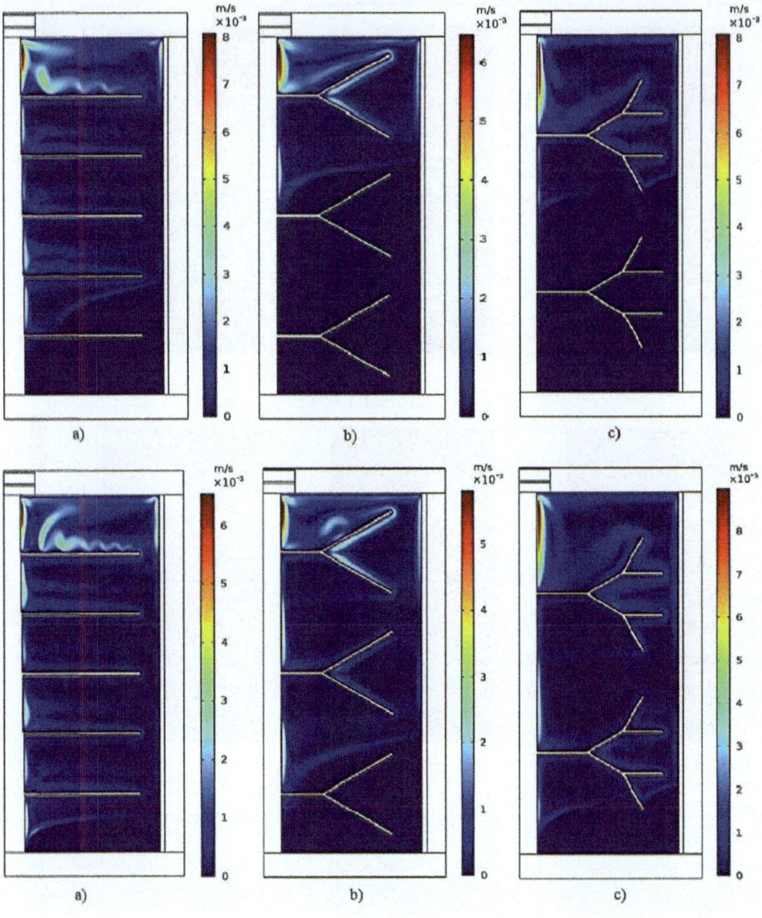

Fig. 6 PCM melted velocity for the three analyzed models: **a** standard, **b** constructal V1, **c** constructal V2, and for 25,800 s and 36,600 s (charging periods), respectively

transfer mechanisms, simulation time, and numerical model). The conclusions are the following:

- The constructal design of the copper heat sink does not positively impact the heat transfer between the TEG's cold side and the LHSCS.
- Modifying the fins design and distribution causes a low performance of the TEG included in the STEG + LHSCS system, observing a reduction in the temperature difference between the TEG hot and cold sides. The average temperature difference during the solar insolation period reaches around 111.9 °C, 89.7 °C, and 89.2 °C for the standard model, V1 constructal and V2 constructal, respectively. Without solar radiation, the average temperature difference reaches around 3.9 °C, 3.2 °C, and 3.4 °C for the standard model, V1 and V2 constructal, respectively.

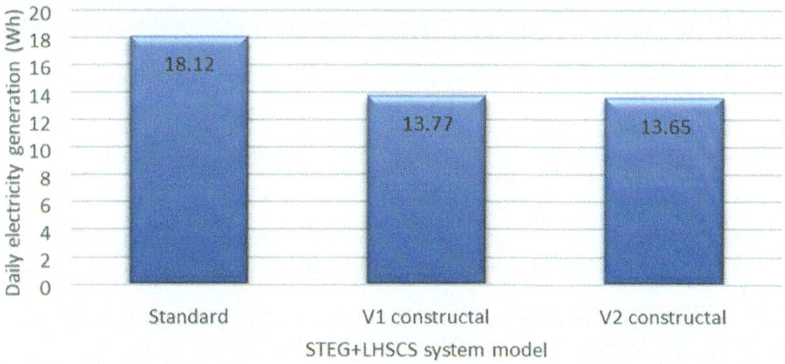

Fig. 7 Daily electricity generation for the three analyzed models: **a** standard, **b** constructal V1, and **c** constructal V2, respectively

- The reduction in the TEG's side temperature difference causes a reduction in the electricity generation. Compared with the standard model, a decrease of 24.0% and 24.7% in electricity generation is observed for the V1 constructal design and V2 constructal design.
- The constructal design of the fins also decreases the heat transfer by conduction (PCM in solid phase) and convection (PCM in liquid phase) if compared to the standard model, which considered radial fins. The reduction in the heat transfer by convection in the liquid PCM causes the reduction of the quantity of melted PCM during the charging time compared with the standard model radial fins design.
- The reduction in the convection heat transfer in the melting PCM is observed as a reduction of the flow velocity inside the liquid PCM.

The use of constructal theory to design efficient solar energy conversion systems and latent heat storage systems needs more future analysis, focusing on particular applications. In this study, including the constructal theory in the design of the copper heat sink embedded in the LHSCS does not positively impact the STEG + LHSCS system performance. Consequently, it is necessary to study other configurations of constructal theory design deeply, such as fins orientation, branching separation, branching angle, and number of branching levels.

References

1. Bejan, A., Lorente, S.: The constructual law of design and evolution in nature. Philosophical Trans. Royal Soc. B, 1335–1347 (2010)
2. Bejan, A.: How nature takes shape: extensions of constructal theory to ducts, rivers, turbulence, cracks, dendritic crystals and spatial economics. Int. J. Therm. Sci. **38**, 653–663 (1999)
3. Bejan, A.: Street network theory of organization in nature. J. Adv. Transp. **30**(2), 85–107 (1996)
4. Reis, H.: Constructal theory: from engineering to physics, and how flow systems develop shape and structure. Appl. Mech. Rev., 269–282 (2006)

5. Bejan, A.: Theory of organization in nature: pulsating physiological processes. Int. J. Heat Mass Transf. **40**(9), 2097–2104 (1997)
6. Bejan, A.: Shape and Structure, from Engineering to Nature. Cambridge University Press, Cambridge (2000)
7. Bejan, A.: Constructal theory: the generation of flow-systems architecture, from engineering and nature. Ingeniería Mecánica, Tecnología y Desarrollo **1**(6), 195–201 (2005)
8. Xu, P., Yu, B., Yun, M., Zou, M.: Heat conduction in fractal tree-like branched networks. Int. J. Heat Mass Transf. **49**, 37463751 (2006)
9. Nakayama, W.: A methodology to work on geometrically complex heat transfer systems: the cases of heat conduction through composite slabs. Int. J. Heat Mass Transf. **46**, 3397–3409 (2003)
10. Rocha, L., Lorente, S., Bejan A.: Conduction tree networks with loops for cooling a heat generating volume. Int. J. Heat Mass Transfer, 2626–2635 (2006)
11. Ziaei, S., Lorente, S., Bejan, A.: Morphing tree structures for latent thermal energy storage. J. Appl. Phys. (2015)
12. Chen, Y., Cheng, P.: Heat transfer and pressure drop in fractal tree-like microchannel nets. Int. J. Heat Mass Transfer, 2643–2648 (2002)
13. Almongbel, M., Bejan, A.: Constructal optimization of nonuniformly distributed tree-shaped flow structures for conduction. Int. J. Heat Mass Transf. **44**, 4185–4194 (2001)
14. Chen, L., You, J., Feng, H., Xie, Z.: Constructal optimization for "discpoint" heat conduction with nonuniform heat generating. Int. J. Heat Mass Transf. **134**, 1191–1198 (2019)
15. Alebrahim, A., Bejan, A.: Constructal trees of circular fins for conductive and convective heat transfer. Int. J. Heat Mass Transf. **42**, 3585–3597 (1999)
16. Ledezma, G., Bejan, A.: Constructal three-dimensional tress for conduction between a volume and one point. J. Heat Transfer **120**, 977–984 (1998)
17. Montero, F., Lamba, R., Ortega, A., Jahn, W., Chen, W., Amador, A., Guzmán, A.: Bidirectional solar thermoelectric generator combining heat storage for daytime and nighttime Powe generation. Appl. Thermal Eng. (2023)
18. Montero, F., Lamba, R., Ortega, A., Jahn, W., Guzmán, A.: A novel 24-h day-night operational solar thermoelectric generator using phase change materials. J. Cleaner Prod., 296 (2021)
19. Lorenzi, B., Chen, G.: Theoretical efficiency of hybrid solar thermoelectric photovoltaic generators. J. Appl. Phys. **124**(2) (2018)
20. Mahan, G.: Inhomogeneous thermoelectrics. J. Appl. Phys., 4551–4554 (1991)
21. Panton, R.: Incompressible Flow. Wiley, Hoboken (1996)
22. Kim, C., Kaviany, M.: A numerical method for phase-change problems. Int. J. Heat Mass Transf. **33**(12), 2721–2734 (1990)
23. Explorador Solar from Universidad de Chile, http://ernc.dgf.uchile.cl:48080/inicio. Accessed 15 July 2023. COMSOL Multiphysics Documentation
24. Montero, F., Lamba, R., Singh, W., Jahn, W., Chen, W.: Energy and exergy of a bidirectional solar thermoelectric generator combining thermal energy storage. Frontiers Energy Res., 11 (2023)

Quantification of the Decrease of Droplets Expelled from the Human Respiratory Tract During Expiratory Activities When Using Masks

Paúl León Benítez, David Loza-Matovelle, Luis Carrión Matamoros, and Santiago Castellanos Villa

Abstract This study aims to assess the effectiveness of three nasal-mouth guards using a novel experimental methodology based on particle quantification. The nasal-mouthpiece protectors under examination include a single-layer protector, a surgical mask, and a KN95 mask. The expiratory phenomena evaluated in this study encompass talking, coughing, and sneezing. The experimental method for quantifying particles involves several steps. Firstly, high-speed videos are recorded. Subsequently, the most representative images illustrating the evolution of the phenomena are saved. Finally, these images are analyzed using a program developed by the author, which quantifies particles through image binarization. The experiment is conducted with three different test individuals to ensure its reliability. This repetition allows for a quantitative analysis of the amount of mucus-saliva mixture expelled during each phenomenon. Moreover, it enables the quantification of such expulsion when using the various nasal-mouth protectors. Additionally, the experimental methodology facilitates a qualitative analysis of each phenomenon, as it allows for observing how expiratory phenomena evolve under different states of nasal-mouth coverage, as depicted in representative images.

Keywords Experimental methodology · Effectiveness quantitative analysis · Expiratory activities · Nasal-mouth guards

P. León Benítez · D. Loza-Matovelle (✉) · L. Carrión Matamoros · S. Castellanos Villa
Departamento de Ciencias de la Energía y Mecánica, Universidad de las Fuerzas Armadas ESPE, Sangolquí 171103, Ecuador
e-mail: dcloza@espe.edu.ec

P. León Benítez
e-mail: pleon@espe.edu.ec

G. F. Olmedo Cifuentes et al. (eds.), *Emerging Research in Intelligent Systems*, Lecture Notes in Networks and Systems 902, https://doi.org/10.1007/978-3-031-52255-0_18

1 Introduction

Three years into the COVID-19 pandemic, its impact has been felt by over 760 million people worldwide, resulting in the tragic loss of nearly 6.9 million lives [1]. Efforts to evaluate the efficacy of face masks have predominantly relied on established standards. These standards involve using machines that simulate particle emissions from the respiratory system, ensuring the repeatability of results. Notable examples of these standards can be found in references [2, 3]. However, these standards fall short of assessing the masks' effectiveness in real-life scenarios where they need to impede the passage of particles expelled from the respiratory tract into the surrounding environment. This limitation is particularly evident during uncontrollable actions like sneezing or other physiological mechanisms [4–6].

Recognizing this limitation, and given the widespread use of face masks as a primary measure for preventing viral diseases [7], this study offers both a qualitative and quantitative evaluation of how effectively three different types of covers reduce the flow of fluids expelled during actual physiological respiratory actions such as talking, coughing, and sneezing. This approach seeks to address the practical shortcomings of existing standards and provides valuable insights into the real-world performance of these masks in mitigating the transmission of infectious agents.

2 Experimental Method

In everyday life, individuals cannot perceive the significant number of particles they expel during ordinary activities like speaking, and the situation worsens when it comes to actions like coughing or sneezing, where these particles become high-velocity flows that are invisible to the naked eye.

The application of high-speed cameras for health-related research dates back to 1941, when scientists began recording anthropogenic natural phenomena, such as sneezing or coughing, at a speed of 1300 frames per second (fps) [8]. This marked the inception of a critical technique: intense lighting to render visible particles that would otherwise remain unseen under normal conditions. This technique is fundamentally known as darkfield imaging.

Darkfield imaging operates on a fundamental principle: capturing an image surrounded by an intense, focused light source while blocking spectral refraction. This effectively illuminates the specific flow or area of interest while isolating it from the background [9]. This technique has been pivotal in enabling researchers to study and visualize the dynamics of particle dispersion during actions like sneezing or coughing, which would otherwise be invisible to human observation.

2.1 Case Study Design

To conduct this case study, high-speed videos are recorded at an approximate rate of 1000 frames per second (fps), a reference value drawn from the research of Turner, Jennison, and Edgerton [8]. Three distinct case studies are considered: the first pertains to voluntary physiological processes of the respiratory system, while the second and third are linked to involuntary physiological processes [10]. Table 1 describes these three case studies, all repeated with three test subjects.

Each expiratory phenomenon underwent examination under four different states of nasal-mouth protection. The first state involved no protection, while the second, third, and fourth states encompassed the use of one-, three-, and five-layer masks, respectively. The nasal-mouth protectors utilized in the study consisted of a single-layer fabric cover, a surgical mask, and a KN95 mask. Consequently, this study encompasses 3 case studies and four states of nasal-mouth protection.

Therefore, the research undertook an extensive analysis of 36 experiments, encompassing the three expiratory actions mentioned earlier and the four distinct nasal-mouth coverage conditions described above.

The individuals participating in this study were 24 years old, male, with an average height of 1.7 m, and had no preexisting respiratory issues. The study is conducted under controlled conditions at 15∘C and 68% relative humidity.

Then comes the description of the three analyzed expiratory phenomena and standardization process (Table 1):

First case. In the first scenario, the test subject spoke with a high-pitched voice. During this exercise, the individual systemically combined all the alphabet's consonants, one by one, with the vowels. This process showed that when the test subject pronounced the consonant 'T' and combined it with the vowel 'A,' a notably higher quantity of droplets was produced.

Second case. The second case involved simulating a coughing event akin to sneezing. This action typically involves an initial irritation, followed by the closure of the glottis (the vocal apparatus in the throat) and rapidly increasing intrathoracic (within the chest) pressure. Air surge is forcefully expelled at high flow upon reopening the glottis, as described in reference [11]. In this scenario, the test subject induced a cough to replicate this phenomenon.

Table 1 Expiratory phenomena provoked in each state of nasal-buccal protection

Case no	Expiratory phenomena	Executed action
1	Speak	Pronouncing 'TA' aloud
2	Cough	Artificial cough provoked by the test person
3	Sneeze	Sneezing provoked by stimulating the mucous membrane of the nasal cavity with a toothpick

Third case. The third case revolved around the deliberate triggering of a sneeze. The test subject stimulated the mucous membrane of the nasal cavity, prompting a sneeze. This stimulation is self-induced using a toothpick.

Additionally, for comparative purposes, exhaled bioaerosols are also considered. The test subject is recorded while exhaling a vaporized water/glycerin mixture. This allowed for the visualization of aerosol dispersion when using nasal-mouth protectors, as outlined in [3].

2.2 *Experimental Darkfield Imaging Setup*

For darkfield imaging in our experiment, this study employs a Phantom v2512 high-speed camera [12]. This camera is strategically positioned perpendicular to the primary direction of particle motion observed during the anthropogenic natural phenomena under investigation.

This study uses an intense light source, a 120 W Cyclops white LED light, to illuminate the particles effectively. This light source is positioned on the side opposite the camera, ensuring that the edge of the cone-shaped beam created by the focused light intersects with the middle of the camera lens, as visually represented in Fig. 1.

Fig. 1 Experimental setup

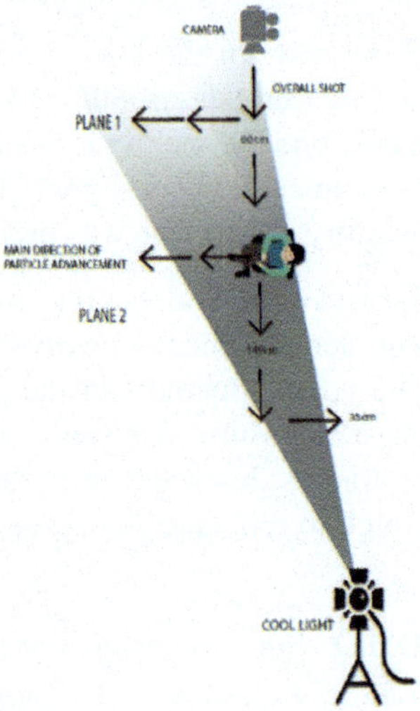

2.3 *Experiment Design*

The primary objective of this study is to quantify the reduction in expelled particles during various physiological processes. A custom-coded program is developed to achieve this, which program calculates the percentage of white pixels relative to the total number of pixels in each image it processes. The percentage of white pixels is interpreted as the total volume of the droplet/bioaerosol mixture expelled during each anthropogenic phenomenon. In contrast, the remaining pixels represent the surrounding air and are interpreted by the program as black pixels.

The program functions through a meticulously defined sequence of steps (Fig. 2):

1. **Image interpretation and transformation to gray-scale**

   Initially, the program reads an image in .jpg format. The program perceives this image as a three-dimensional matrix. The program then eliminates the third dimension of the matrix, which corresponds to color information, thereby reducing the image to its two primary dimensions: position and gray intensity of each pixel.
2. **Binarization of the image (visualizing the droplet/bioaerosol fluid)**

   The objective at this stage is to distinguish the fluid composed of the droplet/bioaerosol mixture from the surrounding air in the image. Therefore, a mask is applied to eliminate the outline of the individual, as the profile could be displayed as white pixels in the picture. Subsequently, a threshold value is defined, ranging from 0 to 1. Pixels above this threshold are assigned the value '1' (white), while those below it are set to '0' (black). Figure 2 of the provided flowchart shows a detailed procedure of this process.

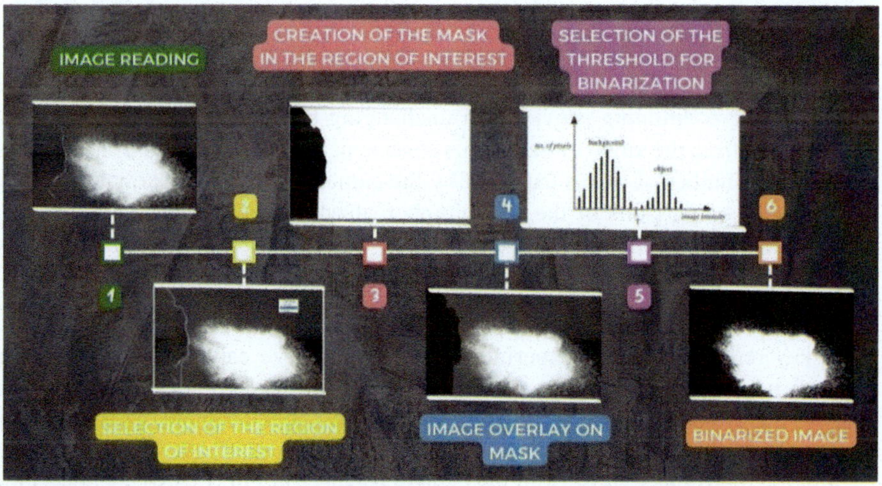

Fig. 2 Image binarization flowchart

3. **Counting white and black pixels in the image (counting droplet/bioaerosol fluid and surrounding air)**

   Once the image is binarized, the program treats it as a two-dimensional matrix with binary values. The subsequent step involves initiating separate counters for each '0' and '1' in the matrix, effectively counting the total number of black pixels (representing surrounding air) and white pixels (representing the droplet/bioaerosol fluid).

4. **Determination of the percentage of white pixels in the image (quantifying the rate of droplet/bioaerosol fluid)**

   The program calculates the image size in terms of pixels by summing the total counts of zeros and ones in the matrix. The number of ones is then divided by the size of the image, resulting in the percentage of white pixels relative to the total number of pixels. This percentage value represents the volume of mucus/saliva or droplet/bioaerosol mixture expelled during each expiratory phenomenon, providing a quantitative measure of the phenomenon's output.

This image processing and binarization approach enables the program to quantify the expelled droplet/bioaerosol mixture while distinguishing it from the surrounding air, providing valuable insights into the dynamics of particle dispersion during the studied physiological processes.

2.4 *Experiment Conditions*

During the execution of the experiment, several critical conditions are observed to ensure the reliability and consistency of the results:

1. **Static position of the setup.**

   The experimental setup is maintained in a static position throughout the study. This static positioning provided an unchanging background for all the images extracted from the video footage. This consistency in the background ensured that the total number of pixels captured by the camera remained invariant, enabling accurate comparisons of the total number of white pixels. These white pixels are interpreted as the percentage amount of fluid expelled during the expiratory phenomena, regarding the total number of pixels acquired.

2. **Main restriction—frame size**

   A key restriction in the experiment is the frame size captured by the camera from its designated shooting position. Moving the subject closer to the camera (plane 1 in Fig. 1) reduced the particles captured by the camera. This reduction is primarily due to the diminished light intensity in that position. Conversely, moving the subject away from the camera (plane 2 in Fig. 1) also decreased the number of particles captured by the camera. This reduction occurred because the plane generated by the primary forward direction of the particles, along with the projection of light, is correspondingly reduced.

These experiment conditions, particularly the static positioning of the setup and the considerations regarding frame size, are crucial in ensuring the consistency and accuracy of the data obtained. They allowed for meaningful comparisons of the expelled fluid across different expiratory phenomena while minimizing potential sources of variability.

3 Results and Discussion

3.1 The Efficiency of Each Mask Type

The images presented in Fig. 3 are captured precisely at the moment of the highest dispersion of droplets/bioaerosols during each of the studied expiratory phenomena under different states of nasal-mouth protection. This visual representation vividly illustrates droplet reduction as a transition to nasal-mouth protection garments with increased layers.

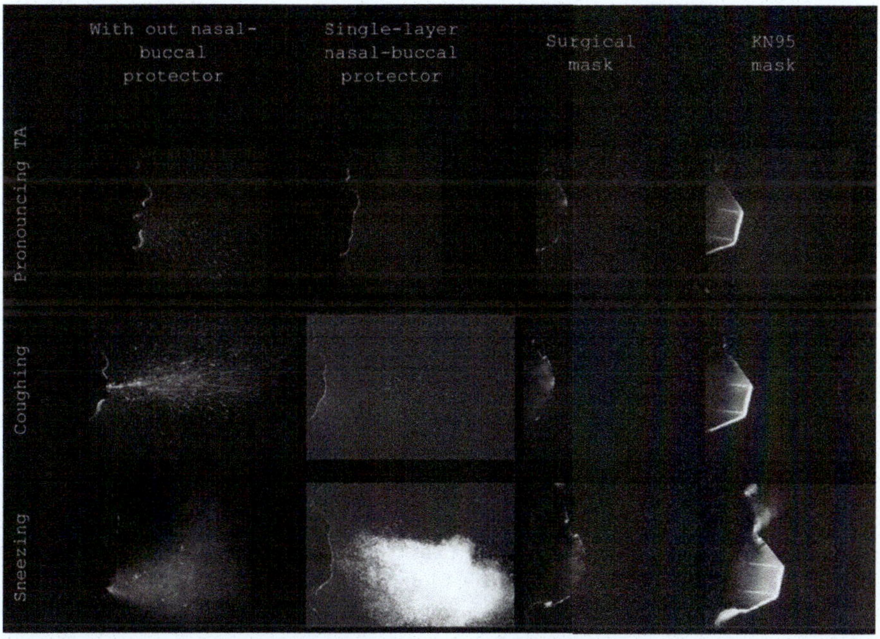

Fig. 3 Droplet and bioaerosol dispersion in three expiratory phenomena in four states of nasal-mouth coverage

(a) (b) (c)

Fig. 4 Maximum bioaerosol dispersion with three different nasal-mouth coverings, using: **a** Single-layer nasal-mouth protector. **b** Three-layer nasal-mouth protector. **c** Five-layer nasal-mouth protector

It is readily apparent that the KN95 mask exhibits the highest efficiency in reducing droplets across all the cases examined. However, even with this considerable reduction, Fig. 4 highlights a noteworthy observation: bioaerosols generated with the glycerin-water mixture [3] tend to disperse beyond the confines of the face shield. This phenomenon underscores the increased risk of potential SARS-CoV-2 virus transmission in enclosed environments.

In contrast, the single-layer nasal-mouth mask depicted in Fig. 4a predominantly disperses aerosols in the direction in front of the subject. Yet, fewer aerosols are expelled upward and downward from the mask. Conversely, three- and five-layer masks, such as the surgical mask and KN95, as shown in Fig. 4b, c, prevent aerosols from escaping to the front but allow for their dispersion toward the upper and lower regions of the mask.

These visual findings shed light on the varying effectiveness of different nasal-mouth protection options in mitigating droplet dispersion. The choice of mask type not only influences the reduction in droplets but also impacts the direction of aerosol dissemination during expiratory events, which carries implications for infection risk management in various settings.

3.2 *Percentage of Saliva-Mucus Mixture*

Following the procedure outlined in Sect. 2.3, independently for each image corresponding to every expiratory phenomenon under different states of nasal-mouth protection and for each of the three test subjects, the data is compiled and presented in Table 2. This table represents the relative percentage of the saliva-mucus mixture in each scenario studied. It is important to note that the data on the number of particles expelled in the various study cases are extracted from the images in Figs. 3 and 4. These are taken precisely at each case's highest droplet/bioaerosol dispersion.

A significant trend becomes evident when observing the data in Table 2 from left to right. In approximately 90% of the data, there is a progressive reduction in the percentage of droplets/aerosols as the number of layers in the nasal-mouth protectors increases. This trend suggests that the use of a five-layer nasal-mouth

Table 2 Percentage amounts of mucus-saliva mixture in the three expiratory phenomena, in four nasal-buccal protection states, on each test person

Expiratory phenomena	Without nasal-mouth protector	Single-layer nasal-mouth protector	Three-layer nasal-mouth protector	Five-layer nasal-mouth protector
First-person case				
Talking	0.8231	0.06318	0.020500	0.002441
Coughing	6.3070	0.43936	0.005957	0.001953
Sneezing	22.820	31.2300	0.299500	2.238400
Second person case				
Talking	0.1682	0.00400	0.007324	0.005664
Coughing	0.1324	0.04267	0.016500	0.014350
Sneezing	42.900	14.0500	0.824800	0.916100
Third person case				
Talking	1.654	0.05195	0.026850	0.0268500
Coughing	4.591	0.08730	0.009765	0.0002929
Sneezing	19.54	14.4500	0.408500	0.9387700

protector (KN95 mask), even in the worst-case scenario such as sneezing, results in a more significant reduction in the number of particles expelled into the environment compared to a three-layer nasal-mouth protector (surgical mask). Furthermore, a surgical mask, in turn, is more effective at reducing particle passage than a single-layer mask.

It is essential to acknowledge the remaining 10% of the data. This deviation can be attributed to the second experimental condition detailed in Sect. 2.4. Due to the experiment's constraint on frame size captured by the camera, some particles may not be captured within the frame during the specific shot, accounting for this small percentage of variance in the data.

3.3 Percentage of Expelled Droplets During Each Phenomenon

Figures 5, 6, and 7 depict the dynamic evolution of the number of particles captured by the camera from the initiation of each physiological phenomenon until its stabilization. The commencement of each phenomenon is marked by the appearance of the first expelled particle from the respiratory tract of each test subject.

These curves are generated by extracting the most representative images from the video footage during the progression of the expiratory phenomena. The program detailed in Sect. 2.3 is then applied to each image.

Fig. 5 Performance curves for various types of protective garments during the initial test on a person. **a** Pronouncing 'TA' aloud. **b** Coughing. **c** Sneezing

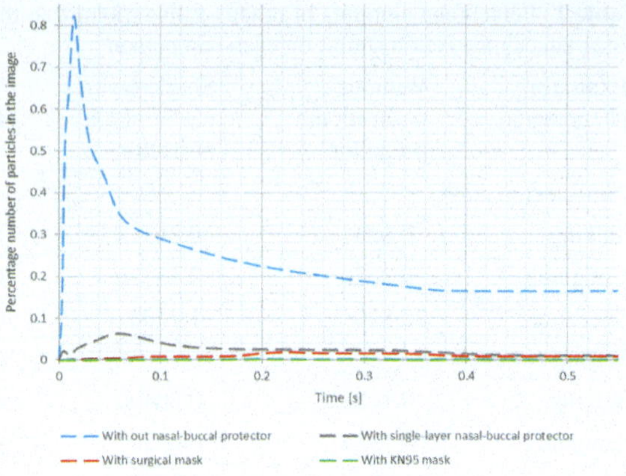

(a)

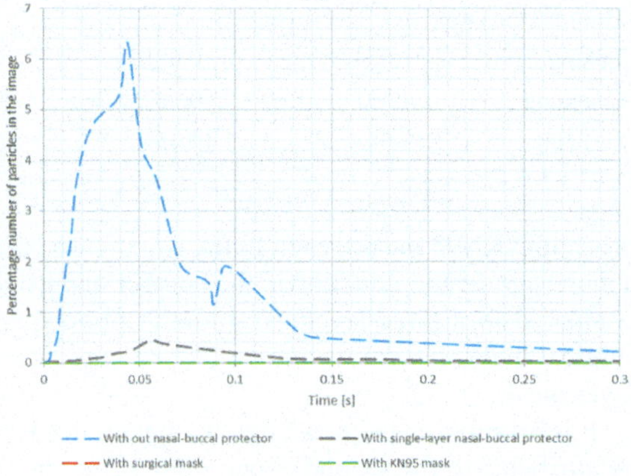

(b)

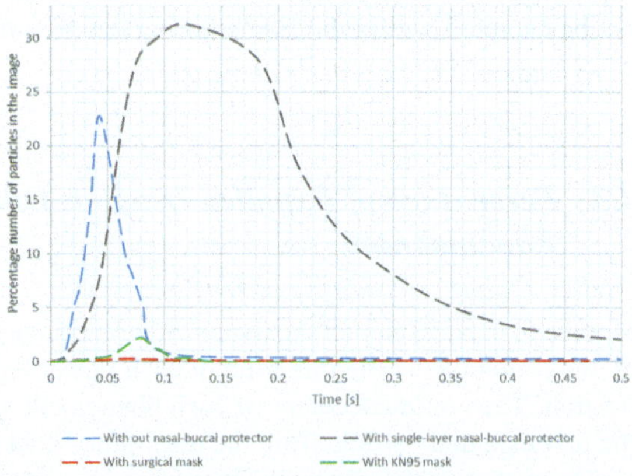

(c)

Fig. 6 Performance curves for various types of protective garments during the third test on a person. **a** Pronouncing 'TA' aloud. **b** Coughing. **c** Sneezing

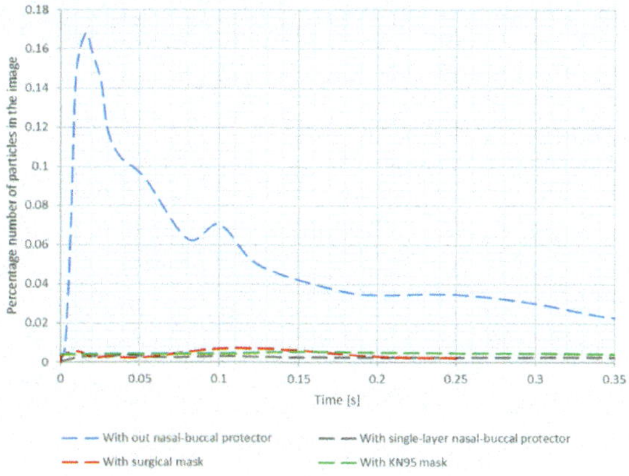

(a)

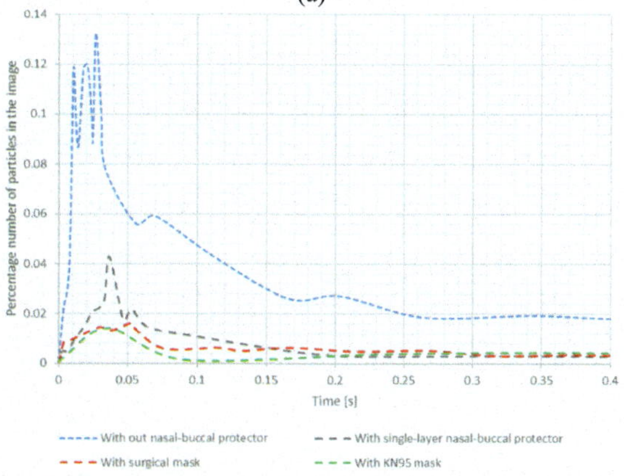

(b)

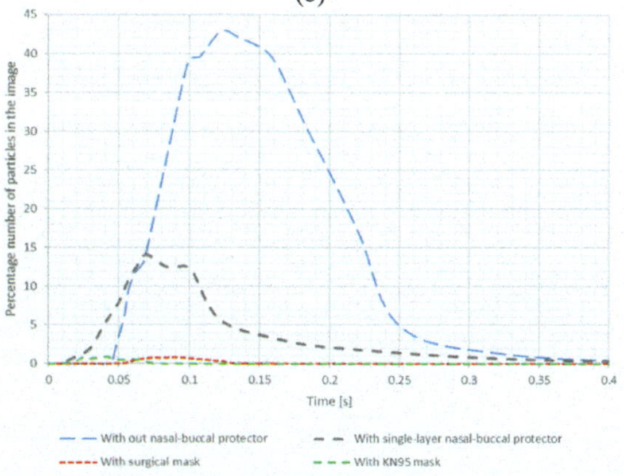

(c)

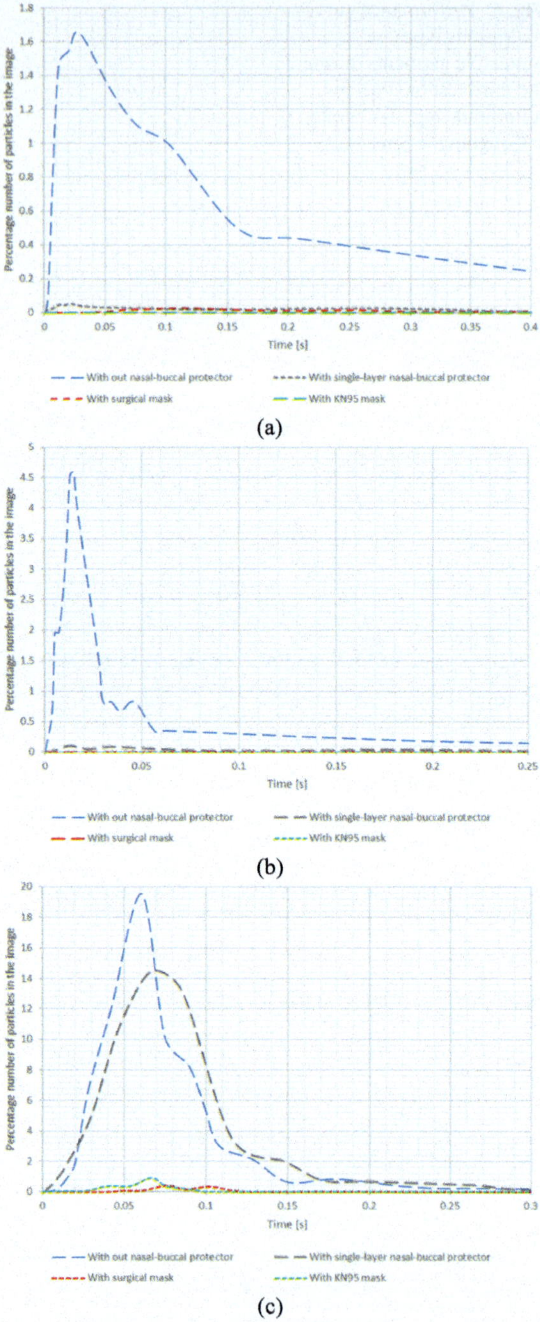

Fig. 7 Performance curves for various types of protective garments during the third test on a person. **a** Pronouncing 'TA' aloud. **b** Coughing. **c** Sneezing

Figures 5, 6, and 7 present the data with the X-axis representing the progression of time during the phenomenon and the Y-axis representing the percentage of the mucus-saliva mixture obtained through the program.

When examining these figures, specific patterns become evident:

1. Speaking ('TA' Pronunciation): Without nasal-mouth protection, there is a rapid surge in the percentage of mucus-saliva mixture, increasing from 0.8 to 1.6% in approximately 20 ms. Subsequently, there is a slight decrease over the next 400 ms. Using any form of nasal-mouth protection significantly reduces the percentage to a maximum of 0.06%.
2. Coughing: Coughing without protection leads to a considerable increase in the percentage of mucus-saliva mixture, rising from 0.13 to 6.3% in roughly 29 ms. When using any nasal-mouth protection, the rate is reduced to a maximum of 0.44%.

Sneezing: Unprotected sneezing generates the highest percentages of the mucus-saliva mixture, ranging from 19.54 to 19.64 in approximately 58 ms. The curve stabilization time varies between test subjects, with 600, 350, and 150 ms observed. Nasal-mouth protection, particularly a single-layer protector, does not significantly reduce the percentage of mucus-saliva mixture. In some cases, there is even an increase.

A total of 36 tests were conducted in the study, involving three expiratory activities (talking, coughing, and sneezing) and four nasal-mouth coverage states (no coverage, single-layer coverage, surgical mask, and KN95 mask).

Table 3 is constructed based on quantifying physiological responses that validate the theorem. Specifically, it demonstrates that 'case 1' consistently substantially reduces exhaled particles more than 'case 2'.

In 'case 1', nine experiments are conducted utilizing a specific type of nasal-mouth covering, as detailed in Table 3. Conversely, 'case 2' corresponds to the same set of 9 experiments but with an alternative nasal-mucosal covering.

Table 3 Effectiveness comparison of three different nasal-buccal protectors in three various expiratory activities of three other test persons

		Case 1		
		Single-layer protector	Three-layer protector	Five-layer protector
Case 2	Without nasal-mouth protector	8	9	9
	Single-layer protector	–	8	8
	Three-layer protector	1	–	6
	Five-layer protector	1	3	–

It is important to note that these nine experiments encompass a combination of 3 distinct physiological actions, each performed by three different test subjects.

To illustrate, the value eight at the upper-left corner of Table 3 represents a theorem derived from these experiments. It suggests that when subjected to 9 physiological actions, 8 of these actions consistently demonstrate that 'case 1', characterized by the utilization of a single-layer nasal-mouth shield, results in a more significant reduction in exhaled particles compared to 'case 2', which involves performing the actions without the use of nasal-mouth protection.

As previously elucidated in the preceding sections, regarding the effectiveness of masks compared to no nasal-mouth protection, Table 3 indicates that in 8 out of 9 tests, a single-layer nasal-mouth protector decreases the amount of expelled droplets from the respiratory tract. In contrast, 100% of the tests show that surgical and KN95 masks effectively reduce the ejected droplets.

Regarding the comparative effectiveness of masks, 8 out of 9 tests indicate that a surgical mask and a KN95 mask are more effective than a single-layer protector in reducing expelled fluid. Additionally, 66.66% of the tests suggest that a KN95 mask is more effective than a surgical mask.

4 Conclusion

The experimental methodology introduced in this study, designed to quantify particles through binarized images captured by a high-speed camera, has proven effective in providing a percentage of white pixels relative to the total number of pixels in each image. This percentage corresponds to the amount of expelled mucus-saliva during various expiratory phenomena.

The novel approach presented for particle quantification and the subsequent comparison of mask effectiveness enables quantitative analysis of different nasal-mouth protection options. It also allows for the visualization of the evolution of expiratory phenomena under various nasal-mouth coverage conditions.

Key findings from this study include:

Timing of Mucus-Saliva Mixture Increase: In expiratory activities such as talking, coughing, and sneezing without any nasal-mouth protection, there is a noticeable increase in the amount of mucus-saliva mixture within the first 20, 29, and 58 ms, respectively, from the onset of each phenomenon. Stabilizing the behavioral curve for each anthropogenic phenomenon takes 400, 150, and 600 ms, respectively.

Comparative Mask Effectiveness: In a comparative analysis of nasal-mouth protectors, all tests indicate that surgical and KN95 masks effectively reduce the number of expelled droplets during talking, coughing, and sneezing. Eight of nine tests suggest that a single-layer nasal-mouth protector reduces droplet discharge outside the mask. Additionally, eight out of nine tests claim that surgical and KN95 masks are more effective than single-layer masks. Furthermore, 66% of the tests indicate that a KN95 mask is more effective than a surgical mask.

Overall, this study provides valuable insights into the quantitative aspects of droplet expulsion during expiratory activities and highlights the varying effectiveness of different nasal-mouth protection options. These findings contribute to our understanding of respiratory particle dynamics and have implications for managing infection risk in various contexts.

The current investigation has primarily centered on establishing a methodology for assessing the effectiveness of three distinct protective measures using particle quantification. Nonetheless, in future research endeavors, it is imperative to incorporate a statistical analysis encompassing diverse participant samples while considering variables such as gender, age, race, and additional factors. This will enable us to ascertain how these variables may influence such physiological phenomena.

Acknowledgements We want to express our heartfelt gratitude to Ph.D. Nicolas Marcelo Paez Flor for his unwavering support within and beyond the academic realm. His dedication to fostering a culture of knowledge creation has been genuinely inspiring. Mr. Paez's guidance and encouragement have played a pivotal role in our research endeavors, and his passion for generating knowledge has been a constant source of motivation. We are deeply thankful for his invaluable contributions to our academic journey.

References

1. Mathieu, E., Ritchie, H., Ortiz-Ospina, E. et al.: A global database of COVID-19 vaccinations. Nature Human Behav. **5**(7), 947–953 (2021), ISSN: 23973374. https://doi.org/10.1038/s41562-021-01122-8. [Online]. Available: https://doi.org/10.1038/s41562-021-01122-8
2. Verma, S., Dhanak, M., Frankenfield, J.: Visualizing the effectiveness of face masks in obstructing respiratory jets. Phys. Fluids **32**(6), 061 708 (2020)
3. Arumuru, V., Pasa, J., Samantaray, S.S.: Experimental visualization of sneezing and efficacy of face masks and shields. Phys. Fluids **32**(11) (2020), ISSN: 10897666. https://doi.org/10.1063/5.0030101. [Online]
4. Archibald, M.S., Amos, J.F.: Auditory response to a photic stimulus. Optometry (St. Louis, Mo.), **74**(1), 50–54 (2003)
5. Garcia Vela, R., Cabanas Valdés, R., Huertas Ríos, S. et al.: Análisis del ajuste de mascarillas autofiltrantes en combinación con las mascarillas quirúrgicas para la protección del profesional sanitario en su atención a pacientes afectos de sars-cov-2 a partir de un estudio experimental. Archivos de Prevención de Riesgos Laborales **24**(2), 154–170 (2021)
6. Grinshpun, S.A., Haruta, H., Eninger, R.M., Reponen, T., McKay, R.T., Lee, S.-A.: Eficacia de la mascarilla facial con filtro de partículas n95 y de la mascarilla quirúrgica durante la respiración humana: Dos vías para la penetración de partículas. J. Occup. Environ. Hyg. **18**(sup1), S1–S14 (2021)
7. Schinder, E.O.: Inocuidad y riesgos del uso de mascarillas naso-oro-faciales para infecciones virales el caso de argentina para la epidemia sars. cov2. breve informe preliminar
8. Turner, C.E., Jennison, M.W., Edgerton, H.E.: Public health applications of high-speed photography. Amer. J. Public Health the Nations Health **31**(4), 319–324 (1941), ISSN: 0002-9572. https://doi.org/10.2105/ajph.31.4.319
9. Nadort, A., Woolthuis, R.G., van Leeuwen, T.G., Faber, D.J.: Quantitative laser speckle flowmetry of the in vivo microcirculation using sidestream dark field microscopy. Biomed. Opt. Express **4**(11), 2347–2361 (2013)

10. Scharfman, B.E., Techet, A.H., Bush, J.W.M., Bourouiba, L.: Visualization of sneeze ejecta: steps of fluid fragmentation leading to respiratory droplets. Exp. Fluids **57**(2), 24 (2016), ISSN: 1432- 1114. [Online]. Available: https://doi.org/10.1007/s00348-015-2078-4
11. Banner, A.S.: Cough: physiology, evaluation, and treatment. Lung **164**(1), 79–92 (1986), ISSN: 03412040. https://doi.org/10.1007/BF02713631
12. Phantom: Ultrahigh-speed 1 mpx phantom v2512., [Online]. Available: https://www.phantomhighspeed.com/products/cameras/ultrahigh1mpx/v2512

Impacts of Park-And-Ride Systems on Vehicle Emissions in Intermediate Cities

Zamir Mera, Fredy Rosero, Jairo Ortega, Ramiro Rosero, and Stefan Hausberger

Abstract The park-and-ride system is a modal interchange point, enabling commuters to switch from conventional vehicles to public transport for their urban journeys. This study explores the potential of the park-and-ride system to reduce fuel consumption and pollution compared to using a private passenger car in Ibarra, Ecuador. The novelty remains in the assessment of the Euro level and the bus occupancy, based on real-world driving profiles and using a detailed micro-level emission model to estimate fuel consumption, carbon dioxide (CO_2), carbon moNO_xide (CO), and nitrogen oxide (NO_x) emissions. The results reveal potential fuel consumption and pollution per passenger reduction when Euro VI buses are fully loaded. CO_2, CO, and NO_x emissions per passenger are reduced by approximately 13 times, eight times, and 1.7 times, respectively, compared to private Euro 3 passenger cars. These findings underscore the importance of optimizing park-and-ride systems' occupancy to substantially reduce fuel consumption and air pollution. The park-and-ride system can promote more sustainable and environmentally friendly urban transport by transitioning from personal vehicles to shared public transportation.

Keywords Park-and-ride · Public transportation · Vehicle emissions

1 Introduction

The park-and-ride (P&R) system is a set of facilities used by conventional vehicle users who travel a certain distance by vehicle and then switch to public transport mode in the P&R system. An evident concept of the P&R system is that it is a modal interchange point between an origin and a destination, thus the idea that users can

Z. Mera (✉) · F. Rosero · J. Ortega · R. Rosero
Universidad Técnica del Norte, 100105 Ibarra, Ecuador
e-mail: zamera@utn.edu.ec

S. Hausberger
Graz University of Technology, 8010 Graz, Austria

G. F. Olmedo Cifuentes et al. (eds.), *Emerging Research in Intelligent Systems*, Lecture Notes in Networks and Systems 902, https://doi.org/10.1007/978-3-031-52255-0_19

travel between the origin and the destination either by using vehicles or the P&R system [1, 2].

Thus, making a frequent commuter who uses the P&R system to go to the city center for work or business purposes would reduce pollution and traffic [3, 4]. Under this preamble, the P&R system could be studied as a tool that minimizes the pollution caused by private vehicles in the city center [5]. The reduction percentage would depend on the number of cars that have stopped circulating and the number of kilometers they have stopped traveling from the P&R system to the city center [6].

There are two routes that users can take, either by conventional vehicle or via the P&R route (using public transportation) from an origin to a destination. Each route produces pollutants, varying depending on various factors such as driving style, fuel type, or road conditions [7]. However, under the same conditions, each route can produce different pollution levels [8]. Usually, the studies showing that the P&R system can reduce pollution have been done theoretically at the city scale [9], and the conventional vehicle-only route has been analyzed through numerous scientific papers using various methods and methodologies [10]. The trip through the P&R system by bus depends on several factors [11], for example, bus type, fuel, driving style, and even passenger occupancy [12]. In fact, unlike private vehicles, passenger load has an essential influence on fuel consumption and bus emissions [13, 14]. For example, some studies, which have mainly used simulations with emission models (MOVES, HBEFA), found that the number of passengers increases the distance emission factors (in g/km) [15, 16]. However, per-passenger fuel consumption and emissions factors, calculated in (g/pp km), decrease as passenger load increases [17]. The latter demonstrates that mass transport systems like buses can improve air quality [18]. The environmental impact of P&R systems has seldom been analyzed in detail. Ortega et al. used a macro simulation model as COPERT to assess the CO_2 and pollutants of P&R systems [19, 20].

Two linked questions remain unanswered. Is a transport route that includes the P&R system capable of reducing, increasing, or equaling the pollution level of the conventional vehicle-only route? What is the level of occupancy that allows urban buses used in P&R systems to reduce emissions?

This article aims to answer this question using a case study. For this purpose, driving profiles are collected in situ. Then, a micro-level emission model is used to calculate in detail the pollution level of a route involving a representative passenger car and an urban bus. Also, the Euro level is assessed in the case of the urban bus. The results show whether the P&R system can reduce pollution with a potentially positive effect.

The main research gaps linked to P&R systems are (a) the need for comprehensive studies that accurately assess the actual fuel consumption, CO_2 emissions, and pollutant levels of vehicles in real-world conditions, (b) the unique characteristics and challenges of intermediate cities have not been evaluated, (c) in-depth evaluations of bus load levels are required to optimize public transportation systems and to evaluate their performance, operational costs, infrastructure requirements, and environmental benefits. Addressing these research gaps will contribute to informed decision-making and the development of sustainable transportation strategies.

The article is composed as follows: Sect. 2 presents and describes the methodology. Section 3 presents the results and the discussion of the results, and finally, in Sect. 5, the conclusions and policy recommendations.

2 Methodology

2.1 *Study Area*

Figure 1 shows the city of Ibarra, located in Ecuador, which serves as the capital of Im-babura province. The selection of Ibarra as a prime location for the study is substantiated by a comprehensive analysis of its mobility scenery and the expanding automotive fleet. This city is selected as the representative intermediate city of Ecuador. Considering the United Nations (UN) definition and different analyses, the intermediate cities' populations vary between 20,000 and 500,000 inhabitants, depending on population density and the respective urban system in a country [21]. Ibarra is a city located in a valley at an altitude of 2,200 m above sea level. It has an estimated population of 221,000 inhabitants, increasing [22]. Over the years, the province has experienced a considerable 55% rise in its automotive fleet from 2013 to 2022. This growth and geographic location make it difficult to expand the road network, which could lead to traffic congestion and air pollution. The city attracts people from surrounding satellite cities such as San Antonio, Atuntaqui Cotacachi, and Otavalo, who come for various daily activities. The city is connected to other provinces and cities by three highways (north, south, and west). However, only the west corridor offers shorter and more direct routes to the most densely populated cities.

2.2 *Park and Ride Facility*

The location of the P&R facilities is recommended to be within a 5–10 km radius of the city center, with exceptions when there are geographic barriers [23]. Usually, the planning starts with improving the coverage of the second phase of the trip through public transport. However, it is certainly a relatively new system in Latin American cities, where planning starts with determining the place of the future facility [23]. In the city of Ibarra, the location of the P&R is assumed to be in the city of San Antonio, 7.4 km from the central business district (CBD). The CBD is considered to be in the hyper center of the city, in the 'Pedro Moncayo' central park, which is close to the city hall, banks, government offices, and shopping business center. The driving route for passenger cars is slightly shorter as they could take alternative routes, often favoring shorter paths to reach the CBD. On average, passenger cars covered 12.6 km for a round trip, while buses traveled 14.1 km for their round trips. The outward

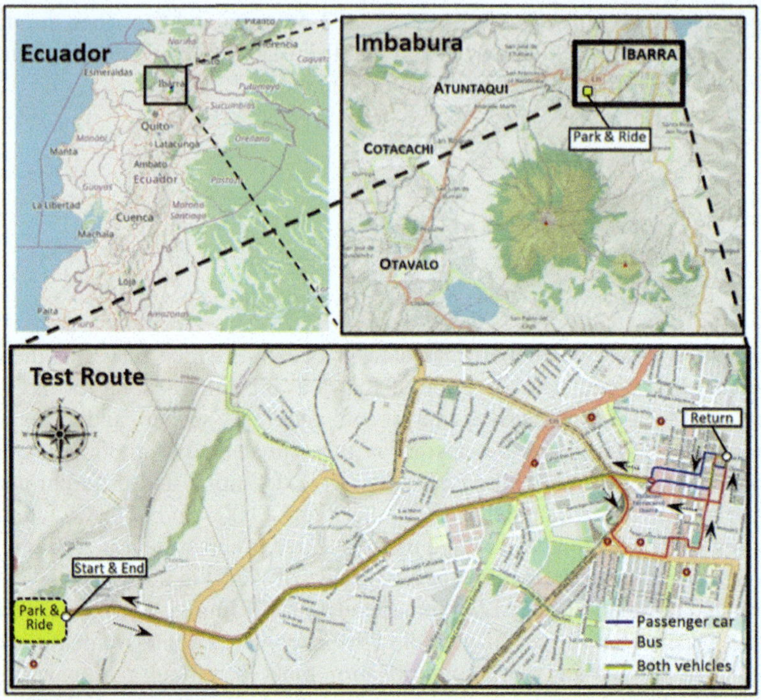

Fig. 1 Location of the park and ride facility for Ibarra city, which is connected to satellite cities by the west corridor

journey resulted in 23.8 min (s.d. 2.4) for the urban bus and 16.6 min (s.d. 1.4) for the passenger car. The return journey is 21.0 min (s.d. 1.5) for the urban bus and 17.1 min (s.d. 2.6) for the passenger car.

2.3 *Vehicles Fleet*

A country's average passenger car and a representative urban bus of the city are used in the P&R evaluation. The average passenger car in Ecuador and its vehicle characteristics are found by weighting the five most-sold passenger cars in 2022 in the automobile and SUV segments, according to the Association of Automotive Companies of Ecuador [24]. The representative urban bus, Chevrolet FTR 32M, was selected because it represents 49% of the current fleet in the city of Ibarra, according to the data provided by the Mobility Public Company in the North of Ecuador [25]. See Table 1. In Ecuador, the sulfur content remains high. Therefore, the homologation process is stacked in Euro II for heavy-duty vehicles and Euro 3 for light-duty cars [26].

Table 1 Average passenger car and representative bus characteristics

Vehicle	Fuel	Displacement [L]	Power [kW]	Emissions level	Curb weight [kg]
Passenger car	Gasoline	1.4	83.4	Euro 3	1214
Bus	Diesel	7.2	170.0	Euro II	7850

2.4 *Driving Profiles*

The driving profiles for passenger cars and buses are obtained from real-world data collected at a sampling rate of 1 s. A GPS + GLONASS logger GL-770 records the profiles [27], with a professional driver simulating the driving behavior for both vehicle types. The speed profile is smoothed, and the road grade is corrected based on altitude profiles according to the procedure established in Regulation EU 2018/1832 [28]. Six round-trips are recorded for each vehicle type, originating from a P&R location and ending at a chosen destination in the central business district. The trips are conducted in the morning and afternoon to capture various traffic conditions. The average speed resulted in 18.0 km/h and 21.4 km/h for the urban bus and the passenger car, respectively.

2.5 *Emissions Modelling*

The vehicles and driving profiles are simulated in the Passenger Car and Heavy Duty Emission Model, PHEM v13 [29]. This micro-level emission model uses longitudinal dynamics and engine emission maps to estimate fuel consumption and emissions to estimate every driving condition [30]. The model is built and calibrated from extensive testing data performed in engine and chassis dynamometer and using on-road emissions tests with Portable Emissions Measurement Systems (PEMS). Thus, PHEM has built-in representative average vehicle models for passenger cars and city buses with different Euro-level configurations. From the wide range of parameters that can be modified in the simulation, the weight of the vehicle, the load, the Euro-level, and the power are adjusted to an average-sales weighted passenger car model and the representative urban bus model of the city. The simulation parameters by occupancy rate and Euro technology are shown in Tables 2 and 3, respectively.

TWC = Three-Way Catalyst, EGR = Exhaust Gas Recirculation, DOC = Diesel Oxidation Catalyst, SCR = Selective Catalytic Reduction.

According to studies conducted in the United States [31] and Europe [32], the average occupancy of passenger cars is estimated to be 1.5 passengers. For urban buses, the minimum occupancy considers the driver, an assisting person, and the average occupancy of passenger cars. Half occupancy is considered when all seats in the bus are occupied, while full occupancy includes standing passengers. The

Table 2 Simulation parameters of vehicles by occupancy rate

Vehicle	Curb weight [kg]	Occupancy	Passengers [#]	Load [kg]
Passenger car	1214	Average	1.5	112.5
Bus	7850	Minimum[a]	1.5	262.5
		Half	36	2850.0
		Full	80	6150.0

[a]Minimum bus occupancy includes the driver, the assisting person, and the average occupancy of a passenger car of 1.5 passengers, 75 kg per person

Table 3 Simulation parameters of vehicles by Euro technology

Vehicle	Euro level	Aftertreatment system	Power [kW]
Passenger car	Euro 4	TWC	83.4
Bus	Euro II	–	170.0
	Euro IV	EGR	185.4
	Euro VI	DOC + EGR + SCR	196.6

Ecuadorian normative NTE INEN 2205 for the specification of urban buses defines the minimum number of seats and the designated space for standing passengers [33].

3 Results and Discussion

3.1 Distance-Specific Emission Factors and Fuel Consumption

Figure 2 shows passenger cars' and urban buses' average distance-specific emission factors. The figure compares the various Euro technologies and considers the occupancy levels for urban buses. The results indicate a clear trend: higher Euro-level powertrains and lower occupancy contribute to reduced fuel consumption (directly linked to CO_2 emissions) and pollutants. Euro VI buses further enhance this reduction in pollutant emissions, as discussed below. When comparing the emissions of current predominant Euro II urban buses to those of Euro IV and VI technologies, it becomes evident that the latter significantly reduces emissions by incorporating additional after-treatment systems.

Fuel consumption and CO_2 emission factors. The bus occupancy level directly influences fuel consumption and resulting CO_2 emissions, leading to varying loads that match previous works' results [15, 16]. A fully loaded bus experiences an increase in fuel consumption and CO_2 emissions by approximately 19–22% compared to an unloaded bus. In the most likely scenario, a half-loaded bus emits

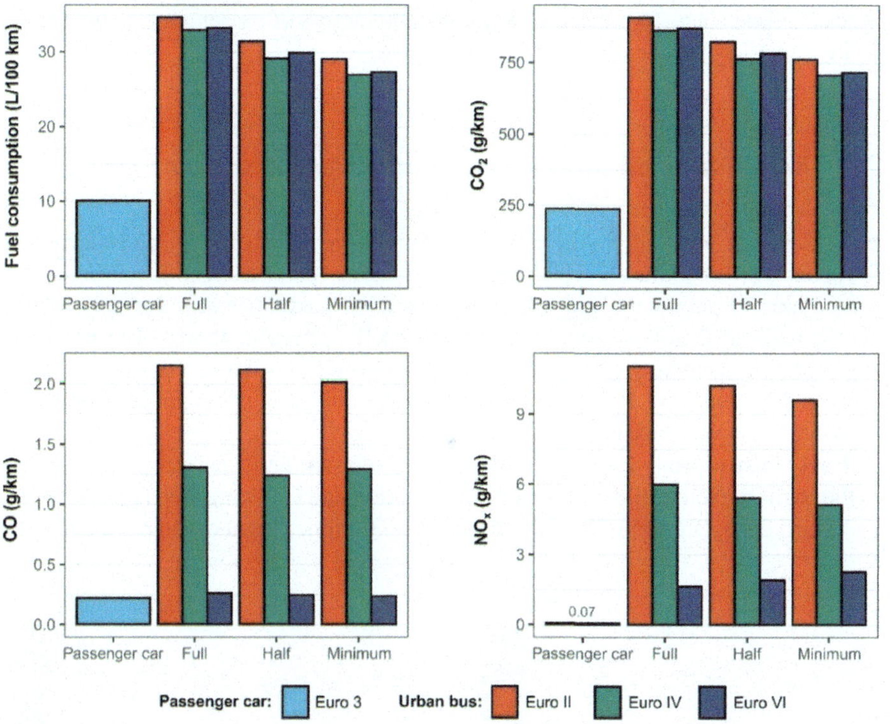

Fig. 2 Fuel consumption and distance-specific emission factors. Fuel consumption is expressed in liters per 100 km of gasoline and diesel for passenger cars and urban buses, respectively

3.3–3.5 times more CO_2 than a passenger car with the average occupancy level. These observations emphasize the importance of considering occupancy levels when assessing the environmental impact of urban buses. The analysis based on the emissions per passenger, which reflects the emissions avoided by the P&R system, is described in Sect. 3.2.

Pollutant emission factors. The CO emissions of spark-ignition vehicles tend to be higher than compression-ignition vehicles due to more rich-mixture combustion; however, the tailpipe emissions finally depend on the effectiveness of the aftertreatment [34]. In Euro II buses, reducing CO emissions as the load decreases is directly linked to the reduction in fuel consumption. However, CO emissions of half-loaded buses are 9.7 times than those of passenger cars. Euro IV buses can reduce emissions by 36%–42% compared to Euro II buses. Furthermore, Euro VI buses demonstrate distance-specific emission factors like passenger cars, primarily due to the Diesel Oxidation Catalyst (DOC) implementation. The CO emissions of half-loaded Euro VI buses are only 11% higher than passenger cars, demonstrating the advantage of using advanced aftertreatments in diesel vehicles.

Euro IV and VI buses significantly reduce NO_x emissions compared to Euro II buses. However, the emissions of half-loaded buses are notably higher, with 139 times, 73 times, and 26 times the NO_x emission factors of passenger cars when Euro II, IV, and VI buses are compared, respectively.

3.2 Emissions and Fuel Consumption Per Passenger

The emissions avoided by the P&R system can be quantified by considering the occupancy of urban buses, which implies replacing the passenger cars used by citizens. Figure 3 shows the average emissions from passenger cars and urban buses due to real-world routes simulated for the P&R system in Ibarra. The emissions depicted represent the cumulative emissions of round trips between the P&R facility and the city center by 1000 passengers. The results show the significant impact of bus occupancy on emissions.

Fuel consumption and CO_2 emission per passenger. Occupancy is recognized as the primary factor influencing fuel consumption and CO_2 emissions per passenger. Unloaded buses present CO_2 emissions of 3.4–3.6 times higher than passenger cars.

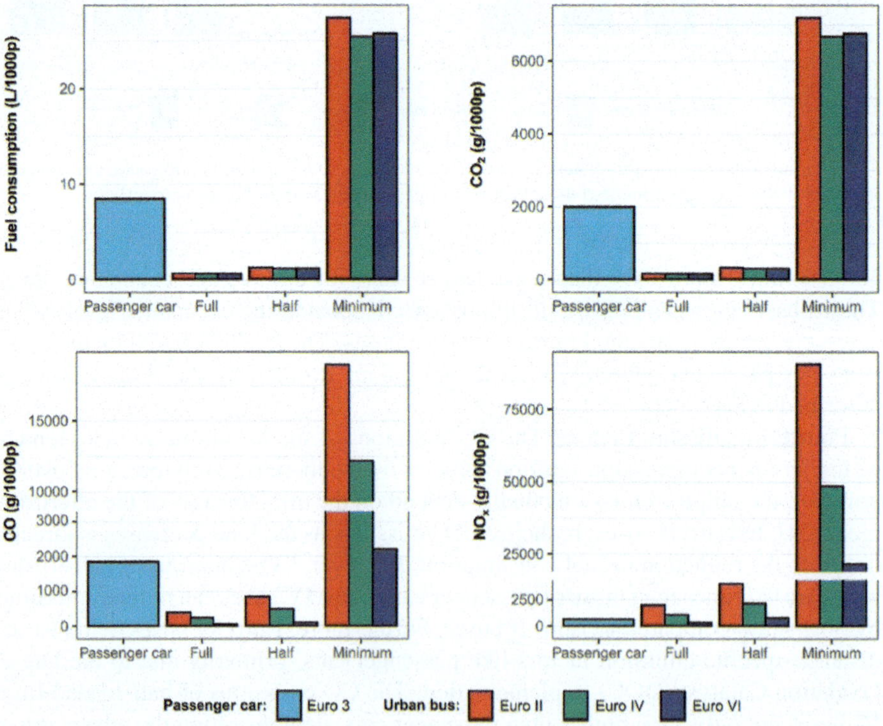

Fig. 3 Comparison of emissions produced by passenger cars and urban buses representing different Euro levels and occupancy. The emissions represent round trips for 1000 passengers between the park-and-ride facility and the city center

In contrast, half- and full-loaded buses demonstrated 6.2–6.6 times and 12.4–13.0 times lower emissions than passenger cars per passenger transported, respectively. These figures underscore the significance of planning bus trips to achieve maximum occupancy. By optimizing bus occupancy, there are two significant advantages: first, to reduce fuel consumption and CO_2 emissions, and second, to decrease the expenses related to subsidies in fuels.

CO emissions per passenger. When the emissions are assigned to each passenger, urban buses take advantage of private passenger cars in case of high occupancy. CO emissions from half- and full-loaded Euro IV buses are 3.8 and 8.0 times lower than using passenger cars, respectively. The emissions are further reduced by Euro VI buses at 20 and 40 times lower than passenger cars for the same occupancy rate.

NO_x emissions per passenger. Due to the inherent challenges in controlling NO_x emissions in compression ignition vehicles, only Euro VI buses demonstrate the potential for reducing NO_x emissions compared to passenger cars. A fully loaded Euro VI bus emits approximately half the NO_x emissions of passenger cars. However, a half-loaded Euro VI bus still exhibits an increase in NO_x emissions by 21% compared to passenger cars. On the other hand, the results indicate that fully-loaded Euro IV and Euro II buses may represent 1.7 times and 3.1 times the NO_x emissions of passenger cars, respectively. This highlights the significance of Euro VI technology in achieving notable reductions in NO_x emissions for urban buses. Also, it remarks that NO_x emissions and other pollutants can be down to zero only by using electric buses.

Ortega et al. found reductions of 20% of CO_2 and 16% of NO_x and CO by a P&R system, modeled with a macro-level emissions model as COPERT [20]. However, the study did not indicate the Euro level and occupancy of the bus, which are important variables when vehicle technology is compared and urban bus trip planning is accounted for. Using electric buses, Ortega et al. found a reduction in 42% of CO_2 and 38–39% of NO_x and CO by a P&R system [20]. These results correspond to tailpipe emissions and could be further analyzed on a life-cycle assessment basis for each powertrain. A report published by the International Council on Clean Transportation (ICCT) demonstrated reductions of 82% of life-cycle greenhouse gas (GHG) emissions when a current diesel urban bus is compared with an electric bus [35].

4 Conclusions

A comprehensive assessment of a park-and-ride facility for an intermediate city in Ecuador was performed. Real-world driving profiles were simulated in a micro-scale emission model, PHEM. The evaluation focused on urban buses, considering the Euro level and the occupancy and fuel consumption, CO_2, CO, and NO_x emissions. The results of the assessment are as follows:

The results highlighted the significance of the Euro level for reducing fuel consumption and pollution. When Euro VI diesel urban buses are used, the emissions are potentially diminished. Moreover, the occupancy played a vital role in quantifying the changes in emissions due to the park-and-ride facility when measured in terms of the number of passengers transported. This implies a transition from private vehicles to shared modes of public transportation when urban buses are half- or full-loaded, with lower fuel consumption and emissions per passenger than private cars, except NO_x emissions, where Euro VI and a full-loaded bus are required.

Evaluating these parameters demonstrates the substantial potential for reducing fuel consumption and air pollution when optimizing the occupancy of Park-and-ride systems. A full-loaded Euro VI bus can decrease by about 13 times, eight times, and 1.7 times the CO_2, CO, and NO_x emissions, respectively. This represents a promising avenue to promote more sustainable and environmentally friendly transportation practices in urban areas.

5 Policy Recommendations

- Implement a Park and Ride (P&R) facility in Ibarra, which can bring potential benefits in CO_2 emissions, air pollution, and traffic reduction, especially when urban buses have high occupancy rates. To ensure high occupancy rates, developing a comprehensive mobility plan and effectively communicating the P&R project to the population is crucial.
- It is advisable to initiate a program for the electrification of urban buses, which could involve fiscal credits and tax reductions. Electric vehicles offer the advantage of zero air pollutant emissions, and their energy efficiency is higher than that of combustion engine vehicles. Although this recommendation is not directly based on our results, it aligns with the current global trend toward transportation electrification.
- For the remaining vehicles, a program should be implemented to renew the urban bus fleet following the highest Euro emission standards, considering the available fuel in the country. This approach will help improve the environmental performance of the overall fleet.

References

1. Zhao, X., Li, Y., Xia, H.: Behavior decision model for park-and-ride facilities utilization. Adv. Mech. Eng. (2017). https://doi.org/10.1177/1687814017708907
2. Ortega, J.: Planning a park and ride system: a literature review, pp. 82–98 (2021).
3. Turnbull, K.F.: Effective Use of Park-and-Ride Facilities. National Academy Press, Washington, D.C (1995)
4. Jairo, O., János, T., Péter, T.: Analyzing pollution reduction by combining the P&R system with electric vehicles. In: Conference: XI Conference on Transport Sciences, GYŐR (2021)

5. Obaid, M., Torok, A., Ortega, J.: A comprehensive emissions model combining autonomous vehicles with park and ride and electric vehicle trans-portation policies. Sustainability **13**, 4653 (2021). https://doi.org/10.3390/su13094653
6. Ortega, J., Hamadneh, J., Esztergár-Kiss, D., Tóth, J.: Simulation of the daily activity plans of travelers using the park-and-ride system and autonomous vehicles: work and shopping trip purposes. Appl. Sci. **10**, 2912 (2020). https://doi.org/10.3390/app10082912
7. Cohen, P., Potchter, O., Schnell, I.: The impact of an urban park on air pollution and noise levels in the Mediterranean city of Tel-Aviv, Israel. Environ. Pollut.Pollut. (2014). https://doi.org/10.1016/j.envpol.2014.08.015
8. Pitsiava-Latinopoulou, M., Basbas, S., Papoutsis, K., Sdoukopoulos, E.: Parking policies for supporting sustainable mobility. Procedia Soc. Behav. Sci.Behav. Sci. **48**, 897–906 (2012). https://doi.org/10.1016/j.sbspro.2012.06.1067
9. Macloszek, E., Kurek, A.: The use of a park and ride system a case study based on the City of Cracow (Poland). Energies (Basel). (2020). https://doi.org/10.3390/en13133473
10. Faghri, A., Lang, A., Hamad, K., Henck, H.: Integrated knowledge-based geographic information system for determining optimal location of park-and-ride facilities. J. Urban Plan. Dev. **128**, 18–41 (2002). https://doi.org/10.1061/(ASCE)0733-9488(2002)128:1(18)
11. Meek, S., Ison, S., Enoch, M.: Role of bus-based park and ride in the UK: a temporal and evaluative review (2008). https://doi.org/10.1080/01441640802059152
12. Wei, T., Christopher Frey, H.: Factors affecting variability in fossil-fueled transit bus emission rates. Atmos. Environ. **233**, 117613 (2020). https://doi.org/10.1016/j.atmosenv.2020.117613
13. Soylu, S.: The effects of urban driving conditions on the operating characteristics of conventional and hybrid electric city buses. Appl. Energy **135**, 472–482 (2014). https://doi.org/10.1016/j.apenergy.2014.08.102
14. Rosero, F., Fonseca, N., López, J.-M., Casanova, J.: Effects of passenger load, road grade, and congestion level on real-world fuel consumption and emissions from compressed natural gas and diesel urban buses. Appl. Energy **282**, 116195 (2021). https://doi.org/10.1016/j.apenergy.2020.116195
15. Alam, A., Hatzopoulou, M.: Investigating the isolated and combined effects of congestion, roadway grade, passenger load, and alternative fuels on transit bus emissions. Transp. Res. D Transp. Environ. **29**, 12–21 (2014). https://doi.org/10.1016/j.trd.2014.03.005
16. Borge, R., de Miguel, I., de la Paz, D., Lumbreras, J., Pérez, J., Rodríguez, E.: Comparison of road traffic emission models in Madrid (Spain). Atmos. Environ. **62**, 461–471 (2012). https://doi.org/10.1016/j.atmosenv.2012.08.073
17. Yu, Q., Li, T., Li, H.: Improving urban bus emission and fuel consumption modeling by incorporating passenger load factor for real world driving. Appl. Energy **161**, 101–111 (2016). https://doi.org/10.1016/j.apenergy.2015.09.096
18. Ibarra-Rojas, O.J., Delgado, F., Giesen, R., Muñoz, J.C.: Planning, operation, and control of bus transport systems: a literature review. Transport. Res. Part B Methodol. **77**, 38–75 (2015). https://doi.org/10.1016/j.trb.2015.03.002
19. Ortega, J., Tóth, J., Péter, T.: A theoretical analysis of car emission levels related to the implementation of a park and ride in an urban environment. In: XXI PANAM Congreso Panamericano de Transporte y Logística, p. 2021.
20. Ortega, J., Tóth, J., Péter, T.: Analyzing pollution reduction by combining the P&R system with electric vehicles. In: XI Conference on Transport Sciences (2021)
21. Bolay, J.C., Kern, A.L.: Intermediate cities. In: The Wiley Blackwell Encyclopedia of Urban and Regional Studies, pp. 1–5 (2019). https://doi.org/10.1002/9781118568446.EURS0163.
22. INEC: Población y Demografía, https://www.ecuadorencifras.gob.ec/censo-de-poblacion-y-vivienda/. Last accessed 14 July 2023
23. Ortega, J., Tóth, J., Péter, T.: Planning a park and ride system: a literature review. Future Transport. **1**, 82–98 (2021). https://doi.org/10.3390/futuretransp1010006
24. AEADE: Anuario 2022 (2023)
25. MOVIDELNOR: Empresa Pública de Movilidad del Norte, https://www.movidelnor.gob.ec/webepm/uniportalepm/. Last accessed 14 July 2023

26. Acevedo, H., Pettigrew, S., Pineda, L., Delgado, O.: Hoja de ruta para descar-bonizar el transporte de carga en América Latina entre 2025 y 2050 (2023)
27. Transystem Inc.: GPS Products, http://www.transystem.com.tw/www/product.php?b=G&m=pe&cid=4&sid=&id=150. Last accessed 19 Sept 2023
28. EC: Commission Regulation (EU) 2018/1832 of 5 November 2018 amending Directive 2007/46/EC of the European Parliament and of the Council, Commission Regulation (EC) No 692/2008 and Commission Regulation (EU) 2017/1151 for the purpose of improving the emission. Official Journal of the European Union 1832, 301 (2018)
29. Hausberger, S.: Simulation of Real World Vehicle Exhaust Emissions. Tech-nische Universität Graz, Graz (2003)
30. Weller, K., Lipp, S., Röck, M., Matzer, C., Bittermann, A., Hausberger, S.: Real world fuel consumption and emissions from LDVs and HDVs. Front. Mech. Eng. **5**, 1–22 (2019). https://doi.org/10.3389/fmech.2019.00045
31. University of Michigan: Personal Transportation Factsheet, https://css.umich.edu/publications/factsheets/mobility/personal-transportation-factsheet, Last accessed 14 July 2023
32. European Environment Agency: Occupancy rates of passenger vehicles, https://www.eea.europa.eu/data-and-maps/indicators/occupancy-rates-of-passenger-vehicles/occupancy-rates-of-passenger-vehicles. Last accessed 14 July 2023
33. INEN: Servicio Ecuatoriano de Normalización, http://apps.normalizacion.gob.ec/descarga/. Last accessed 14 July 2023
34. Heywood, J.B.: Internal Combustion Engine Fundamentals. McGraw-Hill (1988), p. 10987654
35. O'Connell, A., Pavlenko, N., Bieker, G., Searle, S.: A comparison of the life-cycle greenhouse gas emissions of European heavy-duty vehicles and fuels. International Council on Clean Transportation (2023). https://theicct.org/wp-content/uploads/2023/02/lca-ghg-emissions-hdv-fuels-europe-feb23.pdf

Computational Modeling and Artificial Intelligence

Machine Learning Regressors in Forecasting Mechanical Properties in Advanced Manufacturing Processes

Germán Omar Barrionuevo, Jorge Ramos-Grez, and Francisco J. Montero

Abstract This paper presents the advances in forecasting mechanical properties by applying machine learning regressors in advanced manufacturing processes. Computational intelligence is advancing by leaps and bounds in artificial intelligence and data science, which is how data is currently considered the new gold. In addition to advances in information technology, manufacturing processes have also evolved. In that context, additive manufacturing has several advantages over conventional manufacturing processes. On the one hand, it offers the possibility of processing any material: metals, ceramics, polymers, and composite materials. On the other hand, these technologies make it possible to optimize the use of the material, reducing its impact on the environment. Due to its versatility, it is possible to improve design to save material and reduce assembly complexity. This work presents a series of cases of applying machine learning regressors in forecasting mechanical properties in advanced manufacturing processes, from friction welding, laser welding, selective laser melting, wire arc additive manufacturing, and fused filament fabrication. The results show that the gradient-boosting regressors have greater precision in predicting several mechanical properties than conventional artificial neural networks. These algorithms are more effective than using the traditional design of experiments, reducing costs and allowing information about the physical phenomenon involved in each manufacturing process.

Keywords Machine learning · Regressor · Gradient boosting · Manufacturing processes · Additive manufacturing · Welding

G. O. Barrionuevo (✉) · F. J. Montero
Departamento de Ciencias de la Energía y Mecánica, Universidad de las Fuerzas Armadas ESPE, Sangolquí 171103, Ecuador
e-mail: gobarrionuevo@espe.edu.ec

G. O. Barrionuevo · J. Ramos-Grez · F. J. Montero
Department of Mechanical and Metallurgical Engineering, School of Engineering, Pontificia Universidad Católica de Chile, Av. Vicuña Mackenna, 4860 Santiago, Macul, Chile

G. F. Olmedo Cifuentes et al. (eds.), *Emerging Research in Intelligent Systems*, Lecture Notes in Networks and Systems 902, https://doi.org/10.1007/978-3-031-52255-0_20

1 Introduction

Computational intelligence or artificial intelligence (AI) refers to the ability of a computer to learn a specific task from data [1]. In that sense, due to the need to optimize manufacturing processes and the advancement of new AI algorithms, computational intelligence has recently attracted the attention of scientists and decision-makers involved in manufacturing industries. One of the most promising AI techniques is machine learning (ML). ML algorithms use computational methods to learn information directly from data without depending on a programmed model [2]. Among ML methods, it is possible to distinguish between supervised, unsupervised, and reinforcement learning [3].

Supervised ML algorithms map a function from known input–output sets and are primarily used for regression and classification tasks [3]. Regression estimates the relationships between a dependent variable and one or more independent variables [4]. Supervised ML algorithms can solve optimization problems, aiming to find a mapping function that minimizes the loss function described in (1).

$$\min_{\theta} \frac{1}{N} \sum_{i=1}^{N} \mathcal{L}(y^i, f(x^i, \theta)) \tag{1}$$

where $\mathcal{L}$ is the loss function, f is the function to be learned, θ is the mapping function parameter, x^i is the feature vector of the ith samples, y^i is the target value, and N is the number of training samples. The loss function is a method of evaluating how well the algorithm models the dataset, and it is determined as the sum of errors made for each example in training or validation sets. Therefore, minimizing the loss function optimizes the ML accuracy [5].

Consequently, ML is a suitable mathematical tool for predicting mechanical properties in advanced manufacturing processes where data is costly and time-consuming. ML regressors include support vector machines, decision trees, random forests, Gaussian processes, gradient-boosted regressors, artificial neural networks (ANN), and fuzzy logic (FL). An adaptive neuro-fuzzy inference system (ANFIS) is a hybrid algorithm that integrates the ANN and FL principles and can integrate both benefits in a single algorithm [6].

One of the technologies that are revolutionizing the processing of materials is additive manufacturing (AM). AM was created as a rapid prototyping tool [7]. This technology became widespread in the last decade and went from a prototyping technology to a robust process where final functional parts can be achieved directly. AM expands design freedom and allows the processing of almost any material, from metals, ceramics, and polymers to composite materials [8]. Within metal AM, two major technology branches are distinguished, one where there is a bed of metallic powder and the content is fused through a heat source, e.g., laser or electron beam (powder bed fusion—PBF), and the second, where the material is deposited directly in either powder or wires, so-called directed energy deposition (DED) [9].

Selective laser melting (SLM) is a PBF technology designed to use high-power lasers with a power consumption more significant than 100 W to melt and fuse powder regions layer by layer, using computer-aided design (CAD) data [10, 11]. This technique makes it possible to build complex geometries, which are extremely expensive to obtain through conventional processes [12]. However, one of the key challenges to making this technology cost-effective is predicting the resulting mechanical properties [13]. Even using the same manufacturing parameters, the results vary from machine to machine.

Within polymer processing, fused filament fabrication (FFF) technology stands out, where an object is created from a CAD model following the same scheme as PBF but using a polymeric filament as raw material, where the plastic is melted through a nozzle capable of controlling the temperature and speed of displacement [14]. FFF is currently the AM technology that dominates the market due to its flexibility, showing an exponential increase in the last decade [15].

Another promising AM technology is wire arc additive manufacturing (WAAM). WAAM is a DED technology that combines an electric arc as the heat source and a metal wire as feedstock, employing standard welding equipment. WAAM has many advantages compared to PBF processes. It has a higher deposition rate of up to 10 kg/h [16], it does not require a highly inert atmosphere [17], and it is possible to manufacture large and complex metal components [18] and process any alloy used in welding [19].

Metal joining processes are an essential need for various industries. Conventional fusion welding is not feasible for the union of materials with different characteristics due to the difference between their melting points [20]. Solid-state welding is a novel process in which two workpieces are joined under pressure, generating frictional heat but at temperatures below the base materials' melting point [21]. Friction welding (FW) is a solid-state joining method that produces the coalescence of materials under a compressive force when workpieces move in contact, making heat and plastically displacing the material, creating a welding interface [22]. This process does not require filler metal, heat source, or shielding gas. Due to its versatility, FW has promising industrial applicability as a mass-production process for joining dissimilar alloys [23]. Moreover, laser welding is a joining technology characterized by small fusion and heat-affected zones, minimal or non-existent distortions [24], precision, reliability, and the possibility of making dissimilar joints [25].

With the motivation of anticipating the mechanical response of advanced manufacturing process, this work focuses on the applicability of several machine learning regressors for predicting yield strength in FFF, relative density in SLM, the geometry of the deposited layer in WAAM, ultimate tensile strength in FW, and microhardness in laser welding. The data used to predict mechanical properties are part of previous publications in various areas of advanced materials processing.

2 Methodology

The core of the ML algorithms is the data. For this reason, in addition to the data obtained experimentally, it is recommended to use literature data to track their performance. This process is schematized in Fig. 1, where there are three primary activity blocks: (i) data gathering, (ii) training, and (iii) testing [13]. Using suitable keywords, a database can be assembled from literature data extracted from primary scientific databases such as Web of Science and Scopus. Once created, it is recommended to clean data, which means scaling the input parameters, usually applying zero mean and unit variance during training.

ML accuracy is directly related to the error measurement. Low accuracy means a significant difference between the prediction and the actual value. Some of the most employed metrics to evaluate the accuracy of regression algorithms include mean squared error (MSE), root mean squared error (RMSE), mean absolute error (MAE), and the coefficient of determination (R^2) [26]. For MSE, RMSE, and MAE, the lowest value is the highest accuracy, while for R^2, a value close to 1 means a better model.

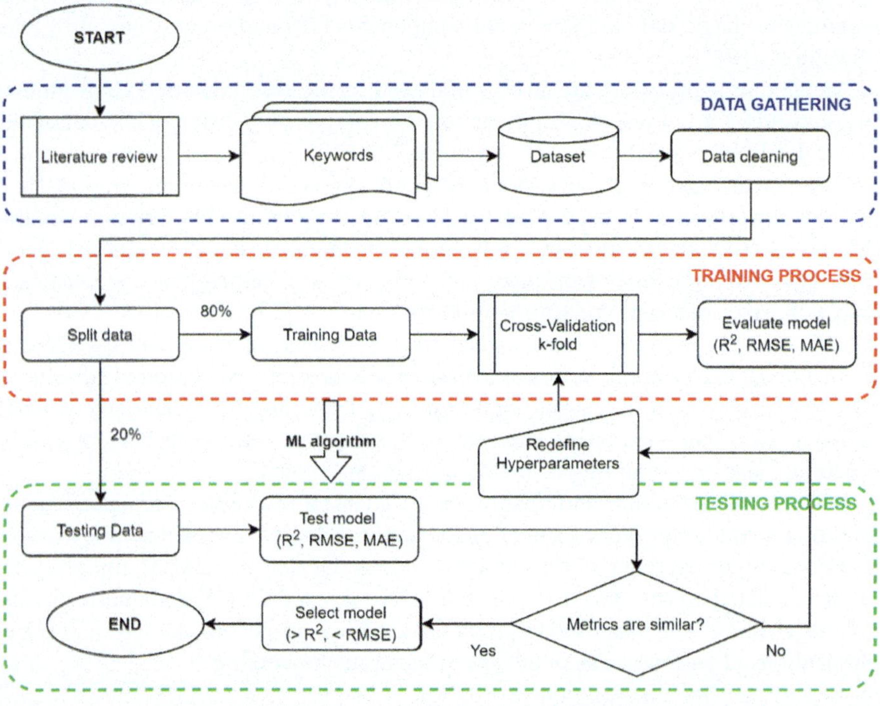

Fig. 1 Schematic process diagram to evaluate the ML algorithm for predicting mechanical properties in advanced manufacturing processes

One of the most recommendable techniques to assess ML accuracy is *k*-fold cross-validation (CV). This method effectively tests the algorithm's ability to predict new data and face common problems such as underfitting or overfitting. In this validation technique, *k* represents the number of parts into which the data is divided; $k-1$ folds are used for training, and the remaining fold is used for testing the model [27]. CV is employed to compare and select a model for a given predictive modeling problem because it is simple to implement, easy to understand, and results in the predictions generally having low biases.

Barrionuevo et al. [28] introduced an index of merit (IM) to assess the prediction accuracy, which combines three metrics (R^2, MSE, MAE) to get a single metric of the algorithms' precision. As the magnitude of the index approaches zero, the lower the error is achieved. The IM is computed in the form of (2), whose components are the coefficient of determination (R^2), the mean squared error (MSE), and the mean absolute error (MAE).

$$\mathrm{IM} = \sqrt{(1 - \mathrm{R}^2)^2 + \mathrm{MSE} + (\mathrm{MAE})^2} \tag{2}$$

The hyperparameters of each ML algorithm must be tuned to achieve the highest accuracy. Random search optimization is one of the most used techniques. This technique automatically examines the hyperparameter search space and attempts to find the optimal values that maximize the coefficient of determination (R^2).

Once identified which algorithm presents the higher accuracy (lower IM), feature importance (FI) analysis can be employed. FI assigns a score to input features based on their usefulness in predicting a target variable. Moreover, FI provides ratings, helping to obtain insight into the data, and the model can improve the efficiency and effectiveness of a predictive model in forecasting.

The property sought varies depending on each manufacturing process. For instance, for SLM, it is crucial to determine the relative density [28]. In WAAM, it is necessary to anticipate the geometry of the deposited layer [29] and select the melting efficiency [30]. In FFF, it is required to know the mechanical response [31], as in friction welding processes [32], and in laser welding, hardness is one of the essential attributes [33]. Therefore, in each cycle, different algorithms can be trained depending on the properties of the material, the processing parameters, or the environmental conditions, see Fig. 2.

3 Case Studies

This section presents the predetermined results of crucial mechanical properties in various advanced manufacturing processes. It starts with 3D printing to predict the composite material's yield strength through ANFIS. Then, the performance of several

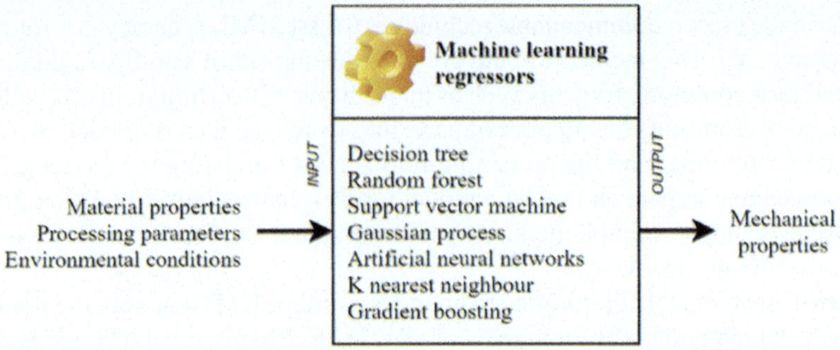

Fig. 2 Input parameters generally employed to train machine learning regressors and anticipate mechanical properties

machine learning algorithms to predict the relative density of stainless steel manufactured by SLM is analyzed. The dimensions of the layer geometry and the melting efficiency in WAAM technology are then evaluated. Finally, advanced welding processes are studied to predict the ultimate tensile strength and the microhardness in friction welding and laser welding, respectively.

3.1 Fused Filament Fabrication

The yield strength of a wood composite filament fabricated by FFF is predicted by applying ANFIS (FIS). A total of 27 samples were employed. The algorithm is trained with the input of layer thickness, fill density, printing speed, and raster angle [31]. Figure 3 illustrates the training, testing, and validation stages.

The forecasted results show a low accuracy. The RMSE shows a value of 3.09, and the mean absolute percentage error (MAPE) as a measure of the prediction accuracy of a forecasting method is 10.65. The low precision of the algorithm is attributed to the small amount of data used during training.

3.2 Selective Laser Melting

Seven of the most utilized ML algorithms predicted the relative density (RD) of 316L stainless steel. The employed regressors are (1) Support vector machines (SVR), (2) Decision trees (DTR), (3) Random forests (RFR), (4) Gradient boosted regressors (GBR), (5) Gaussian process (GPR), (6) k-nearest neighbors (KNN), and (7) artificial neural networks (ANN) or multi-layer perceptron (MLP). A total of 112 datasets are used, where the input parameters are laser power, scanning speed, layer thickness, hatch spacing, and particle size [28].

(a)

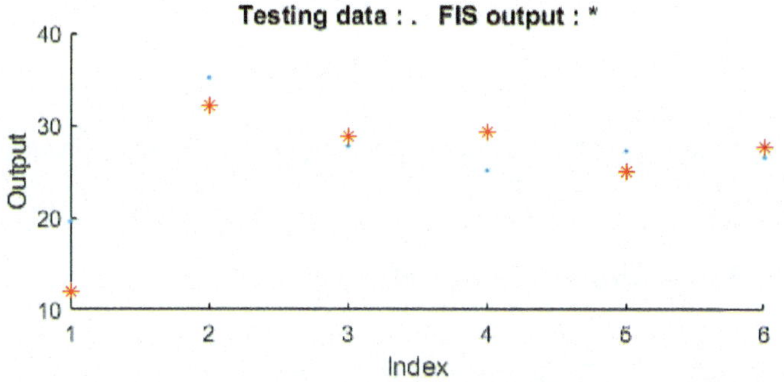

(b)

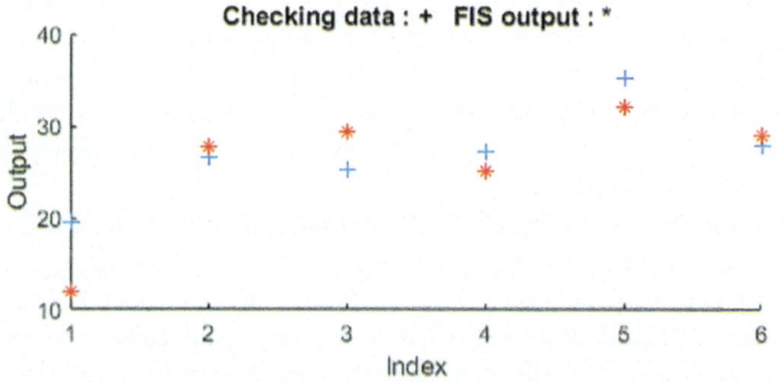

(c)

Fig. 3 Fuzzy inference system (ANFIS) for predicting yield strength. Response during **a** training, **b** testing, and **c** checking [31]

Figure 4 illustrates the behavior of each algorithm compared to the accurate RD. GBR best fits the actual value and shows the highest R^2 and the lowest RMSE. Although MLP also presents good accuracy, presenting the lowest IM (0.5797) versus IM (0.6089) of the GBR. Nevertheless, MLP shows the highest computing time, reducing its applicability.

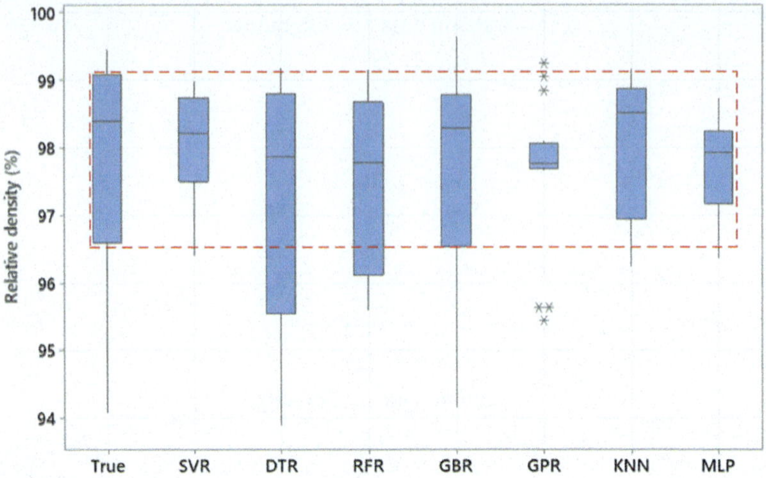

Fig. 4 Boxplot predictions of ML regressor while testing the relative density in samples produced by SLM [28]

3.3 Wire Arc Additive Manufacturing

In WAAM, two different case studies are presented: (1) layer geometry prediction of Ti-6Al-4V alloy [29] and (2) melting efficiency evaluation of low-alloy carbon steel [30].

For the wall width forecasting, twenty-five samples of Ti-6Al-4V are used. The input parameters are arc current, voltage, and plasma gas flow rate. Figure 5 displays the algorithms that better predicted the wall width. The better algorithm is MLP, which shows the highest R^2, the lowest RMSE, and the lowest MAE, followed by linear SVR and extreme GBR regressor.

In Fig. 6, GPR, extreme-GBR (XGBR), and MLP are utilized to predict the melting efficiency in carbon steel processed by WAAM. The input parameters are wire diameter (WD), wire feed speed (WFS), travel speed (TS), and power (P) [30]. All the algorithms show good accuracy ($R^2 > 0.86$). A feature importance analysis is carried out to determine what parameter determines the predictability, indicating that power (P) is the only significant factor (Fig. 7).

3.4 Friction Welding

Determining the maximum ultimate tensile strength (UTS) is critical for joining dissimilar materials. In the study by Barrionuevo et al. [32], machine-learning predictions are compared to response surface methodology (RSM) predictions. The GBR algorithm shows greater precision with an R^2 of 0.909 against 0.8816 of the RSM.

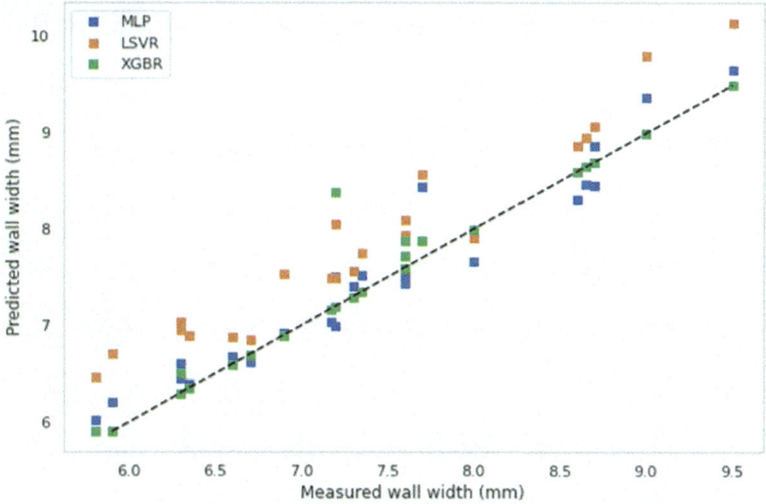

Fig. 5 Scatter plot for comparing supervised ML algorithms' accuracy in the wall width prediction [29]

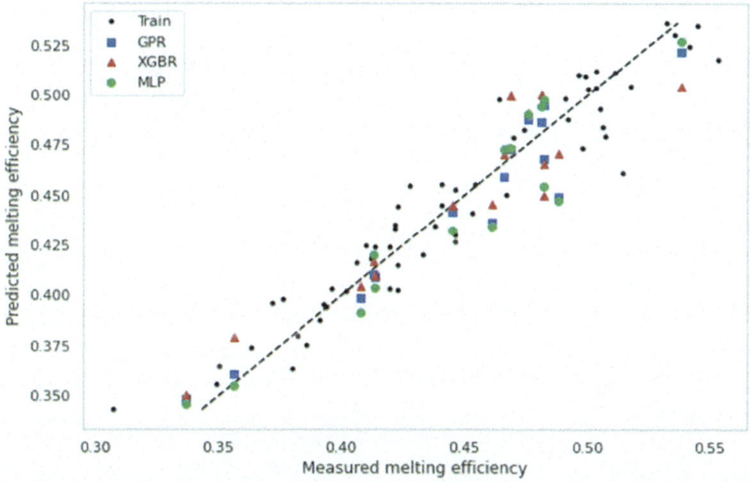

Fig. 6 Scatter plot to evaluate the ML algorithms' performance in melting efficiency prediction during WAAM [30]

The algorithm is trained with 48 datasets. The input parameters are the rotation speed, friction pressure, and laser power. Figure 8 shows the algorithm's performance in predicting the UTS in the entire range of values obtained experimentally [23].

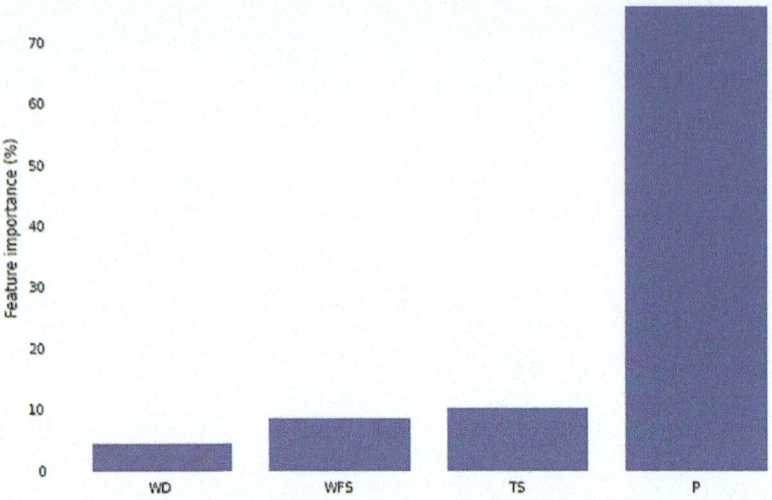

Fig. 7 Feature importance analysis of the input parameters in melting efficiency forecasting [30]

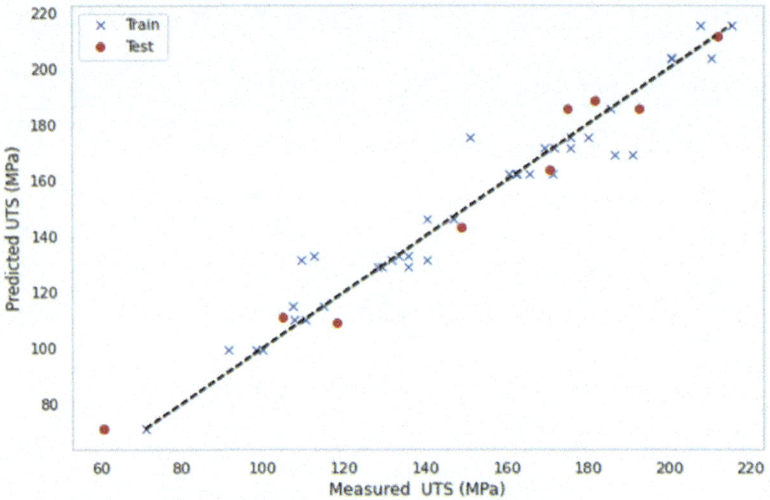

Fig. 8 Scatter plot for accuracy evaluation of the gradient boosting regressor performance in the UTS prediction during rotary friction welding of dissimilar alloys [32]

3.5 *Laser Welding*

Hardness is one of the most widely used mechanical properties that determine the suitability of a welded structure. Vidal et al. [33] determined microhardness by applying Adaboost and RF regressors. Adaboost creates multiple models and combines them to produce improved results, reducing overfitting. The algorithms are trained with

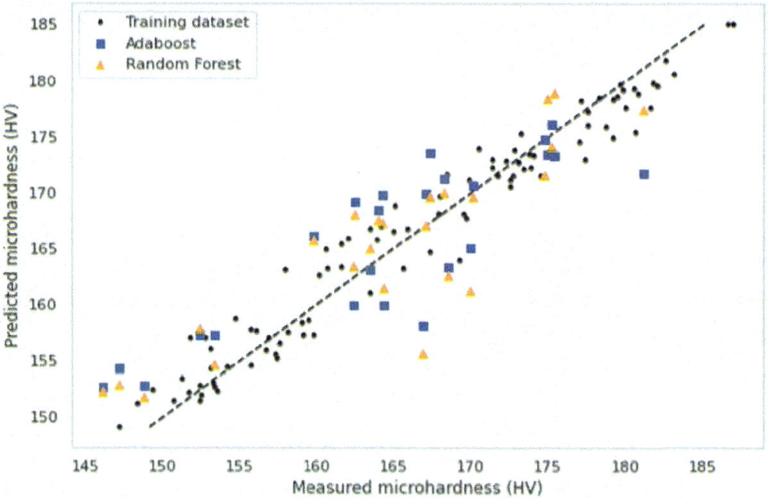

Fig. 9 Scatter plot for accuracy evaluation of the AdaBoost and Random Forest algorithms in the microhardness prediction during laser welding [33]

120 datasets. The input parameters are the vibration amplitude, frequency, and indentation position. Figure 9 shows the evaluation of the ML algorithms. RF offers better accuracy with an R^2 of 0.75, while Adaboost has a value of 0.7058.

4 Discussion

The no-free lunch theorem states that no machine learning algorithm is universally better than any other [2]. According to the reported results, the larger the dataset, the better the predictions. Moreover, as more training parameters are used, the algorithms can better understand how each parameter affects the physical process involved. In that sense, feature-important analysis is highly recommended to determine which parameters influence the predictability of the algorithms.

The metrics to evaluate the accuracy play a significant role in determining which algorithm is suitable depending on the task. It is recommended that several assessment metrics be used to ensure the accuracy of predictions. The index of merit (IM) provides a robust indicator for evaluating multiple metrics [28].

5 Conclusions

In this work, the following conclusions can be drawn from the applicability of several machine learning algorithms to predict properties of interest in advanced manufacturing processes such as laser welding and additive manufacturing.

- The yield strength can be forecasted by applying ANFIS. It is useful when the amount of data is small. Nevertheless, an old-fashioned algorithm limits its application. It is not possible to redefine all the hyperparameters, and it is not possible to define evaluation metrics. Therefore, its applicability in advanced manufacturing processes is not recommended.
- The algorithm that better predicts the relative density of stainless steel fabricated by SLM was the gradient boosting regressor (GBR). This algorithm achieves the highest accuracy and the lowest standard deviation. Multi-layer-perceptron (MLP) also responds well to predicting relative density during SLM. However, significantly more computing time is required for its processing, limiting its applicability. Nevertheless, the relevance of these algorithms is applicable only in cases with a relative density greater than 97%.
- Two essential features were forecasted in WAAM: (1) the wall width, where the algorithm better performed was multi-layer perceptron, which shows the highest accuracy, and (2) melting efficiency, where it was possible to predict the melting efficiency without using intricate measurement systems. The Gaussian process regressor was the algorithm that better predicted melting efficiency in WAAM.
- For the case of welding processes, the algorithm that better performs predicting the UTS in rotary friction welding was the gradient boosting regressor, which showed the lowest error. The average microhardness prediction was fulfilled in laser welding by applying AdaBoost and Random Forest algorithms, which agreed well with the measured values. The obtained relative errors (± 3%) denote the usefulness of the models.

References

1. Siddique, N., Adeli, H.: Computational Intelligence: Synergies of Fuzzy Logic (2013)
2. Goodfellow, I., Bengio, Y., Courville, A.: Deep Learning Adaptive Computation and Machine Learning, vol. 1 (2016). https://doi.org/10.2172/1462436
3. Gianey, H.K., Choudhary, R.: Comprehensive review on supervised machine learning algorithms. In: Proceedings of the 2017 International Conference on Machine Learning and Data Science, MLDS 2017, vol. 2018-Janua, pp. 38–43 (2018). https://doi.org/10.1109/MLDS.2017.11
4. Kostopoulos, G., Karlos, S., Kotsiantis, S., Ragos, O.: Semi-supervised regression: a recent review. J. Intell. Fuzzy Syst. **35**(2), 1483–1500 (2018). https://doi.org/10.3233/JIFS-169689
5. Yang, N., Zheng, Z., Wang, T.: Model loss and distribution analysis of regression problems in machine learning. In: ACM International Conference Proceeding Series, vol. Part F 1481(1), pp. 1–5 (2019). https://doi.org/10.1145/3318299.3318367

6. Mathew, J., Griffin, J., Alamaniotis, M., Kanarachos, S., Fitzpatrick, M.E.: Prediction of welding residual stresses using machine learning: comparison between neural networks and neuro-fuzzy systems. Appl. Soft Comput. J. **70**, 131–146 (2018). https://doi.org/10.1016/j.asoc.2018.05.017
7. Alafaghani, A., Qattawi, A., Alrawi, B., Guzman, A.: Experimental optimization of fused deposition modelling processing parameters: a design-for-manufacturing approach. Procedia Manuf. **10**, 791–803 (2017). https://doi.org/10.1016/j.promfg.2017.07.079
8. Rinaldi, M., Ghidini, T., Cecchini, F., Brandao, A., Nanni, F.: Additive layer manufacturing of poly (ether ether ketone) via FDM. Compos. B Eng. **145**(December), 162–172 (2018). https://doi.org/10.1016/j.compositesb.2018.03.029
9. Li, N., Huang, S., Zhang, G., Qin, R., Liu, W., Xiong, H., Shi, G., Blackburn, J.: Progress in additive manufacturing on new materials: a review. J. Mater. Sci. Technol. **35**(2), 242–269 (2019). https://doi.org/10.1016/j.jmst.2018.09.002
10. Gong, G., Ye, J., Chi, Y., Zhao, Z., Wang, Z., Xia, G., Du, X., Tian, H., Yu, H., Chen, C.: Research status of laser additive manufacturing for metal: a review. J. Mater. Res. Technol. **15**., 855–884. Elsevier Editora Ltda (2021). https://doi.org/10.1016/j.jmrt.2021.08.050
11. Haghdadi, N., Laleh, M., Moyle, M., Primig, S.: Additive manufacturing of steels: a review of achievements and challenges. J. Mater. Sci. **56**(1), 64–107 (2021). https://doi.org/10.1007/s10853-020-05109-0
12. . Yap, C.Y., Chua, C.K., Dong, Z.L., Liu, Z.H., Zhang, D.Q., Loh, L.E., Sing, S.L.: Review of selective laser melting: materials and applications. Appl. Phys. Rev. **2**(4) (2015). https://doi.org/10.1063/1.4935926
13. Barrionuevo, G.O., Walczak, M., Ramos-grez, J., Sánchez-sánchez, X.: Microhardness and wear resistance in materials manufactured by laser powder bed fusion: machine learning approach for property prediction. CIRP J. Manuf. Sci. Technol. **43**, 106–114 (2023). https://doi.org/10.1016/j.cirpj.2023.03.002
14. Kozior, T., Kundera, C.: Evaluation of the Influence of Parameters of FDM Technology on the Selected Mechanical Properties of Models. Procedia Eng. **192**, 463–468 (2017). https://doi.org/10.1016/j.proeng.2017.06.080
15. Gordelier, T.J., Thies, P.R., Turner, L., Johanning, L.: Optimising the FDM additive manufacturing process to achieve maximum tensile strength: a state-of-the-art review. Rapid Prototyp. J. **25**(6), 953–971 (2019). https://doi.org/10.1108/rpj-07-2018-0183
16. Karmuhilan, M., Sood, A.K.: Intelligent process model for bead geometry prediction in WAAM. Mater. Today Proc. **5**(11), 24005–24013 (2018). https://doi.org/10.1016/j.matpr.2018.10.193
17. Li, J.L.Z., Alkahari, M.R., Rosli, N.A.B., Hasan, R., Sudin, M.N., Ramli, F.R.: Review of wire arc additive manufacturing for 3d metal printing. Int. J. Autom. Technol. **13**(3), 346–353 (2019). https://doi.org/10.20965/ijat.2019.p0346
18. Liu, J., Xu, Y., Ge, Y., Hou, Z., Chen, S.: Wire and arc additive manufacturing of metal components: a review of recent research developments. Int. J. Adv. Manuf. Technol. **111**(1–2), 149–198 (2020). https://doi.org/10.1007/s00170-020-05966-8
19. Xiong, J., Yin, Z., Zhang, W.: Closed-loop control of variable layer width for thin-walled parts in wire and arc additive manufacturing. J. Mater. Process. Technol. **233**, 100–106 (2016). https://doi.org/10.1016/j.jmatprotec.2016.02.021
20. Mehta, K.P.: A review on friction-based joining of dissimilar aluminum-steel joints. J. Mater. Res. **34**(1), 78–96 (2019). https://doi.org/10.1557/jmr.2018.332
21. Cai, W., et al.: A state-of-the-art review on solid-state metal joining. J. Manuf. Sci. Eng. Trans. ASME **141**(3), 1–35 (2019). https://doi.org/10.1115/1.4041182
22. Maalekian, M.: Friction welding—critical assessment of literature. Sci. Technol. Weld. Joining **12**(8), 738–759 (2007). https://doi.org/10.1179/174329307X249333
23. Mullo, J.L., Ramos-Grez, J., Barrionuevo, G.O.: Effect of laser heat treatment on the mechanical performance and microstructural evolution of AISI 1045 steel-2017-T4 aluminum alloy joints during rotary friction welding. J. Mater. Eng. Perform. (2021). https://doi.org/10.1007/s11665-021-05614-6

24. Kurc-Lisiecka, A., Lisiecki, A.: Laser welding of stainless steel. J. Achiev. Mater. Manuf. Eng. **98**(1), 32–40 (2020). https://doi.org/10.5604/01.3001.0014.0815
25. Yang, J., et al.: Laser techniques for dissimilar joining of aluminum alloys to steels: a critical review. J. Mater. Process. Technol. **301**(November), 117443 (2022). https://doi.org/10.1016/j.jmatprotec.2021.117443
26. Hyndman, R.J., Koehler, A.B.: Another look at measures of forecast accuracy. Int. J. Forecast. **22**(November), 679–688 (2005). https://doi.org/10.1016/j.ijforecast.2006.03.001
27. Baturynska, I., Martinsen, K.: Prediction of geometry deviations in additive manufactured parts: comparison of linear regression with machine learning algorithms. J. Intell. Manuf. (2020). https://doi.org/10.1007/s10845-020-01567-0
28. Barrionuevo, G., Ramos-Grez, J., Walczak, M., Betancourt, C.: Comparative evaluation of supervised machine learning algorithms in the prediction of the relative density of 316L stainless steel fabricated by selective laser melting. Int. J. Adv. Manuf. Technol. (2021). https://doi.org/10.1007/s00170-021-06596-4
29. Barrionuevo, G.O., Ríos, S., Williams, S.W., Ramos-Grez, J.A.: Comparative evaluation of machine learning regressors for the layer geometry prediction in wire arc additive manufacturing. In: 12th International Conference on Mechanical and Intelligent Manufacturing Technologies, pp. 186–190 (2021). https://doi.org/10.1109/ICMIMT52186.2021.9476168.
30. Barrionuevo, G.O., Sequeira-Almeida, P.M., Ríos, S., Ramos-Grez, J.A., Williams, S.W.: A machine learning approach for the prediction of melting efficiency in wire arc additive manufacturing. Int. J. Adv. Manuf. Technol. 0123456789 (2022). https://doi.org/10.1007/s00170-022-08966-y
31. Barrionuevo, G.O., Ramos-Grez, J.A.: Machine learning for optimizing technological properties of wood composite filament-timberfill fabricated by fused deposition modeling BT. In: Botto-Tobar, M., Zambrano Vizuete, M., Torres-Carrión, P., Montes León, S., Pizarro Vásquez, G., Durakovic, B. (eds.) Applied Technologies, pp. 119–132. Springer International Publishing, Cham (2020)
32. Barrionuevo, G.O., Mullo, J.L., Ramos-Grez, J.A.: Predicting the ultimate tensile strength of AISI 1045 steel and 2017–T4 aluminum alloy joints in a laser-assisted rotary friction welding process using machine learning: a comparison with response surface methodology. Int. J. Adv. Manuf. Technol. **116**(3–4), 1247–1257 (2021). https://doi.org/10.1007/s00170-021-07469-6
33. Vidal, J.G., Barrionuevo, G.O., Ramos-grez, J.A., Vergara, J., Mullo, J.L.: Vibration-assisted laser welding : Frequency and amplitude vibration effect on the microhardness of A36 steel. Proc. Instit. Mech. Eng. Part L J. Mater. Des. Appl. **2022**. https://doi.org/10.1177/14644207221102659

Comparative Evaluation of Performance Efficiency in Terms of Temporal Behavior and Resource Utilization, According to the ISO/IEC 25,010 Model, in a Web Application Developed with Angular, React.js, and Vue.js

Angel Loja and Tania Maita

Abstract Choosing a web application framework is crucial for achieving efficient performance in the current software development landscape. This study focuses on evaluating and comparing the performance efficiency of web applications developed using three popular frameworks: Angular, React.js, and Vue.js. The evaluation is based on the ISO/IEC-25010 model, which defines this set of quality characteristics for software systems. The performance of web applications is essential to ensure a positive user experience and maximize resource utilization. However, there is a lack of comparative studies that assess performance efficiency in ISO/IEC-25010 for web applications. This gap creates uncertainty for developers when selecting the most suitable framework for their requirements. This paper compared a web application developed with Angular, React.js, and Vue.js to address this issue. Also, the ISO/IEC-25010 model is adopted, focusing on temporal behavior and resource utilization sub-characteristics. The evaluation uncovered significant insights into the performance efficiency of web applications developed with Angular, React.js, and Vue.js. It is found that each framework exhibited distinct strengths and weaknesses in temporal behavior and resource utilization. The results of this assessment have significant implications for developers and organizations involved in web application development. Firstly, the choice of framework should align with the specific performance requirements of the application.

Keywords JavaScript · Frameworks · Vue.js · React.js · Angular

A. Loja (✉) · T. Maita
Faculty of Computer Science and Electronics, Escuela Superior Politécnica de Chimborazo, Orellana, Ecuador
e-mail: angel.loja@espoch.edu.ec

T. Maita
e-mail: tania.maita@espoch.edu.ec

G. F. Olmedo Cifuentes et al. (eds.), *Emerging Research in Intelligent Systems*, Lecture Notes in Networks and Systems 902, https://doi.org/10.1007/978-3-031-52255-0_21

1 Introduction

Currently, the development of web applications is an increasingly relevant task in the field of information technology. An organization needs to compete and provide quality services. In this context, the comparative evaluation of performance efficiency becomes crucial to ensure an optimal user experience and overall project success. Loading speed, responsiveness, resource consumption, and stability are fundamental aspects that define the quality of a web application. A careful and exhaustive evaluation ensures users enjoy a smooth experience without delays or interruptions.

Furthermore, efficient performance contributes to user retention and enhances the organization's image and reputation in the market. It is important to note that selecting the appropriate JavaScript framework for a web application plays a pivotal role in achieving high performance and a satisfactory user experience. In this regard, conducting a comparative evaluation of frameworks regarding temporal behavior and resource utilization becomes a critical step for making informed decisions [1]. The research problem consists of evaluating and comparing JavaScript frameworks in terms of their temporal behavior and resource utilization, using the criteria established by the ISO/IEC 25,010 model within the context of a web application.

The main objective is to provide a comprehensive and objective overview of the different frameworks, analyzing their ability to meet a web application's performance and efficiency requirements. Tests and measurements are carried out on the mentioned frameworks to achieve the stated objective. These tests and measurements allow for the analysis of temporal behavior and resource utilization in different situations. Special attention is given to ensuring compliance with the quality criteria established by the ISO/IEC 25,010 model, ensuring that the development of the web application follows standards and best practices to guarantee the delivery of a high-quality final product. Subsequently, performance tests are conducted on the developed web application. The expectation is to obtain results that allow for a detailed comparison of the evaluated JavaScript frameworks (Angular, React.js, and Vue.js) regarding temporal behavior and resource utilization. Through the analysis of these results, it is anticipated that it will be possible to determine which of the frameworks offers the best performance and whether there are significant differences in performance efficiency according to the criteria established by the ISO/IEC 25,010 model.

2 Methodology

Performance efficiency is one of the attributes outlined in the ISO 25010 standard, with the primary objective of evaluating software performance. This assessment relies on two sub-characteristics: temporal behavior analysis to gauge system response times and resource utilization examination to assess memory consumption. Temporal behavior analysis is instrumental in determining the extent to which response times, processing durations, and overall system performance manifest [2].

Defining benchmark testing scenarios is necessary for measuring and evaluating performance, as they establish the metrics and tools to consider in web frameworks. This allows comparing them through experiments consistently executing and analyzing tests, as described in the proposed research methodology. By validating said scenarios through testing, the performance of frameworks is measured in terms of time behavior and resource utilization. In this way, the scenarios facilitate quantifying the attributes required by the ISO 25010 standard and determining the degree to which frameworks meet performance objectives [3].

2.1 *Stack*

The selected technological stack for web application development comprises several technologies and tools that work together to achieve complete and efficient functioning. A framework like React.js, Vue.js, or Angular on the front layer creates an interactive and appealing user interface. On the backend layer, Node.js is employed, along with a framework like Nest.js, to develop the logic and data management on the server side. A database like MongoDB is used on the storage layer for storing and retrieving data. Together, this technological stack allows for the creation of a complete and functional web application, providing a structured and efficient approach to development (Fig. 1).

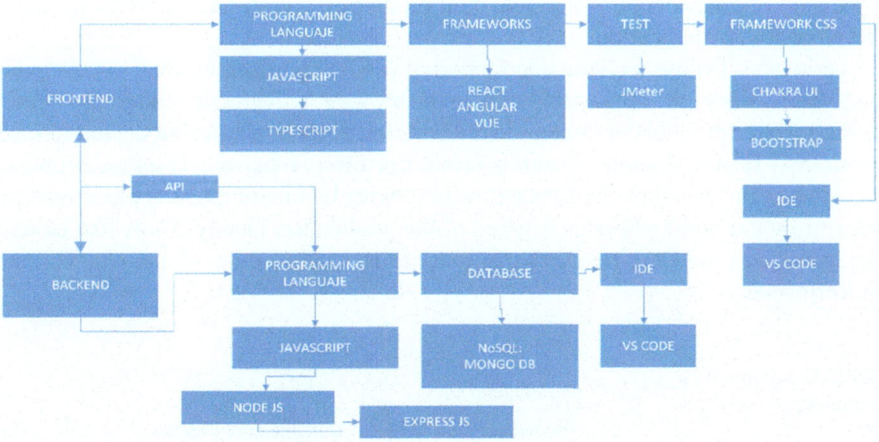

Fig. 1 Stack to develop a web application

2.2 Test Equipment

A comparative evaluation is conducted using practical criteria. This analysis involves comparing the results of different scenarios concerning specific research criteria. The equipment parameters used in creating the test environment are also considered, as seen in Table 1.

2.3 Determination of Performance Effectiveness

As mentioned, the sub characteristics for performance efficiency are measured: temporal behavior, which involves measuring and calculating the time it takes for the system to respond, and resource utilization, which analyzes the quantity and time required for the test.

Practical tests are performed using testing applications such as JMeter and Chrome DevTools toolkit, all executed locally. Tests are conducted for 3 cases in each scenario:

- Adding 500 rows
- Updating 500 rows
- Loading 500 rows.

Following the ISO/IEC 25,010 model, performance efficiency is evaluated regarding temporal behavior and resource utilization. An evaluation experiment is proposed for each framework. The collected data compares Angular and other considered frameworks to determine their relative efficiency and performance in the developed web application.

Fariz et al. [4] use a client-side rendering (CSR) approach by React.js, Angular, and Vue.js, where the application runs in the browser. Client-side rendering refers to using JavaScript to render content in the browser. As a result, during the initial load, a minimal HTML document with a JavaScript file is received. Tests are proposed to measure the speed of the selected technologies by creating test applications and measuring the build and launch speed of the application locally. Tools like Google Analytics and the Chrome DevTools toolkit are used to monitor the application's performance.

Table 1 Equipment parameters

Parameters	Value
Processor	Inside core i7 vpro
Ram	16 Gb
Graphics card	Intel UHD 5000
Solid-state drive	SSD de 256 GB
Operative system	Windows Pro version 22H2

Although they describe the technical characteristics and evaluation criteria, privacy, data security, and user impact are not addressed. This can be problematic, as it does not consider the potential ethical risks or harm that the software may cause, which could lead to unethical decisions and developments that are detrimental to users or society.

A single test setup with specific characteristics and capabilities may not accurately reflect performance in broader environments and configurations. These limitations restrict the generalization of results and the applicability of conclusions to other cases and situations, which limits a comprehensive understanding of performance effectiveness and its relevance in a broader context.

3 Background

In this context, the growing popularity of JavaScript has led to the emergence of various JavaScript frameworks that aim to assist developers in tackling programming tasks. However, the number of JavaScript frameworks has rapidly increased to thousands of versions. It is a challenge for professionals to identify the frameworks that best suit their needs and develop new ones that meet them. Additionally, there is a lack of understanding about what drives developers to make that choice. This explores the factors and actors that influence the option of a JavaScript framework [5].

3.1 Frameworks

In the work of Bielak et al. [6], they mention the comparison of frameworks where they explain their definition and some of their characteristics:

Vue.js. It is a progressive JavaScript framework for building user interfaces. It was developed by Evan You and is designed to be easy to learn and use. It uses a compo-nent-based architecture and a virtual DOM to manage the representation of UI elements. It is commonly used in single-page web applications and integrates well with other libraries and frameworks. Vue.js is based on the concept of components, which are reusable code blocks that encapsulate the logic and user interface of a specific part of the application. Components in Vue.js allow for creating a modular and scalable structure for building applications.

Angular. Angular is an open-source web application framework based on TypeScript developed by Google. It is a complete rewrite of AngularJS and is used for building dynamic single-page web applications (SPAs). Angular uses a declarative approach to building user interfaces and provides various features, such as two-way data binding, component-based architecture, and dependency injection.

In Angular, modules are fundamental building blocks of an application. They consolidate associated components, services, and resources into a cohesive framework, facilitating organization and structure within the application.

Angular uses the dependency injection pattern to manage the dependencies of components and services. Dependency injection allows for the modular and decoupled creation of components, facilitating code reuse and maintenance.

React.js. Does Facebook develop a JavaScript library for building dynamic and high-performance user interfaces? It uses a component-based architecture and a virtual DOM to manage the representation of UI elements. It is often combined with other libraries and frameworks to create complete web applications.

React is based on the concept of reusable components. React components can be functional (Function Components) or class-based (Class Components). State and props are two fundamental concepts in React. The state represents the internal data of a component and can change over time. Props are values passed from a parent to a child component and communicate information between components.

Frontend applications (client-side applications) are executed directly in the user's web browser. They are developed using JavaScript, TypeScript, and frameworks such as Vue.js, React.js, and Angular [7].

Speed of response and security are crucial factors when choosing a front-end framework. Security refers to the protection of the application against potential hacker attacks. If a framework that is not secure or does not have good performance is chosen, it can negatively affect the user experience [8].

A secure and performant framework protects the application against potential attacks and ensures a positive user experience. Performance unit testing is a practical methodology for evaluating the performance of web applications and identifying issues during development. Although this approach has limitations, such as not mentioning how to increase the workload and automatically generate test cases, it allows for fast response times and ensures optimal application performance [9].

The performance efficiency of the software is crucial for user satisfaction. Madelyn et al. [10] analyze internationally recognized standards and methodologies to identify quality activities that help understand and improve performance efficiency. These activities should be an integral part of the quality control process in software development institutions and companies. It is recommended to incorporate them from the beginning of development and apply them consistently throughout all stages.

3.2 Selection Criteria

The article by Vyas [11] identifies some criteria for selecting a specific framework. These criteria include:

Popularity and community.

The popularity of a framework can be an indicator of its quality and usefulness. Additionally, an active community can provide support and valuable resources for developers.

Features and functionalities. Frameworks can have different parts and functionalities better suited for other projects or specific needs.

Learning curve. Some frameworks may be easier or harder for developers to learn, which can affect the speed and efficiency of development.

Performance. Performance is vital to ensure that a web application is fast and responsive. Some frameworks may be quicker or slower than others in different situations.

Compatibility with other technologies. Frameworks should be compatible with other technologies used in the project, such as databases or web servers.

3.3 Architecture

Client–server architecture involves a software design where system tasks are distributed between two main components: the client and the server. In this model, the client is the software running on the end user's computer, while the server runs on a remote machine and provides services to the client [12].

The client and the server can communicate through different network protocols, such as TCP/IP and HTTP. Pereira et al. [13], They mention that in client–server architecture, scalability is possible as more clients and servers can be added according to the system's needs. It facilitates component reuse, as multiple clients can utilize the server's services (Fig. 2).

As in the research conducted by Ollila et al. [14], this is a desirable property to achieve a fast response when using browser APIs directly or through a wrapper. However, when DOM API calls are dynamically generated by a framework, such as Nest, in our case, for building the API, the framework must perform additional work to determine which calls are necessary for each specific transition, which results in additional overhead.

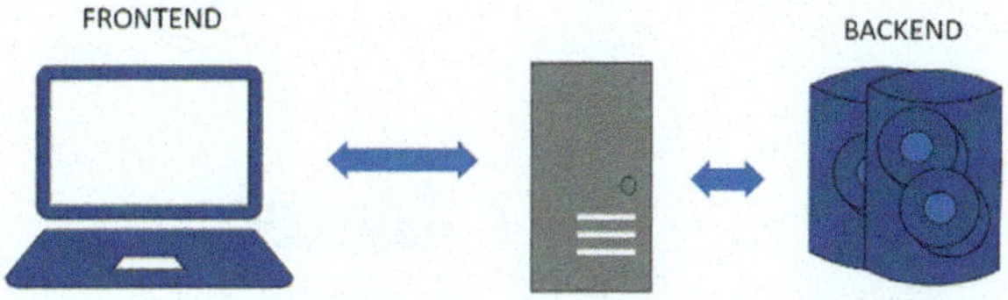

Fig. 2 Client–server architecture

4 Results

4.1 Experiment 1: Test Scenario with React.js

Before the experiment, specific criteria and test scenarios are established to evaluate the performance and efficiency of React.js in the context of the web application. The main objective of this experiment is to analyze how React.js performs in the web application in terms of temporal behavior and resource utilization.

Data is collected on the loading time and responsiveness of the user interface, as observed in Table 2 (Fig. 3).

In Experiment 1, a test scenario is conducted using the React.js framework. In this scenario, various tools are utilized, and pre-defined tests are performed to evaluate the performance and efficiency of React.js (Figs. 4, 5 and 6).

Temporal Behavior in Response Times.

In the React.js experiment, tests are conducted to measure the response times and resource utilization of three transactions: "save", "update", and "load". The results showed that the "save" transaction had an average execution time of 44,049.9 ms, with a response time ranging from 8,139 to 50,451 ms. On the other hand, the "update" transaction had an average execution time of 5,929.49 ms. The throughput

Table 2 Response times recorded in the React.js experiment

Requests	Executions		Response times						Throughput
Label	Samples	Average	Min	Max	Median	90th pct	95th pct	99th pct	Transaction
Save	500	44,049.9	8139	50,451	48,498	49,302	49,427,9	49,966.7	9.73
Update	500	5929.49	1030	8660	6734	7939.8	8582	8647.98	51.98
Load	1	2493	2493	2493	2493	2493	2493	2493	0.4

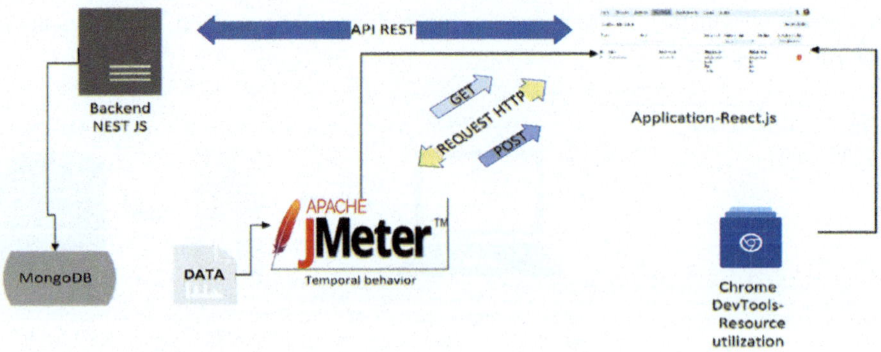

Fig. 3 Test scenario with React.js

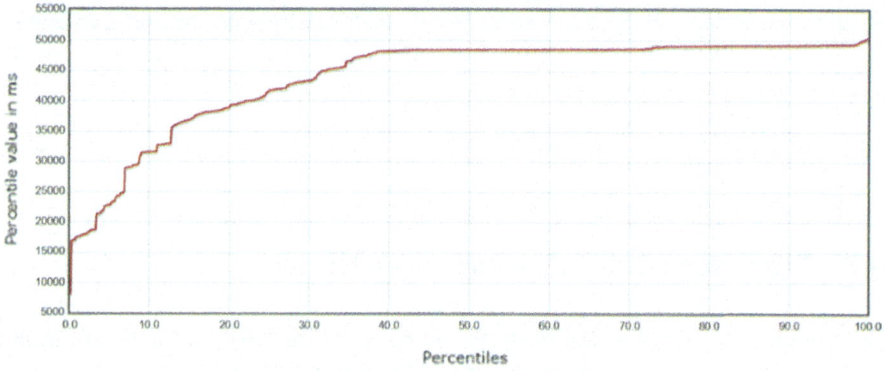

Fig. 4 Results for test scenario with React.js when adding 500 rows

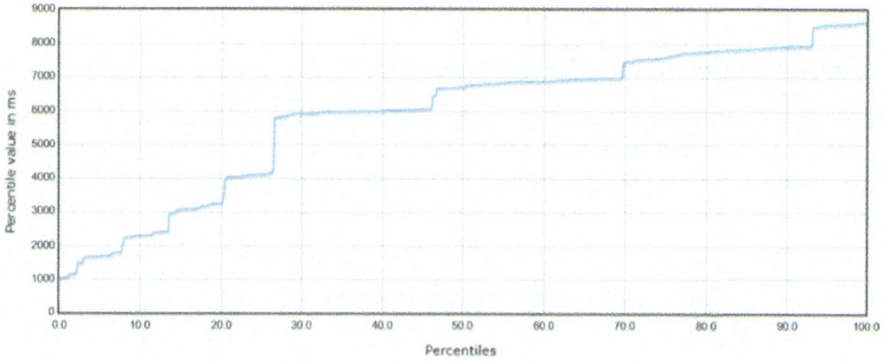

Fig. 5 Update 500 rows for test scenario with React.js

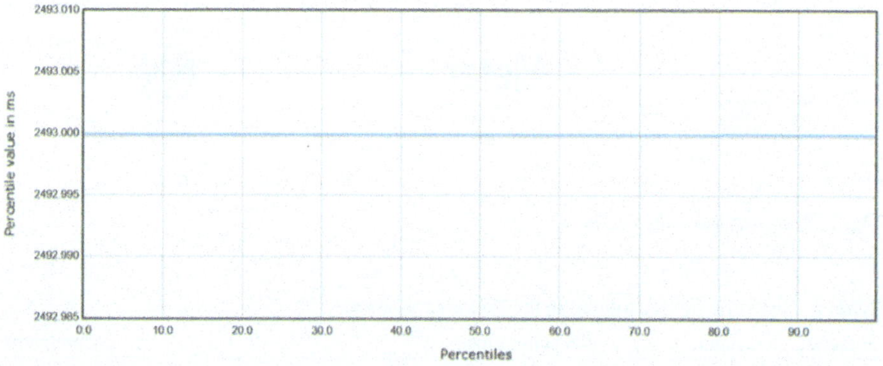

Fig. 6 Load 500 rows for test scenario with React.js

for this transaction is 51.98 transactions per second. Additionally, resource utilization is recorded, with the React.js application consuming 51,152 KB of memory and the GPU memory being 5,760 KB, as shown in the data. The results of the performance tests conducted on the React.js, Vue.js, and Angular frameworks are summarized in Tables 5 and 6.

4.2 Experiment 2: Test Scenario with Vue.js

In Experiment 2, a test scenario is conducted using the Vue.js framework. In this scenario, various tools are used, and pre-defined tests are executed to evaluate the performance and efficiency of Vue.js in the web application (Fig. 7).

The data related to the loading time and responsiveness of the web application is shown in Table 3 (Figs. 8, 9 and 10).

Temporal Behavior in Response Times.

In the Vue.js experiment, three types of requests are performed: "update","save", and "load". Each submission was executed 500 times, and the corresponding response times were recorded. On average, the "update" request had a response time of 11,635.1 ms. The performance, measured in transactions per second, is 26.69. The

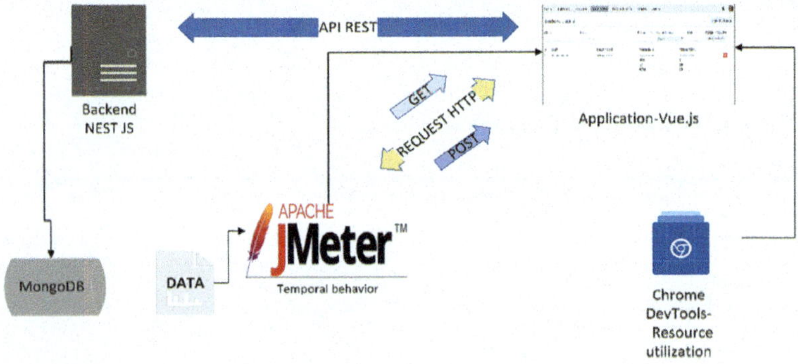

Fig. 7 Test scenario with Vue.js

Table 3 Response times captured in the Vue.js experiment

Requests	Executions		Response times						Throughput
Label	Samples	Average	Min	Max	Median	90th pct	95th pct	99th pct	Transaction
Update	500	11,635.1	1014	17,820	12,517	16,193	16,660.5	17,043.6	26.69
Save	500	10,110	1170	12,278	10,207	11,664	11,928.8	12,016.8	36.44
Load	1	837	837	837	837	837	837	837	1.19

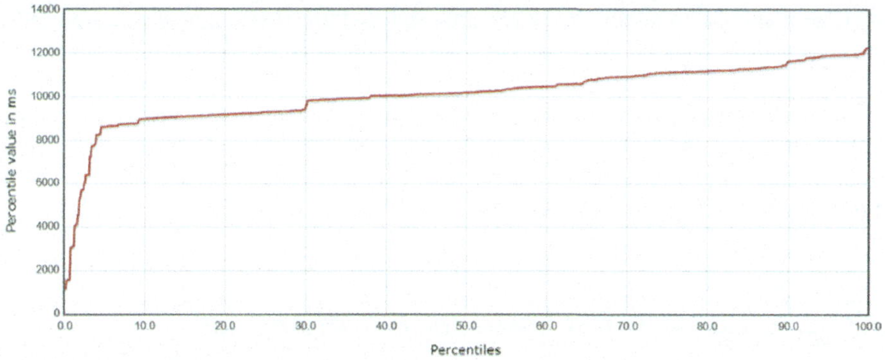

Fig. 8 Update 500 rows for test scenario with Vue.js

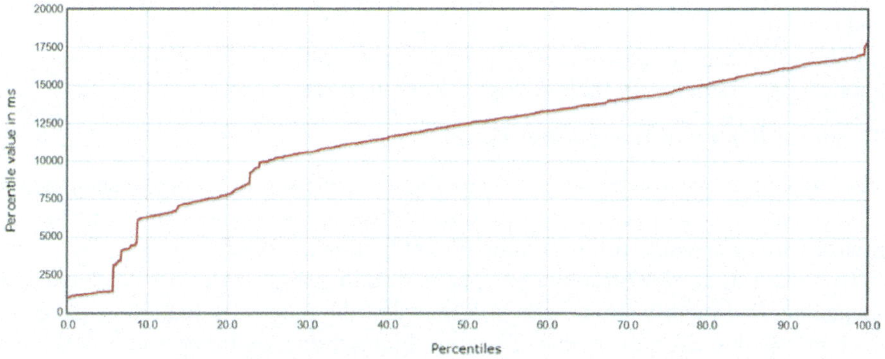

Fig. 9 Save 500 rows for test scenario with Vue.js

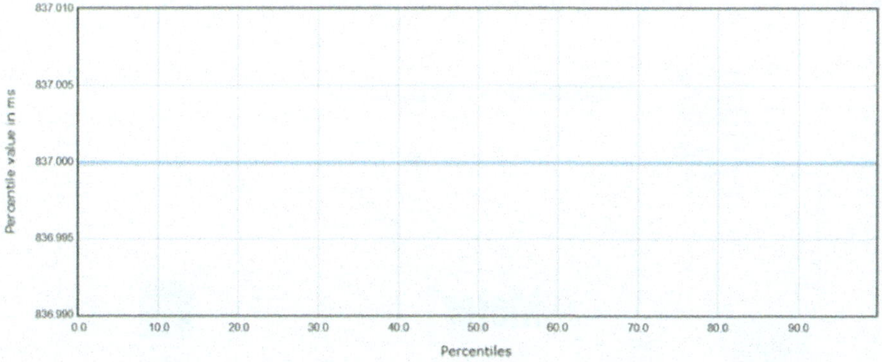

Fig. 10 Load 500 rows for test scenario with Vue.js

average response time for the "save" request is 10,110 ms, and the performance is 36.44 transactions per second. The "load" request is only performed once, with a response time of 837 ms. In addition to performance data, information about the resource utilization of the Vue.js application is provided. The application used 54,416 KB of memory and 8,041 KB of GPU memory. The results of the performance tests conducted on the React.js, Vue.js, and Angular frameworks are summarized in Tables 5 and 6.

4.3 Experiment 3: Test Environment with Angular

This experiment aims to evaluate the performance of Angular in the web application, focusing on key aspects such as loading speed, responsiveness of the user interface, and efficiency in resource utilization, such as memory (Fig. 11).

The information regarding the loading time and responsiveness of the web application is detailed in Table 4 (Figs. 12, 13 and 14).

Temporal Behavior in Response Times

The "update" request in the Angular experiment displayed an average response time of 11,474.86 ms, whereas the "save" request demonstrated significantly faster performance with an average response time of 4,405.53 ms. The impressive speed of the "save" request, at 62.12 transactions per second, highlights Angular proficiency in managing data modification tasks. Additionally, the request to load data, executed once, resulted in a response time of 843 ms, highlighting the framework's capability to retrieve information efficiently. The application used 88,612 KB of memory and 7,934 KB of GPU memory. The results of the performance tests conducted on the React.js, Vue.js, and Angular frameworks are summarized in Tables 5 and 6.

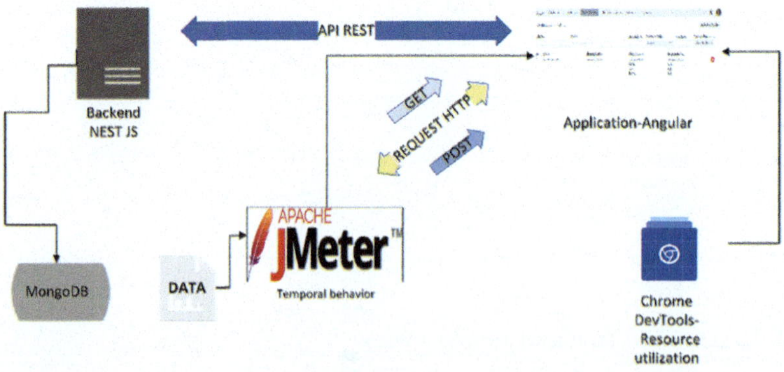

Fig. 11 Angular scenario

Table 4 Response times measured in the Angular experiment

Requests	Executions		Response times						Throughput
Label	Samples	Average	Min	Max	Median	90th pct	95th pct	99th pct	Transaction
Update	500	11,474.86	1231	17,328	12,351.5	15,761.8	16,146.6	16,558.46	26.58
Save	500	4405.53	536	7541	4554	6385.8	6907.5	7121.69	62.12
Load	1	843	843	843	843	843	843	843	1.19

Table 5 Resource utilization

Framework	Memory usage (kB)	GPU memory (kB)
React.js	51,152	5760
Vue.js	54,416	8041
Angular	88,612	7934

Table 6 Performance efficiency of each framework based on average performance

Framework	Update (ms)	Add (ms)	Load (ms)
React.js	44,049,9	5929,49	2493
Vue.js	11,635,1	10,110	837
Angular	11,474,86	4405,53	

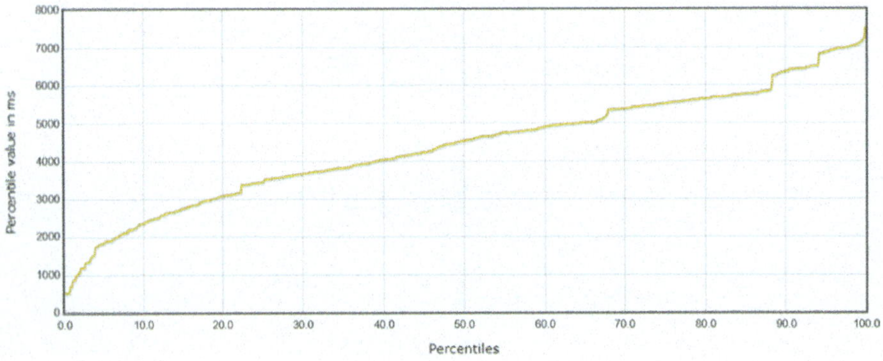

Fig. 12 Update 500 rows for Angular scenario

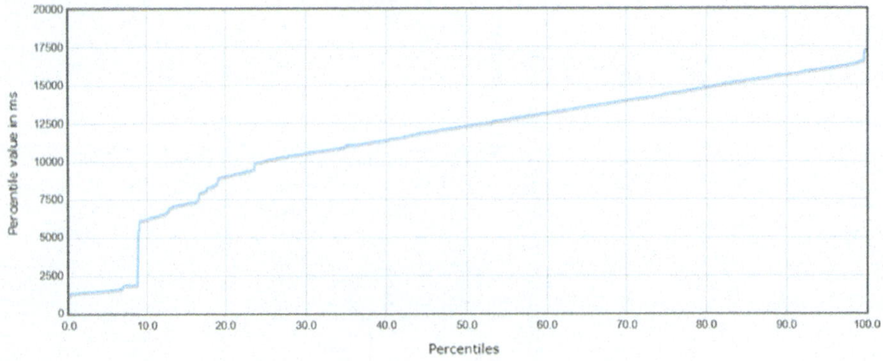

Fig. 13 Save 500 rows for Angular scenario

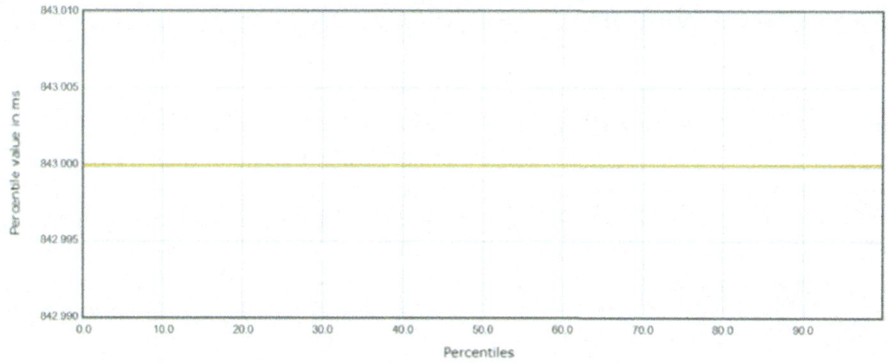

Fig. 14 Load 500 rows for Angular scenario

5 Summary of Results

The results of the performance tests conducted on the React.js, Vue.js, and Angular frameworks are summarized in Tables 5 and 6.

6 Conclusions

During the study, the performance efficiency of each framework was evaluated using the ISO/IEC-25010 model. The results revealed that each framework had strengths and weaknesses regarding temporal behavior and resource utilization. In some transactions, Angular exhibited a higher average response time than React.js and Vue.js, suggesting that it may have slightly lower performance in specific scenarios. However, Angular demonstrated higher resource utilization efficiency than the other two frameworks, indicating that it can manage system resources more efficiently. This study contributes to the field of research by filling a gap in knowledge regarding the performance efficiency of web application frameworks according to the ISO/IEC-25010 model. The comparative evaluation of Angular, React.js, and Vue.js using specific criteria provides an objective insight into the different frameworks and their ability to meet performance and efficiency requirements in web application development. The results of this study have significant practical implications for developers and organizations involved in web application development. By understanding the strengths and weaknesses of each framework, developers can make informed decisions when selecting the most suitable one for their projects. Furthermore, optimizing performance and resource utilization can enhance user experience and improve efficiency in utilizing system resources.

7 Recommendations

To validate and compare frameworks, it is recommended to perform evaluation experiments using specific test scenarios and relevant data. This will allow for obtaining more accurate and comparable results among different technologies. To ensure the reliability of the results, it is essential to make multiple requests for measurements and consider consistent parameters of the equipment used in the test environment. When evaluating the performance of a framework in a specific context, it is suggested to gather data on the responsiveness of the user interface and resource consumption. These data will provide critical information about the framework's performance and efficiency.

References

1. Lizcano, D., Martínez-Ortíz, A.L., López, G., Grignard, A.: End-user modeling of quality for web components. J. Softw. Evol. Process. (2020). https://doi.org/10.1002/smr.2256
2. Yuniasri, D., Damayanti, P., Rochimah, S.: Performance efficiency evaluation frameworks based on ISO 25010. In: EECCIS 2020—2020 10th Electrical Power, Electronics, Communications, Controls, and Informatics Seminar, pp. 254–258. Institute of Electrical and Electronics Engineers Inc (2020). https://doi.org/10.1109/EECCIS49483.2020.9263432
3. Medin, H.F., Vieyra, A.N., Robles, K.A.A., Lagunas, J.J.R., Jacuinde, J.M.C., Vaca, H.Z.: How to improve web application development using a layered reference model and quality indicators. J. Eng. Res. **3**, 2–16 (2023). https://doi.org/10.22533/at.ed.3173102330032
4. Fariz, M., Lazuardy, S., Anggraini, D.: Modern Front End Web Architectures with React.Js and Next.Js. Int. Res. J. Adv. Eng. Sci. **7**, 132–141 (2022)
5. Pano, A., Graziotin, D., Abrahamsson, P.: Factors and actors leading to the adoption of a JavaScript framework. Empir. Softw. Eng.. Softw. Eng. **23**, 3503–3534 (2018). https://doi.org/10.1007/s10664-018-9613-x
6. Bielak, K., Borek, B., Plechawska-Wójcik, M.: Web application performance analysis using angular, react and vue frameworks. J. Comput. Sci. Instit., 77–83 (2021)
7. Kornienko, D.V., Mishina, S.V., Melnikov, M.O.: The single page application architecture when developing secure web services. J. Phys. Conf. Ser. IOP Publishing Ltd (2021). https://doi.org/10.1088/1742-6596/2091/1/012065
8. Negara, B., Eko Indrajit, R., Dazky, E.: Frontend framework consideration for IT developer (2022)
9. Amira, A., Maghawry, H.A., Badr, N.: Automation of performance testing: a review. Int. J. Intell. Comput. Inf. Sci. **22**, 35–50 (2022). https://doi.org/10.21608/ijicis.2022.161846.1219.
10. Maidelyn, P., Marin Diaz, A., Trujillo Casañola, Y., Buedo Hidalgo, D., Llopiz, R.P.: Quality activities for performance efficiency from early software stages. Revista Cubana de Ciencias Informáticas. **15**, 281–296 (2021)
11. Vyas, R.: Comparative analysis on front-end frameworks for web applications. Int. J. Res. Appl. Sci. Eng. Technol. **10**, 298–307 (2022). https://doi.org/10.22214/ijraset.2022.45260.
12. Darbord, G., Verhaeghe, B., Etien, A., Anquetil, N.: Migrating the communication protocol of client-server applications (2022)
13. Pereira, D.S., Bezerra, L.F.V, Nunes, J.S., Filho, I.M.B., Lopes, F.A.S.: Performance efficiency evaluation based on ISO/IEC 25010:2011 applied to a case study on load balance and resilient (2023). https://doi.org/10.5753/wtf.2023.787
14. Ollila, R., Mäkitalo, N., Mikkonen, T.: Rendering performance. J. Web Eng. **21**, 789–814 (2022). https://doi.org/10.13052/jwe15

Prediction Models of Oral Diseases: A Systematic Review of the Literature

Mayra Alejandra Dávila Olivos and Félix Melchor Santos López

Abstract Oral diseases impose a significant burden on many countries, affecting individuals throughout their lives and causing pain, disfigurement, and even death. These diseases share similar risk factors with other important non-communicable diseases. In high-income countries, dental treatment accounts for 5% of healthcare expenditures and 20% of patient expenses. Unfortunately, low- and middle-income countries often struggle to afford preventive and treatment services for oral health disorders. Prediction models are crucial in optimizing resource allocation, particularly in environments where advanced sensing technologies foster healthier living conditions. In this context, this study conducts a systematic review to explore the applications and potential of artificial intelligence in addressing these challenges. This study aims to identify the solutions employed and the performance metrics used to assess their impact on public health, ultimately striving for improved outcomes.

Keywords Artificial intelligence · Oral disease · Model · Prediction · Public health

1 Introduction

Oral diseases have been documented as far back as 2500 BC. Unfortunately, these conditions are irreversible; they cannot self-correct once they begin. Only specific oral tissues have the capacity for autonomous recovery [1]. Enhancing our ability to predict, detect, and diagnose various oral cavity conditions can significantly increase survival rates and improve the overall well-being of individuals at the highest risk. One crucial initial defense is orofacial self-examination or noninvasive examinations and interventions conducted at healthcare centers by specialized personnel [2, 3]. Emerging technologies like artificial intelligence (AI), machine learning (ML), deep learning (DL), and big data, collectively referred to as the "fourth industrial

M. A. Dávila Olivos (✉) · F. M. Santos López
Faculty of System Engineering and Informatics, Universidad Nacional Mayor de San Marcos, Lima, Perú
e-mail: 12200133@unmsm.edu.pe

G. F. Olmedo Cifuentes et al. (eds.), *Emerging Research in Intelligent Systems*, Lecture Notes in Networks and Systems 902, https://doi.org/10.1007/978-3-031-52255-0_22

revolution", span various disciplines and hold immense promise. Models rooted in ML and DL techniques can predict and make decisions that often outperform traditional or alternative solutions [4]. Developing efficient systems that enhance public health under favorable conditions for holistic well-being is imperative. This concern is relevant to both developed and developing nations. This paper investigates the application of AI in predicting oral diseases and improving public health. The primary objective is to conduct a comprehensive systematic review of the literature (RSL) on AI models aimed at developing intelligent systems for enhancing oral health. This study performs an in-depth analysis of oral disease prediction systems, following the Preferred Reporting Items for Systematic Reviews and Meta-Analyses (PRISMA) methodology. In addition, this work considers the characteristics used to evaluate their performance. The structure of this review article is as follows: Sect. 2 outlines the method, including the search strategy, inclusion and exclusion criteria, research questions (RQs), and the study selection process. Section 3 presents the results and discussions, addressing the RQs. Finally, Sect. 4 provides the conclusion of the investigation.

2 Materials and Methods

The systematic review in this study is conducted following the PRISMA methodology, which comprises a set of evidence-based elements designed to facilitate the presentation of systematic review and meta-analysis reports. PRISMA is also a valuable tool for critically evaluating published systematic reviews. The process can be broken down into well-defined steps, outlined in the following sections.

2.1 Search Strategy

To address the search criteria and keywords, the authors employed four databases: Scopus, IEEE, Web of Science, and PubMed. The search for relevant publications commenced on October 18, 2021, and a filter was applied to include studies from 2016 onwards. Two search queries are utilized, each with the following keyword combinations: "(diagnosis OR detection OR prediction) AND (oral diseases) AND (deep learning)" and "(diagnosis OR detection OR prediction) AND (oral diseases) AND (machine learning)". In total, the first search string yielded 172 scientific articles. These are distributed as follows: 55 from Scopus, 13 from IEEE, 64 from Web of Science, and 40 from PubMed. For the second search string, a total of 341 articles are identified. The breakdown by database is as follows: 113 from Scopus, 16 from IEEE, 118 from Web of Science, and 94 from PubMed.

2.2 *Inclusion and Exclusion Criteria*

The authors thoroughly assessed all the articles from the search and selected those that met the study's criteria, organizing them into personal folders. These documents are carefully reviewed to identify various methods related to the application of AI models for diagnosing oral diseases. Inclusion in this review is contingent upon fulfilling specific eligibility criteria within the research articles. Inclusion criteria: (1) The presence of "Diagnosis/Detection/Prediction of oral diseases" in the title and abstract; (2) At least one ML and DL algorithm included in the study; (3) Availability of data related to oral diseases for analysis; (4) Inclusion of images depicting oral diseases for study; (5) Presentation of at least one diagnostic/detection/prediction metric within the study; (6) The article falls within the first and second quartiles in SJR (SCImago Journal Rank) ranking. Exclusion criteria: (1) Duplicate documents; (2) Publications categorized as general or secondary studies, such as review articles or demonstration articles; (3) Articles lacking explicit information regarding the prediction of oral diseases and their performance indicators.

2.3 *Research Questions*

The number of individuals at risk of developing oral diseases is on the rise, drawing attention from government agencies and healthcare centers. Leveraging disruptive technologies, such as AI, holds significant promise in addressing the challenges within the dental field. The authors of this systematic review have identified key research questions (RQs) and embarked on this comprehensive study to uncover pertinent insights. The RQs guiding this systematic review are as follows:

1. (RQ1) What models are available for predicting oral diseases?
2. (RQ2) What metrics are employed to assess the performance of models predicting oral diseases?
3. (RQ3) Which oral diseases can be predicted using existing models?
4. (RQ4) What types of data/images are utilized in models predicting oral diseases?

The authors formulated these questions to present a systematic review of AI models to predict oral diseases. Additionally, this article aims to deliver a comprehensive overview of the preliminary evaluation measures employed, the precision achieved, the scope of oral diseases studied, and elucidate the challenges and prospects for future research in this field.

2.4 Selection of Studies

The documents obtained by applying the search queries are systematically analyzed following the PRISMA guidelines. Initially, the publications are scrutinized to identify and eliminate duplicate investigations, excluding 227 articles. Subsequently, the remaining 286 articles underwent a second round of evaluation. The assessment of article importance involved considering their titles, abstracts, and content. During this phase, 222 papers were excluded due to failing to meet the predetermined inclusion and exclusion criteria. These articles primarily focused on various aspects of the health of individuals with oral diseases, exploring connections with different disease types and investigating models and methods for prediction, diagnosis, detection, classification, evaluation, segmentation, comparison, and identification. These approaches either incorporated or did not incorporate AI models and are applied in diverse areas of dentistry. Following the specified eligibility criteria, the authors retained 64 articles for a more in-depth examination in the third stage. Within this list, five articles either lacked specific AI prediction algorithms or consisted solely of literature reviews within the field of oral pathology. One article exclusively provided an overview of the broader landscape of ML in orthodontics without implementing a diagnostic system. Some researchers failed to define the parameters for evaluating model predictions.

In contrast, others had valuable content but were published in journals lacking an adequate impact index or did not fit into any category. Consequently, these studies did not meet the inclusion criteria and were excluded. In the end, this systematic review meticulously analyzed 42 pivotal articles addressing the RQs identified in the preceding section. Figure 1 illustrates the selection process following the PRISMA guidelines.

3 Results and Discussions

This systematic review, sourced from three databases, encompasses 42 articles concerning diagnostic models of oral diseases utilizing AI. Among these, nine articles (21.42%) originated from Scopus, 18 (42.85%) from Web of Science, and 15 (35.71%) were retrieved from PubMed. It is important to note that these databases are exclusively the result of applying the guidelines outlined in the PRISMA methodology for the RSL.

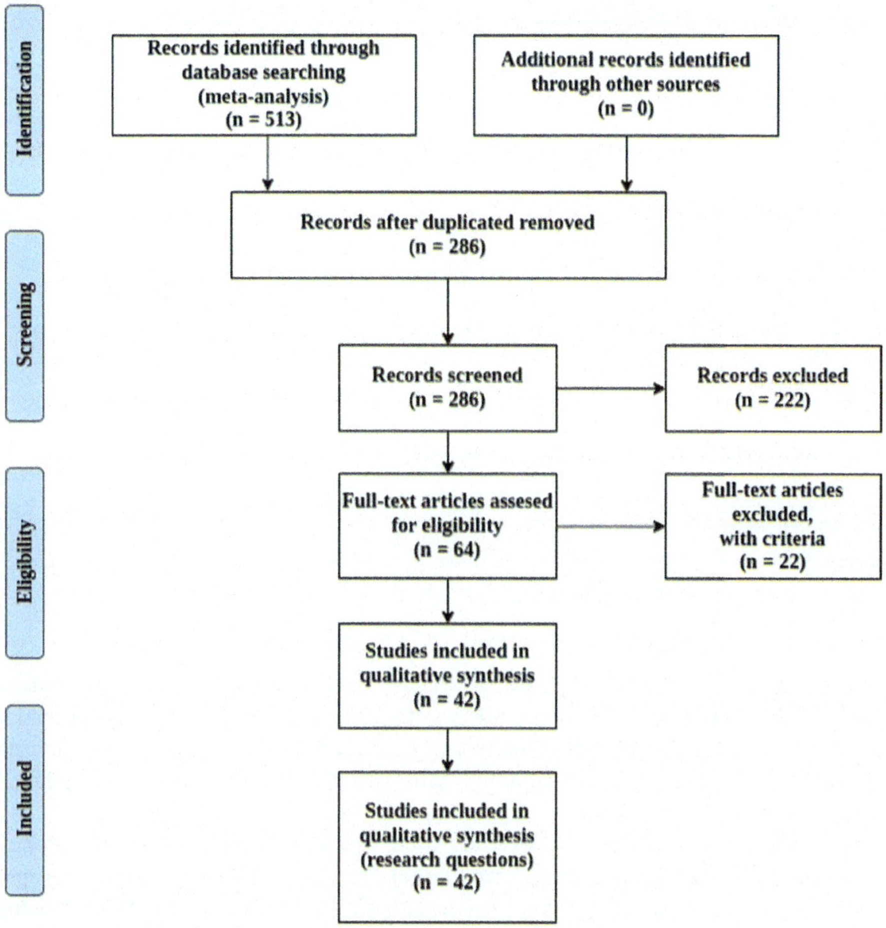

Fig. 1 PRISMA flowchart illustrating the selection of studies included in this systematic review

3.1 Response to RQ1

The first RQ pertains to the models utilized for diagnosing oral diseases, and the results are summarized in Table 1. Most studies employ prediction methods developed by the authors or commercially available. Notably, 11.90% ($N = 5$) of the analyzed articles [5]-[6], which are systematic reviews, highlight the utilization of convolutional neural networks (CNN) in their research endeavors. These publications analyze the application of CNNs across different dental disciplines, underscoring their significant relevance to oral health-related issues. Regarding the most frequently employed AI models in oral disease prediction systems, most researchers concentrated on different neural network architectures. Specifically, 34 studies (80.95%) harnessed meth-ods based on CNNs. Additionally, six studies focused on support

Table 1 Models used in different studies

Model	Reference
Convolutional neural network	[6, 8–10, 15–44]
Support vector machine	[6–11]
Random forest	[5, 11]
Decision forest	[12]
Decision tree	[5]
Extreme gradient boosting trees	[13]
Deep artificial neural networks	[5, 6, 14]
Hybrid graph-cut technique	[15]
Fuzzy inference	[5]
Fuzzy neural network	[8]
Optimal long and short-term memory with dragonfly optimization algorithm	[16]
Model actively	[45]
Single shot multiple box detector	[44]
ASDL-TCI	[46]
TensorFlow lite (RMSProp optimization algorithm)	[37]

vector machines [6–11], random forests [5, 11], decision forests [12], decision trees [5], and extreme gradient augmentation trees [13]. Furthermore, three studies [5, 6, 14] explored artificial neural networks, while one study employed the hybrid graph-cut technique [15] in conjunction with two-dimensional radiographic dental images. Fuzzy logic, such as fuzzy inference [5] and fuzzy neural networks [8], appeared in certain studies, showcasing their potential in health problem prediction. Additionally, long and short-term memory emerged as a viable solution, especially when combined with CNNs, to extract characteristics and facilitate diagnosing dental caries in periapical dental images [16]. CNNs predominate among other AI models due to their superior performance when processing image inputs. Furthermore, their layered architecture enables CNNs to handle increasing complexity by identifying larger image sections. CNNs excel in image recognition and artificial vision, providing valuable information to enhance decision-making processes in oral health diagnostics.

3.2 Response to RQ2

Regarding the metrics used to evaluate prediction models, researchers focused on conventional indicators commonly employed to assess prediction method performance, as detailed in Table 2. Notably, 52.38% ($N = 22$) of the analyzed studies utilized the area under the curve (AUC) metric, while 59.52% ($N = 25$) employed the specificity metric. Additionally, 64.28% ($N = 27$) of studies relied on sensitivity

Table 4 Oral diseases diagnosed by existing models

Oral disease	References
Oral cancer	[5–8, 11, 12, 17, 19, 20, 24–27, 29–31, 33, 34, 37, 39, 40, 42, 45]
Decalcification, dental plaque, and deformities	[6, 8, 15, 28, 31, 43, 45]
Cavities	[6, 9, 10, 14, 16, 17, 23, 28, 31, 44]
Oral dysplasia, Sjögren's syndrome, oral leukoplakia, edentulism, lichen planus, and solar cheilosis	[5, 8, 13, 21, 31, 38, 45]
Periodontitis	[6, 22, 23, 31]
Periapical pathology, lesions and pathologies	[6, 17, 31, 35, 36, 45]
Gingivitis	[18, 27]
Root fractures	[6, 31, 45]
Oral/mouth injuries	[8, 41]
Other oral diseases	[46]

dysplasia, while two analyzed Sjögren's syndrome and oral leukoplakia. Additionally, three articles explored edentulism, lichen planus, and solar cheilosis. These studies collectively account for 16.67% of the selected articles. It is worth noting that oral cancer receives significant attention in research due to its higher incidence in adults, its association with factors like tobacco and alcohol consumption, and human papillomavirus infection. Moreover, its rapid progression underscores the importance of early detection.

3.4 Response to RQ4

The fourth RQ pertains to the data and/or images employed in developing predictive models. Accordingly, the studies can be categorized into four major groups: (1) Studies utilizing mucosa/cavity oral imaging; (2) Investigations relying on panoramic radiographs; (3) Studies dependent on clinicopathological data; and (4) Publications utilizing histopathological data. The results of this categorization are presented in Tables 5 and 6. Notably, 45.23% ($N = 19$) of the analyzed articles incorporate images of the entire oral cavity or specific areas, including the tongue, maxillary and mandibular teeth, and lips. Clinicopathological data encompass personal data [3], clinical and pathological data [1, 3, 6, 33, 35], injury conditions [3], treatment details [1], chronic medical conditions [15], and dental records [34]. This category represents 16.66% ($N = 7$) of the publications examined. Images of the oral cavity are the most frequently utilized data type compared to other datasets. This preference stems from the enhanced precision of CNNs in detecting anomalies, surpassing the human eye's capabilities through image recognition with artificial vision, as explored in the third research question. It is imperative to conduct a meticulous collection and analysis of relevant datasets to enable the development of effective and efficient oral disease prediction systems, thereby ensuring improved solution performance.

Table 5 Images used with existing prediction models

Images	Reference
Mucosa/cavity oral	[6, 8, 17, 18, 21, 25–28, 32, 33, 37, 39–41, 43–46]
Panoramic radiographs	[6, 9, 15, 20, 22, 29, 31, 45]
Periapical radiographs	[6, 16, 23, 31, 35, 45]
Cone beam computed tomography scans	[6, 17, 30, 31, 34, 36, 45]
Computed tomography scans	[6, 17, 19, 31, 42, 45]
Histopathological	[24]
Autofluorescent and white light/ immunofluorescent	[8, 38]
Contrast enhanced computed tomography scans	[10, 31]
Face scan	[6, 10, 17]
Other images	[7]

Table 6 Data used with existing prediction models

Data	Reference
Clinicopathological	[5–7, 11, 12, 17]
Laboratory (Histopathological)	[6, 7, 10]
Demographic and social	[7, 10, 14]
Risk habits	[7]
Of specters	[7]
Nutritional	[10, 14]
Lifestyle/socioeconomic	[5, 8, 10]
Oral exam	[5, 7, 10]
Autofluorescence	[7]
Other data	[7, 26, 31]

4 Conclusion

This study conducted an RSL focusing on prediction models of oral diseases based on AI methods. The review encompassed the research and analysis of scientific articles published in Scopus, IEEE, Web of Science, and PubMed databases. The most significant papers were reviewed following predefined RQs and eligibility criteria, adhering to the PRISMA methodology guidelines. This approach effectively underscored AI's potential in addressing oral health-related issues. Researchers demonstrated a keen interest in investigating various oral diseases. Among these, oral cancer emerged as the most extensively studied. Most studies incorporated data and images from the mucosa/oral cavity to construct the datasets used for implementing their models. The reviewed literature highlights the potential of ML and DL, mainly through the utilization of CNNs. These models were evaluated using performance metrics such as AUC, sensitivity, specificity, accuracy, precision, recall, and F1 score to enhance oral health. While access to the datasets remains restricted, requiring formal requests to the researchers or research centers, this study provides valuable insights into crucial opportunities within the dental field and public health management. Moreover, future research should ensure that the developed systems practically apply in real-time scenarios.

References

1. Chitnis, G., Bhanushali, V., Ranade, A., Khadase, T., Pelagade, V., Chavan, J.: A review of machine learning methodologies for dental disease detection. In: 2020 IEEE India Council International Subsections Conference (INDISCON), pp. 63–65. IEEE (2020)
2. Anantharaman, R., Anantharaman, V., Lee, Y.: Oro vision: Deep learning for classifying orofacial diseases. In: 2017 IEEE International Conference on Healthcare Informatics (ICHI), pp. 39–45. IEEE (2017)
3. Ehtesham, H., Safdari, R., Mansourian, A., Tahmasebian, S., Mohammadzadeh, N., Ghazisaeedi, M., Bashiri, A.: Clinical decision support system, a potential solution for diagnostic accuracy improvement in oral squamous cell carcinoma: a systematic review. J. Oral Health Oral Epidemiol. **6**(4), 187–195 (2017)
4. Sun, M.L., Liu, Y., Liu, G., Cui, D., Heidari, A.A., Jia, W.Y., Ji, X., Chen, H., Luo, Y.: Application of machine learning to stomatology: a comprehensive review. IEEE Access **8**, 184360–184374 (2020)
5. Adeoye, J., Tan, J.Y., Choi, S.W., Thomson, P.: Prediction models applying machine learning to oral cavity cancer outcomes: a systematic review. Int. J. Med. Inf. **154**, 104557 (2021)
6. Reyes, L.T., Knorst, J.K., Ortiz, F.R., Ardenghi, T.M.: Scope and challenges of machine learning-based diagnosis and prognosis in clinical dentistry: A literature review. J. Clin. Transl. Res. **7**(4), 523 (2021)
7. Alabi, R.O., Youssef, O., Pirinen, M., Elmusrati, M., Makitie, A.A., Leivo, I., Al-mangush, A.: Machine learning in oral squamous cell carcinoma: current status, clinical concerns and prospects for future—a systematic review. Artif. Intell. Med. **115**, 102060 (2021)
8. García-Pola, M., Pons-Fuster, E., Suárez-Fernández, C., Seoane-Romero, J., Romero-Méndez, A., Lopez-Jornet, P.: Role of artificial intelligence in the early diagnosis of oral cancer. a scoping review. Cancers **13**(18), 4600 (2021)

9. Bui, T.H., Hamamoto, K., Paing, M.P.: Deep fusion feature extraction for caries detection on dental panoramic radiographs. Appl. Sci. **11**(5), 2005 (2021)
10. Hung, M., Voss, M.W., Rosales, M.N., Li, W., Su, W., Xu, J., Bounsanga, J., Ruiz-Negron, B., Lauren, E., Licari, F.W.: Application of machine learning for diagnostic prediction of root caries. Gerodontology **36**(4), 395–404 (2019)
11. Shan, J., Jiang, R., Chen, X., Zhong, Y., Zhang, W., Xie, L., Cheng, J., Jiang, H.: Machine learning predicts lymph node metastasis in early-stage oral tongue squamous cell carcinoma. J. Oral Maxillofac. Surg.Maxillofac. Surg. **78**(12), 2208–2218 (2020)
12. Bur, A.M., Holcomb, A., Goodwin, S., Woodroof, J., Karadaghy, O., Shnayder, Y., Kakarala, K., Brant, J., Shew, M.: Machine learning to predict occult nodal metastasis in early oral squamous cell carcinoma. Oral Oncol. **92**, 20–25 (2019)
13. Elani, H.W., Batista, A.F., Thomson, W.M., Kawachi, I., Chiavegatto Filho, A.D.: Predictors of tooth loss: a machine learning approach. PLoS ONE **16**(6), e0252873 (2021)
14. Zanella-Calzada, L.A., Galvan-Tejada, C.E., Ch'avez-Lamas, N.M., Rivas-Gutierrez, J., Magallanes-Quintanar, R., Celaya-Padilla, J.M., Galv'an-Tejada, J.I., Gamboa-Rosales, H.: Deep artificial neural networks for the diagnostic of caries using socioeconomic and nutritional features as determinants: data from NHANES 2013–2014. Bioengineering **5**(2), 47 (2018)
15. Al Kheraif, A.A., Wahba, A.A., Fouad, H.: Detection of dental diseases from radiographic 2D dental image using hybrid graph-cut technique and convolutional neural network. Measurement **146**, 333–342 (2019)
16. Singh, P., Sehgal, P.: GV black dental caries classification and preparation technique using optimal CNN-LSTM classifier. Multimed. Tools Appl. **80**, 5255–5272 (2021)
17. Ren, R., Luo, H., Su, C., Yao, Y., Liao, W.: Machine learning in dental, oral and craniofacial imaging: a review of recent progress. PeerJ **9**, e11451 (2021)
18. Alalharith, D.M., Alharthi, H.M., Alghamdi, W.M., Alsenbel, Y.M., Aslam, N., Khan, I.U., Shahin, S.Y., Diani˘skov'a, S., Alhareky, M.S., Barouch, K.K.: A deep learning-based approach for the detection of early signs of gingivitis in orthodontic patients using faster region-based convolutional neural networks. Int. J. Environ. Res. Public Health **17**(22), 8447 (2020)
19. Ariji, Y., Fukuda, M., Kise, Y., Nozawa, M., Yanashita, Y., Fujita, H., Katsumata, A., Ariji, E.: Contrast-enhanced computed tomography image assessment of cervical lymph node metastasis in patients with oral cancer by using a deep learning system of artificial intelligence. Oral Surg Oral Med Oral Pathol Oral RadiolPathol. Oral Radiol. **127**(5), 458–463 (2019)
20. Ariji, Y., Yanashita, Y., Kutsuna, S., Muramatsu, C., Fukuda, M., Kise, Y., Nozawa, M., Kuwada, C., Fujita, H., Katsumata, A., et al.: Automatic detection and classification of radiolucent lesions in the mandible on panoramic radiographs using a deep learning object detection technique. Oral Surg Oral Med Oral Pathol Oral RadiolPathol. Oral Radiol. **128**(4), 424–430 (2019)
21. Camalan, S., Mahmood, H., Binol, H., Araujo, A.L.D., Santos-Silva, A.R., Vargas, P.A., Lopes, M.A., Khurram, S.A., Gurcan, M.N.: Convolutional neural network- based clinical predictors of oral dysplasia: class activation map analysis of deep learning results. Cancers **13**(6), 1291 (2021)
22. Chang, H.J., Lee, S.J., Yong, T.H., Shin, N.Y., Jang, B.G., Kim, J.E., Huh, K.H., Lee, S.S., Heo, M.S., Choi, S.C., et al.: Deep learning hybrid method to automatically diagnose periodontal bone loss and stage periodontitis. Sci. Rep. **10**(1), 7531 (2020)
23. Chen, H., Li, H., Zhao, Y., Zhao, J., Wang, Y.: Dental disease detection on periapical radiographs based on deep convolutional neural networks. Int. J. Comput. Assist. Radiol. Surg.Comput. Assist. Radiol. Surg. **16**, 649–661 (2021)
24. Das, N., Hussain, E., Mahanta, L.B.: Automated classification of cells into multiple classes in epithelial tissue of oral squamous cell carcinoma using transfer learning and convolutional neural network. Neural Netw.Netw. **128**, 47–60 (2020)
25. Fu, Q., Chen, Y., Li, Z., Jing, Q., Hu, C., Liu, H., Bao, J., Hong, Y., Shi, T., Li, K., et al.: A deep learning algorithm for detection of oral cavity squamous cell carcinoma from photographic images: a retrospective study. EClinicalMedicine **27** (2020)
26. Guo, J., Wang, H., Xue, X., Li, M., Ma, Z.: Real-time classification on oral ulcer images with residual network and image enhancement. IET Image Proc. **16**(3), 641–646 (2022)

27. Jubair, F., Al-karadsheh, O., Malamos, D., Al Mahdi, S., Saad, Y., Hassona, Y.: A novel lightweight deep convolutional neural network for early detection of oral cancer. Oral Dis. **28**(4), 1123–1130 (2022)
28. Kim, D., Choi, J., Ahn, S., Park, E.: A smart home dental care system: integration of deep learning, image sensors, and mobile controller. J. Ambient Intell. Humaniz. Comput., 1–9 (2021)
29. Kwon, O., Yong, T.H., Kang, S.R., Kim, J.E., Huh, K.H., Heo, M.S., Lee, S.S., Choi, S.C., Yi, W.J.: Automatic diagnosis for cysts and tumors of both jaws on panoramic radiographs using a deep convolution neural network. Dentomaxillofacial Radiol. **49**(8), 20200185 (2020)
30. Lee, J.H., Kim, D.H., Jeong, S.N.: Diagnosis of cystic lesions using panoramic and cone beam computed tomographic images based on deep learning neural network. Oral Dis. **26**(1), 152–158 (2020)
31. Leite, A.F., Vasconcelos, K.F., Willems, H., Jacobs, R.: Radiomics and machine learning in oral healthcare. Proteomics–Clin. Appl. **14**(3), 1900040 (2020)
32. Li, W., Liang, Y., Zhang, X., Liu, C., He, L., Miao, L., Sun, W.: A deep learning approach to automatic gingivitis screening based on classification and localization in RGB photos. Sci. Rep. **11**(1), 16831 (2021)
33. Lin, H., Chen, H., Weng, L., Shao, J., Lin, J.: Automatic detection of oral cancer in smartphone-based images using deep learning for early diagnosis. J. Biomed. Opt. **26**(8), 086007–086007 (2021)
34. Lin, Y., He, M.: Deep learning-based three-dimensional oral conical beam computed tomography for diagnosis. J. Healthc. Eng. 2021 (2021)
35. Moidu, N.P., Sharma, S., Chawla, A., Kumar, V., Logani, A.: Deep learning for categorization of endodontic lesion based on radiographic periapical index scoring system. Clin. Oral Invest. **26**(1), 651–658 (2022)
36. Orhan, K., Bayrakdar, I., Ezhov, M., Kravtsov, A., Özyürek, T.: Evaluation of artificial intelligence for detecting periapical pathosis on cone-beam computed tomography scans. Int. Endod. J. **53**(5), 680–689 (2020)
37. Song, B., Sunny, S., Li, S., Gurushanth, K., Mendonca, P., Mukhia, N., Patrick, S., Gurudath, S., Raghavan, S., Imchen, T., et al.: Mobile-based oral cancer classification for point-of-care screening. J. Biomed. Opt. **26**(6), 065003–065003 (2021)
38. Song, B., Sunny, S., Uthoff, R.D., Patrick, S., Suresh, A., Kolur, T., Keerthi, G., Anbarani, A., Wilder-Smith, P., Kuriakose, M.A., et al.: Automatic classification of dual-modalilty, smartphone-based oral dysplasia and malignancy images using deep learning. Biomed. Opt. Express **9**(11), 5318–5329 (2018)
39. Tanriver, G., SolukTekkesin, M., Ergen, O.: Automated detection and classification of oral lesions using deep learning to detect oral potentially malignant disorders. Cancers **13**(11), 2766 (2021)
40. Warin, K., Limprasert, W., Suebnukarn, S., Jinaporntham, S., Jantana, P.: Automatic classification and detection of oral cancer in photographic images using deep learning algorithms. J. Oral Pathol. Med.Pathol. Med. **50**(9), 911–918 (2021)
41. Welikala, R.A., Remagnino, P., Lim, J.H., Chan, C.S., Rajendran, S., Kallarakkal, T.G., Zain, R.B., Jayasinghe, R.D., Rimal, J., Kerr, A.R., et al.: Automated detection and classification of oral lesions using deep learning for early detection of oral cancer. IEEE Access **8**, 132677–132693 (2020)
42. Xu, S., Liu, C., Zong, Y., Chen, S., Lu, Y., Yang, L., Ng, E.Y., Wang, Y., Wang, Y., Liu, Y., et al.: An early diagnosis of oral cancer based on three-dimensional convolutional neural networks. IEEE Access **7**, 158603–158611 (2019)
43. You, W., Hao, A., Li, S., Wang, Y., Xia, B.: Deep learning-based dental plaque detection on primary teeth: a comparison with clinical assessments. BMC Oral Health **20**, 1–7 (2020)
44. Zhang, X., Liang, Y., Li, W., Liu, C., Gu, D., Sun, W., Miao, L.: Development and evaluation of deep learning for screening dental caries from oral photographs. Oral Dis. **28**(1), 173–181 (2022)

45. Hung, K., Yeung, A.W.K., Tanaka, R., Bornstein, M.M.: Current applications, opportunities, and limitations of AI for 3D imaging in dental research and practice. Int. J. Environ. Res. Public Health **17**(12), 4424 (2020)
46. Mansour, R.F., Althobaiti, M.M., Ashour, A.A.: Internet of things and synergic deep learning based biomedical tongue color image analysis for disease diagnosis and classification. IEEE Access **9**, 94769–94779 (2021)

How Can Personal Knowledge Graphs Contribute to Precision Nutrition?

Mauricio Espinoza-Mejía, Víctor Saquicela, and Victoria Abril-Ulloa

Abstract Nutrition is undoubtedly a significant factor in disease prevention, and keeping track of it is incredibly important for our overall health. One of the latest technologies that has emerged to help healthcare professionals and users make informed decisions is the Knowledge Graph (KG). These models have been used to implement various knowledge-based systems related to nutrition, taking advantage of the vast amount of information about foods, nutrients, dietary recommendations, and healthy eating patterns available in multiple media. However, a KG needs to incorporate individual information, represented by a personal knowledge graph (PKG), to facilitate reasoning and decision-making about an individual's diet or nutrition plan. This article introduces the concept of a Personal Nutritional Knowledge Graph (PNKG), a structured information resource containing data related to an individual's specific nutritional aspects. This study analyzes the concepts and requirements and proposes a generic architecture to assist developers in building adaptive personalized nutrition advice systems.

Keywords Intelligent systems · Personal Knowledge Graph · Precision Nutrition

1 Introduction

Nutrition, food, and diet prevent disease and maintain good health [1]. According to the World Health Organization (WHO), better nutrition is associated with improved child health, more robust immune systems, safer pregnancy and childbirth, reduced risk of noncommunicable diseases (such as diabetes and cardiovascular disease), and increased longevity [2].

M. Espinoza-Mejía (✉) · V. Saquicela
Computer Science Department, University of Cuenca, Cuenca, Ecuador
e-mail: mauricio.espinoza@ucuenca.edu.ec

V. Abril-Ulloa
Technologies Applied to Health Research Group, Faculty of Medical Sciences, University of Cuenca, Cuenca, Ecuador

G. F. Olmedo Cifuentes et al. (eds.), *Emerging Research in Intelligent Systems*, Lecture Notes in Networks and Systems 902, https://doi.org/10.1007/978-3-031-52255-0_23

Despite constant efforts to innovate and incorporate digital technology to improve patients' diagnosis, prognosis, and treatment, the healthcare domain still faces numerous challenges due to manual processes that slow the work of doctors and healthcare professionals [3]. Unfortunately, nutrition is not immune to these problems, as tasks such as assessing nutritional status, correlating nutrient data, reviewing personal/family history, suggesting personalized meal plans, or searching for patterns in a patient's eating habits can be incredibly time-consuming and require valuable resources.

KGs are one of the emerging technologies that are helping to improve medical care in nutrition. A KG encodes structured information about entities and their relationships by capturing information retrieved from various resources [4]. KGs can enhance the efficiency of information retrieval and facilitate identifying and applying patterns in data to improve clinical outcomes [5]. In addition to the potential benefits in healthcare, KGs can be a valuable tool for those seeking a healthy lifestyle, as these tools could integrate and relate scattered Web information about nutrients, food groups, recipes, dietary recommendations, and guidelines for healthy eating. This way, the barriers of fragmented and disorganized information access can be overcome, allowing for better information management and more informed decision-making. Without these knowledge models, it would be challenging to recognize how food affects their health. Understanding this relationship, however, has significant benefits for consumers, such as identifying which foods can improve their chances of preventing heart disease or diabetes, the world's leading causes of death.

In recent years, general knowledge graphs have been widely used in the field of nutrition in a variety of applications, including the development of new recipes, food question-answering systems, the discovery of links between diet and disease, the visual analysis of food, dietary recommendations, food supply chain management, and the design of intelligent food machines. A comprehensive description of the KGs' contribution to these applications is presented in [6].

Although KGs in the nutrition domain are well established and widely used to represent relevant entities or topics in fields such as recipes, nutrients and health, food safety, and general knowledge about food, they do not directly contain personal nutrition information. They are generally not designed to incorporate an individual's nutrition data. However, as argued in several recent papers, collecting and analyzing personal information in a PKG is critical to capturing the nuances of an individual's health and addressing the need for personalized care for an aging global population [5, 7–9].

PKGs, proposed by [10], fill this gap by providing structured information resources about entities related to specific aspects of an individual, their attributes, and the relationships between them.

This article analyzes the current situation of PKGs as tools that can improve the management and organization of information related to an individual's nutrition. It is argued that these personal nutritional knowledge models, along with other advances in Artificial Intelligence (AI), are crucial to strengthen one of the trends gaining the most strength in food and health, precision nutrition or personalized nutrition (PN).

The remainder of this article is organized as follows. Section 2 describes the characteristics of PN and its relationship to PKGs. Section 3 defines the concept of a PKG specifically designed to support personalized nutrition. Section 4 describes a generic architecture that guides the design of different adaptive personalized nutrition advice (APNA) systems. Section 5 describes the scenario in which the generic architecture is tested. Finally, Sect. 6 presents the conclusions of this work.

2 Precision Nutrition

In recent decades, there has been a shift in the approach to nutrition. In the 1980s, evidence-based nutrition became popular, suggesting that what worked for a large group of people would work for everyone. However, genetics has shown that each individual is unique and responds differently to treatments and diets. A new trend in food and health has emerged from this premise: precision nutrition or personalized nutrition. This trend argues that diets and treatments should be tailored to the specific characteristics of each person based on individual variables that include genetics, microbiome, metabolic profile, health status, physical activity, dietary patterns, food environment, and socioeconomic and psychosocial characteristics [11]. Rather than following a flashy diet that has worked for others, precision nutrition focuses on gathering data to assess health and tailor dietary recommendations to individual needs.

Precision nutrition research is growing exponentially with advances in various techniques and technologies. This has significant global implications for future research priorities, commercialization of products and services, and implementation in health services and public health policy [12]. This paper aims to analyze how PKGs can comprehensively and structurally represent all information related to an individual's integral nutritional status, aiming to develop personalized solutions that promote healthy lifestyles.

Nutritionists consider factors such as body mass index, dietary habits, and a patient's personal and family health history. However, by combining this information with genetic knowledge and other relevant data, innovative nutritional strategies can be developed to treat obesity, diabetes, hypertension, and other diet-related diseases. It is critical to note that generating an individual's nutritional information is not a task that a single individual can accomplish. It requires the collaborative efforts of researchers, laboratory technicians, nutritionists, clinicians, and the individual.

Through nutrigenetic and nutrigenomic studies, researchers are identifying genetic variants and bioactive compounds that influence dietary response. Nutrigenetic tests developed by laboratory technicians identify the population subgroup to which each person belongs. Then, nutritionists and clinicians collect patient data and interpret nutrigenetic reports to provide personalized nutritional intervention. Finally, people collect data to set goals and objectives for a healthier life.

The heterogeneity of information needed to implement a personalized nutrition strategy is evident. The information generated by different stakeholders is diverse in terms of data types, sources, and formats, making integration and analysis difficult. A PKG can address this heterogeneity by creating a unified data structure that integrates information from different sources and forms into a single entity. In addition, by establishing semantic relationships between other graph nodes, it is possible to gain new insights and make inferences that can improve dietary personalization.

The food industry has the potential to use the nutritional data gathered through a PKG to introduce products tailored to groups of individuals with similar metabolisms, following the trend of group precision nutrition [13]. While creating fully personalized products and services tailored to each customer's unique needs may not be financially viable, some companies already promote this idea for certain products [14].

In addition, biotech companies can use the information stored in a PKG, such as genetic profiles and dietary habits, to develop personalized treatments for specific health conditions.

The growing interest of IT companies in precision nutrition is evident in the wide range of mobile apps and wearables that encourage behavior change and monitor lifestyle. The information collected in PKGs could be precious in developing applications tailored to users' specific needs, emphasizing the importance of personalized nutrition in improving health outcomes.

3 Personal Nutrition Knowledge Graph

Although research on PKGs is still in its early stages, work from several fields, including information retrieval, natural language processing, and the semantic web, has paved the way for their development and future advancement. There is no clear consensus on the meaning of a PKG. Still, the term has been widely used to refer to the structured representation of information about a specific user, mainly in entities personally related to the user.

Based on an extensive analysis of the literature, it can be distinguished that these knowledge domain models have been studied from at least two perspectives:

(i) their ability to manage and organize personal data [10], and (ii) their potential as a knowledge base source to provide personalized services [15]. In addition, in recent years, there has been a growing interest in PKG research, which has moved from extracting personal information to construct or populate them [16–18] to exploring different ways of using this information [7–9, 19, 20]. However, several open questions remain about PKG's definition, construction, use, and practical implementation.

This article defines a PKG specifically designed to support personalized nutrition based on the statements above. This type of PKG is referred to as a Personal Nutrition Knowledge Graph. The information captured by a PNKG is limited to the stakeholders involved in precision nutrition (described in the previous section) rather than encompassing their broader personal sphere. Undoubtedly, the implementation of a PNKG faces several challenges. Firstly, it is necessary to address knowledge management, which includes acquiring personal information, recognizing and linking entities, storing individual nutritional knowledge packages, and connecting with external KGs. Second, it is necessary to consider privacy and user consent to share their data with doctors, laboratory personnel, and other stakeholders to provide more complete information. Finally, the maintenance of acquired knowledge in a PNKG must remain up-to-date, which implies the propagation of changes made in the PNKG to external knowledge graphs and vice versa, the control of possible inconsistencies in the linked knowledge, and the most efficient ways to activate these synchronizations.

Experts agree that social and organizational difficulties, rather than technological issues, are the main challenges for the development and application of PKG [15]. The architecture proposed in this paper demonstrates that, given the current state of AI technology, it is feasible to design applications tailored to collect personal nutritional information.

4 Proposed Approach

The development of PN systems is based on some novel elements and possibilities. To date, the current problem of PN has not been effectively addressed using PNKG. This approach can influence how the nutrition community uses accessible KG and PNKG for their applications. However, there are several obstacles that this research must overcome.

- Cost and Accessibility: PN can be expensive and requires specialized resources such as genetic testing and individualized assessments. As a result, its use may be restricted to privileged groups, limiting its accessibility to the broader population and exacerbating health inequalities.
- Ethical and privacy considerations: Implementing precision nutrition involves handling sensitive genetic and personal data. This raises ethical and privacy concerns about collecting, storing, and using these data. It is necessary to establish appropriate policies and safeguards to protect individuals' privacy and ensure the ethical use of information.
- Challenges in Interpreting and Communicating Results: Interpretation and communication of PN results can be complex. Translating genetic and other personalized data into understandable and practical dietary recommendations for the general population can be challenging. Lack of education and public awareness about PN can also limit its acceptance and adoption.

The authors agree with Renner et al. [21] that two key issues need to be tackled to address the problems above and achieve effective implementation of PN in the population: (i) broadening the current PN objectives to include individual nutritional goal preferences beyond the currently recommended biomedical objectives; and (ii) creating adaptive personalized nutrition advice (APNA) systems that are in line with the needs, abilities, and receptivity of each individual.

This section proposes a generic architecture to guide the development of APNA systems, revealing crucial aspects of structure, user interface, functionality, interaction, and personalization that are considered important in designing these applications. The focus of the architecture is on (i) the collection of individual nutritional data that provides in-situ and timely information for decision-making in natural food environments, (ii) the establishment of goals that lead to behavioral changes through participatory dialog between users and experts, (e.g., a nutritionist or a real or virtual advisor), and (iii) the explanation of the results achieved in a way understandable to the stakeholders.

The AI contributions to the proposed architecture can be summarized in two main aspects: (i) collecting and improving the quality of health, diet, and behavior data using different modalities (e.g., wearables, chatbots, and applications), and (ii) developing algorithms and machine learning techniques to extract relevant information from the collected data.

The architecture comprises User Front-End, Expert Front-End, and Back-End (see Fig. 1).

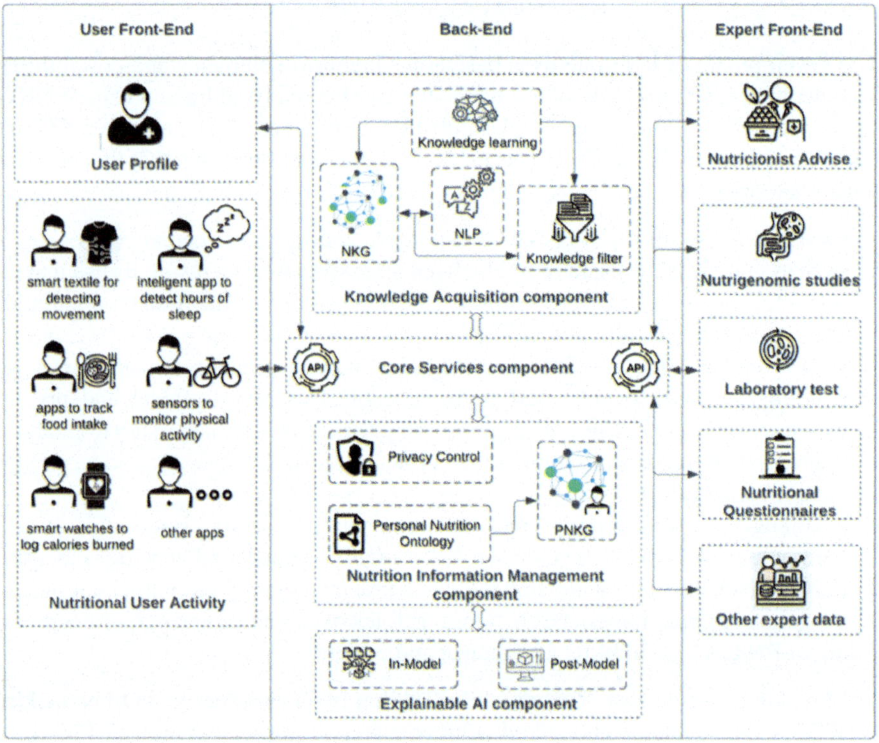

Fig. 1 Generic architecture to support adaptive, personalized nutrition advice systems

4.1 User Front-End

The user front-end is responsible for interacting with the user to obtain the input data and display the results. The input data can be explicitly entered by the user and updated in his profile or collected through various applications, allowing the nutritional user activity to be ordered in situ. The user's profile represents what the user likes and dislikes. It helps to understand the context better and provide more accurate recommendations.

The amount of information that needs to be collected depends on the specific goals the user wants to achieve, such as preventing the onset of chronic diseases, ensuring healthy aging, or achieving goals such as losing weight or increasing muscle mass, among others. These goals are monitored and reviewed by nutritionists and other experts.

The ubiquity of smartphones and the emergence of wearable devices have created the opportunity to collect nutritional data automatically. Below are various proposals (recent or underway) that exemplify the advances in AI to collect data on physical activity levels, daily food intake, human emotions, or sleep quality. It is worth noting that these data are not the only ones needed to complete a person's nutritional profile, but they are perhaps the most complex to collect. The proposals are organized according to the main components they use to collect the data:

- Wearables. The study in [22] analyzes commonly used wearable activity trackers that provide information about a person's physical activity, including the number of steps taken, time spent inactive, vigorous vs. moderate exercise, and even energy expenditure. These devices are typically linked to smartphones or computers to display this information in meaningful graphs and statistics. Although this work focuses on rheumatic and musculoskeletal diseases, this approach can be applied to other purposes. In the work described in [23], the GoBe2 wristband (Healbe Corp) is designed to automatically track users' daily energy intake (calories) and macronutrient intake (grams of protein, fat, and carbohydrate). The technology uses computational algorithms to convert bioimpedance signals into measured patterns of extracellular and intracellular fluids associated with the input of glucose and essential nutrients into the body. By analyzing changes in fluid concentration, the technology estimates calories consistent with glucose absorption into the bloodstream.
- Intelligent Applications. The application described in [24] outlines artificial intelligence that can detect with 77% accuracy whether people have slept at least 7 h or only 3 h, based on the tone and pace of their speech. The goal of the DietBot project [25] is to automate the measurement of food intake and provide personalized nutritional advice. By reading receipts and calculating food intake, DietBot will generate a customized diet plan tailored to the individual's health status, dietary habits, and preferences. The project will use modular methods to continuously improve data accuracy and advice quality by adding new functionalities.

- The review article described in [26] examines how the implementation of mobile applications and wearable technologies has provided opportunities for the real-time collection of granular data related to nutrition. The authors conclude from their analysis that digital technology for clinical nutrition is still young, although there is much promise for growth and disruption in the future.
- Smart Textiles. Researchers at the MIT Media Lab propose using smart textiles [27] that conform to the body to obtain precise data from a pressure sensor to analyze and classify a user's posture and movements. An advantage of this technology is that this approach is less intrusive than camera-based systems, which can raise privacy issues due to their continuous and potentially invasive detection and recognition.

4.2 Expert Front-End

The Expert graphical user interface allows other stakeholders in the precision nutrition process to interact with the system. In its most basic form, a nutritionist can create a dietary guide based on the user's goals. Of course, some nutritional goals will require nutrigenomic studies, laboratory tests, or the completion of dietary questionnaires. Some examples are:

- Vitamin D Deficiency: To optimize vitamin D levels, a blood test is needed to measure current levels and determine the need to increase vitamin D intake through sun exposure, diet, or supplementation.
- Disease Prevention: Some genetic variants are associated with a higher risk of developing certain diseases, such as cardiovascular disease or type 2 diabetes. A nutrigenomic study can identify these variants and set nutritional goals to reduce risk or improve disease management.
- Identifying nutritional imbalances: A nutritional questionnaire can provide information about possible deficiencies or excesses of certain nutrients in the diet. The proposal of a conversational agent[1] for conducting health questionnaires aims to improve communication between patients and healthcare institutions. The idea is to move from a static questionnaire where users provide health information to more interactive digital applications.

4.3 Back-End

Depending on the logic of the application to be developed, the back-end can take many different configurations. Four components are suggested in the generic architecture:

Knowledge acquisition component. This component collects pertinent user data, including demographics, medical history, and dietary activity, to generate adaptive

[1] A project of the HumanE AI Net initiative (https://www.humane-ai.eu/), which aims to facilitate AI systems that augment human capabilities.

and tailored nutritional recommendations based on the learned information. Knowledge Learning is responsible for learning new information from data that comes from users, experts, and other sources. The system collects relevant data, finds patterns, and updates its knowledge base using machine learning techniques and data analysis to improve the quality and accuracy of recommendations.

Once the data is collected, automatic techniques based on named entity recognition (NER) and natural language processing (NLP) can be used to identify entities and relationships, enabling the population of Nutritional Knowledge Graphs (NKG). Extracting entities and relationships from unstructured data is an active area of research [28, 29].

It is essential to integrate nutrition terms and relationships into ontologies and knowledge resources to gain a broader understanding of how nutrition and food influence aspects such as health. In [30], the authors conclude that it is vital to expand the representation of nutrition in ontologies and knowledge resources in areas that are already modeled, such as food, as well as in unexplored areas, such as nutritional biomarkers, behavioral nutrition counseling, cultural beliefs, and food processing. The same work highlights the need for collaboration and continuous communication across the field of nutrition to achieve a complete representation of the different domains of nutrition through existing ontologies.

To this end, efforts such as the Working Group, composed of curators of various ontologies and representatives from the US Department of Agriculture, meet regularly to discuss how to represent nutrition in the current ontological landscape. The working group operates through a GitHub page[2] and is open to participation by interested individuals or groups.

The proposed models in the area of personalized nutrition provide a solid foundation for further development toward a more comprehensive and guiding proposal. However, many areas still need to be explored, such as the integration of genetic information and microbiome analysis. Due to the evolving nature of user behavior over time, a knowledge filter algorithm is used to dynamically recalculate the user's main points of interest within each period. This element calculates the weights and applies a filter to the recognized entities from the NKG, which are then passed to the PKG. This ensures the adaptation and refinement of the user's preferences and interests.

Core services component. This component provides intuitive application programming interfaces (APIs) to (i) retrieve data or information from the environment (user activity or expert advice) and (ii) continuously monitor, adjust, and send recommendations to the front end based on user feedback, expert advice, and changes in user data, ensuring adaptive and evolving nutrition advice.

Nutrition information management component. The role of this component within the APNA system is twofold: (i) to collect and organize a comprehensive knowledge base of personal nutrition information, and (ii) to integrate and update the personal nutrition information knowledge base to ensure the accuracy and relevance of the data used to generate personalized recommendations.

[2] https://github.com/FoodOntology/joint-food-ontology-wg.

Implementation of this component requires overcoming the challenges described in Sect. 3. Fortunately, the community is working to address many of these issues by implementing supporting architectures for PKG through research and technology development (see, for example, [15, 20, 31]).

Depending on the scope of the APNA system, the Personal Nutrition Ontology will need to be designed to annotate the PNKG that captures and links information from precision nutrition stakeholders (described in Sect. 2). The ontology defines how the facts and statements that make up the content of the PNKG are logically represented, structured, and expressed.

Some models for representing personal nutrition information can be found in the literature: Food4healthKG [32], which integrates ontologies and heterogeneous data on food, gut microbiota, and human health for precision nutrition applications. The work proposed by Gyrard et al. describes a health PKG, including using existing medical ontologies. They mention some challenges in ontology reuse but do not share how they are being addressed in practice [33]. The work in [8] proposes a knowledge model to capture dietary preferences and personal context to provide personalized nutritional recommendations. The authors present the creation of a new ontology called "Personal Health Ontology," which is based on semantic technologies and aims to integrate the patient's medical information, as well as the social determinants of health and daily life observations obtained from interviews with diabetic patients.

To ensure the protection and control of the personal data and privacy of the users within the APNA system, the Privacy Control must be appropriately configured.

Explainable AI component. This component is critical in fostering trust and transparency by assisting nutrition professionals and users in decision-making. Knowledge graphs used in Explainable AI (XAI)[3] provide explainability by organizing information, extracting relevant features and relationships, and performing logical reasoning.

As described in [34], KGs have proven valuable in improving explainability in healthcare. They have been used to detect healthcare misinformation, identify adverse drug reactions and drug-drug interactions, and bridge the knowledge gap between healthcare experts and AI-based models.

It is envisioned that KG explainability can be used at different stages in the APNA system development pipeline: during the development of models and algorithms (in-model explainability), knowledge-based explainability can be used to understand and visualize how decisions are made within the nutrition system, or during the phase of interpretation and communication of results (post-model explainability), knowledge-based explainability can help present and effectively communicate the nutrition recommendations generated by the system.

[3] The term"Explainable AI" was first coined by the Defense Advanced Research Projects Agency (DARPA) in 2016 to address the need for increased transparency and accountability in AI systems, particularly in safety–critical domains such as defense and healthcare.

5 Use Case: Healthy Substitutes for Processed Products

To assess the suitability and effectiveness of the suggested framework, a system has been implemented to propose alternative food options that offer superior nutritional value to processed items [35]. The aim is also to check how this type of application can benefit from using a PNKG, improving its capacity for personalization.

The front end is built with React Native and uses Firebase for authentication. A mobile app has also been developed to allow users to browse and search for specific processed products, find information about their nutritional content, ingredients, and portion sizes, and get personalized recommendations. The Express framework and Node.js runtime environment are used to power the back-end functionality of the application. RESTful APIs are used to communicate between the front end and back end.

The work incorporated dietary guidelines from three primary sources: the ADA Standards of Care in Diabetes, the "Diabetes mellitus tipo 2 Guía de Práctica Clínica (GPC)" from the "Ministerio de Salud Pública del Ecuador" (MSP), and the "Documento Técnico de las Guías Alimentarias Basadas en Alimentos (GABA) del Ecuador" from the MSP and the Food and Agriculture Organization (FAO). This information is modeled and stored in a KG. In addition, a transformation process of semi-structured data is implemented to create a KG containing information on processed and ultra-processed products and their chemical composition. Finally, a PNKG is modeled to register user information and the log of requested substitutions. All knowledge graph models are stored in the Neo4j database.

The system uses a self-organizing map and K-means clustering to recommend appropriate food substitutions based on their nutritional characteristics. It is worth noting that this version of the system does not implement any privacy mechanisms over the PNKG, nor does it explain the recommended substitutions.

5.1 Results

The current line of research is still in its early stages. In the initial attempt to implement an APNA system, a fundamental ontology has been designed for personalized recommendations of processed products. Further work is needed to extend the ontology to other precision nutrition topics.

Nevertheless, the system's back end suggests that the recommendation of possible substitutes can be personalized for each user using the knowledge graph information. Unwanted replacements can be filtered for a given user using the Restricts relationship between User nodes and Food, Category, or Tag nodes modeled in the ontology. Figure 2 shows an example of food substitutions for an ultra-processed product, Tuna Tenderloin Pieces in Soybean Oil.

When a user requires a food substitution, appropriate foods and foods that match the user's dietary preferences are recommended. To help users make better decisions,

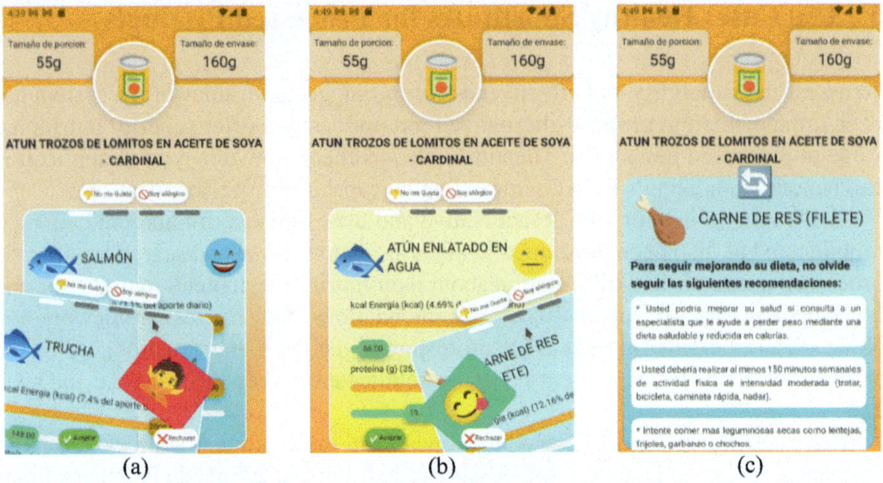

Fig. 2 An example of food substitutes for an ultra-processed product. **a** Food substitutions without dietary preference information. **b** Food substitutions with dietary preference information. **c** Food substitution accepted by the user

reference values for the intake contributions to their daily requirements are displayed for each recommended food. In Fig. 2a, the system has not yet registered that this user does not like salmon and suggests it as a possible substitute food. Still, when the user registers this preference in the PNKG, the system will avoid recommending this food, see Fig. 2b. In addition, if the user accepts a substitute suggestion, some general and dietary advice messages are displayed to improve their diet, see Fig. 2c. Given the individual's dislike of salmon, the PNKG can infer that alternative fish options would be preferable. This inference is derived from the connections within the graph, which establish relationships between food preferences, specific attributes, and viable substitute choices.

6 Conclusions

This work has proposed using PNKGs as an organized knowledge structure to support personalized nutrition. To this end, an analysis of the characteristics that define precision nutrition has been conducted, and the potential contribution of PNKGs in integrating information about an individual's nutritional status has been explained. A generic framework has been proposed to guide the design of personalized nutrition advice systems that adapt to individual needs. The framework has been applied in the implementation of the first system. This study has aimed to establish a new line of research in developing knowledge-based intelligent applications, such as conversational AI agents and recommendation systems, that can provide the best dietary advice for disease prevention or treatment based on a person's characteristics. Initially, this research can explore the knowledge acquisition processes and the maintenance, creation, and updating factors of PNKGs. Also, privacy is a constant concern with personalized features.

In future work, there is a desire to implement an AI chatbot that utilizes the knowledge of a PNKG. Hopefully, this article will stimulate discussion on applying PNKGs to personalized nutrition.

Acknowledgements This work is part of the research project "2021 030 UPS ESPINOZA MAURICIO", supported by the Vice-Rectorate for Research of the University of Cuenca.

References

1. Carbajal Azcona, A.: Manual de Nutrición y Dietética. Departamento deNutrición. Facultad de Farmacia. Universidad Complutense de Madrid
2. World Heath Organization: Nutrition. https://www.who.int/es/health-topics/nutrition (2023). Last accessed 05 July 2023
3. Patra, R.: Introduction to knowledge graphs in healthcare. https://blogs.oracle.com/ai-and-datascience/post/introduction-to-knowledge-graphs-in-healthcare (2019). Last accessed 05 July 2023
4. Chen, X., Jia, S., Xiang, Y.: A review: knowledge reasoning over knowledge graph. Expert Syst. Appl. 141, 112948 (2020). https://doi.org/10.1016/j.eswa.2019.112948
5. Rastogi, N., Zaki, M.J.: Personal health knowledge graphs for patients (2020)
6. Min, W., Liu, C., Xu, L., Jiang, S.: Applications of knowledge graphs for food science and industry. Patterns **3**(5) (2022)
7. Dragoni, M., Bailoni, T., Donadello, I., Martin, J.C., Lindgren, H.: Integrating functional status information into knowledge graphs to support self-health management. Data Intell., 1–22 (2023)
8. Seneviratne, O.W., Harris, J., Chen, C.H., McGuinness, D.L.: Personal health knowledge graph for clinically relevant diet recommendations. ArXiv abs/2110.10131 (2021)
9. Shirai, S., Seneviratne, O., McGuinness, D.L.: Applying personal knowledge graphs to health (2021)
10. Balog, K., Kenter, T.: Personal knowledge graphs: A research agenda. In: Proceedings of the ACM SIGIR International Conference on the Theory of Information Retrieval (ICTIR) (2019)
11. Berciano, S., Figueiredo, J., Brisbois, T.D., Alford, S., Koecher, K., Eckhouse, S., Ciati, R., Kussmann, M., Ordovas, J.M., Stebbins, K., Blumberg, J.B.: Precision nutrition: maintaining scientific integrity while realizing market potential. Front. Nutr. **9**(2022). https://doi.org/10.3389/fnut.2022.979665
12. Livingstone, K.M., Ramos-Lopez, O., P'erusse, L., Kato, H., Ordovas, J.M., Martinez, J.A.: Precision nutrition: a review of current approaches and future endeavors. Trends Food Sci. Technol. **128**, 253– 264 (2022). https://doi.org/10.1016/j.tifs.2022.08.017
13. Kirk, D., Catal, C., Tekinerdogan, B.: Precision nutrition: a systematic literature review. Comput. Biol. Med. **133**, 104365 (2021). https://doi.org/10.1016/j.compbiomed.2021.104365
14. Revista Lideres.: Comida sana y empacada es su fórmula emprendedora. https://www.revistalideres.ec/lideres/comida-sana-empacada-formula-emprendedora.html (2021). Last accessed 05 July 2023
15. Skjæveland, M.G., Balog, K., Bernard, N., Lajewska, W., Linjordet, T.: An ecosystem for personal knowledge graphs: a survey and research roadmap. SSRN (2023). https://ssrn.com/abstract=4432100

16. Tigunova, A., Yates, A., Mirza, P., Weikum, G.: CHARM: inferring per- sonal attributes from conversations. In: Proceedings of the 2020 Conference on Empirical Methods in Natural Language Processing (EMNLP), pp. 5391–5404. Association for Computational Linguistics (2020). https://doi.org/10.18653/v1/2020.emnlp-main.434
17. Yu, D., Sun, K., Cardie, C., Yu, D.: Dialogue-based relation extraction. In: Proceedings of the 58th Annual Meeting of the Association for Computational Linguistics, pp. 4927–4940. Association for Computational Linguistics (2020). https://doi.org/10.18653/v1/2020.acl-main.444
18. Safavi, T., Fourney, A., Sim, R., Juraszek, M., Williams, S., Friend, N., Koutra, D., Bennett, P.N.: Toward activity discovery in the personal web. In: Proceedings of the 13th International Conference on Web Search and Data Mining (WSDM '20). Association for Computing Machinery, pp. 492–500. New York, NY, USA (2020). https://doi.org/10.1145/3336191.3371828
19. Chakraborty, P., Dutta, S., Sanyal, D.K.: Personal research knowledge graphs. In: Companion Proceedings of the Web Conference 2022 (WWW '22). Association for Computing Machinery, pp. 763–768. New York, NY, USA (2022). https://doi.org/10.1145/3487553.3524654
20. Ilkou, E.: Personal knowledge graphs: use cases in e-learning plat- forms. In: Companion Proceedings of the Web Conference 2022 (WWW '22). Association for Computing Machinery, pp. 344–348. New York, NY, USA (2022). https://doi.org/10.1145/3487553.3524196
21. Renner, B., Buyken, A.E., Gedrich, K., Lorkowski, S., Watzl, B., Linseisen, J., Daniel, H., Conrad, J., Ferrario, P.G., Holzapfel, C., Leitzmann, M., Richter, M., Simon, M.C., Sina, C., Wirsam, J.: Perspective: a conceptual framework for adaptive personalized nutrition advice systems (apnass). Adv. Nutr. (2023). https://doi.org/10.1016/j.advnut.2023.06.009
22. Davergne, T., Kedra, J., Gossec, L.: Wearable activity trackers and artificial intelligence in the management of rheumatic diseases. Zeitschrift fur Rheumatologie **80**(2021). https://doi.org/10.1007/s00393-021-01100-5
23. Dimitratos, S., German, J., Schaefer, S.: Wearable technology to quantify the nutritional intake of adults: validation study. JMIR Mhealth Uhealth 8(7) (2020). https://doi.org/10.2196/16405
24. Wong, C.: AI can tell if you are sleep deprived by listening to your voice. https://www.newscientist.com/article/2368838-ai-can-tell-if-you-are-sleep-deprived-by-listening-to-your-voice/ (2023). Last accessed 05 July 2023
25. Wageningen University: DietBot. https://www.wur.nl/en/project/DietBot-AI-based-personalized-nutritional-advice-generated-from-individual-health-parameters-and-food-purchase-behavior.htm (2023). Last accessed 05 July 2023
26. Limketkai, B., Mauldin, K., Manitius, N., Jalilian, L., Salonenet, B.R.: The age of artificial intelligence: use of digital technology in clinical nutrition. Nutr. Metab. Surg. 9(20) (2021). https://doi.org/10.1007/s40137-021-00297-3
27. MIT Media Lab: 3DKnITS: Three-dimensional Knitted Intelligent Textile Sensor. https://www.media.mit.edu/projects/3dknits/overview/ (2022). Last accessed 05 July 2023
28. Li, J., Sun, A., Han, J., Li, C.: A survey on deep learning for named entity recognition. IEEE Trans. Knowl. Data Eng. **34**(1), 50–70 (2022). https://doi.org/10.1109/TKDE.2020.2981314
29. Pawar, S., Palshikar, G.K., Bhattacharyya, P.: Relation extraction: a survey (2017)
30. Chan, L., Vasilevsky, N., Thessen, A., McMurry, J., Haendel, M.: The landscape of nutri-informatics: a review of current resources and challenges for integrative nutrition research. Database (2021). https://doi.org/10.1093/database/baab003
31. Bosanska, D.C., Huptych, M., Lhotska, L.: A pipeline for population and analysis of personal health knowledge graphs (phkgs). In: 18th International Conference on Semantic Systems SEMANTICS 2022, CEUR Workshop Proceedings (2022). https://ceur-ws.org/Vol-3235/paper8.pdf
32. Fu, C., Huang, Z., van Harmelen, F., He, T., Jiang, X.: Food4healthkg: knowledge graphs for food recommendations based on gut microbiota and mental health. SSRN (2022)
33. Gyrard, A., Gaur, M., Shekarpour, S., Thirunarayan, K., Sheth, A.: Personalized Health Knowledge Graph. Kno.e.sis Publications (2018). https://corescholar.libraries.wright.edu/knoesis/1173, zSCC: 0000014

34. Rajabi, E., Kafaie, S.: Knowledge graphs and explainable ai in healthcare. Information **13**(10), 459 (2022). https://doi.org/10.3390/info13100459
35. Solano, P., Vıctor, H.: Sistema de informacion nutricional y recomendación de sustituciones saludables de productos procesados para la prevención de DMT2 en Cuenca. Tesis Pregrado, Universidad de Cuenca, Facultad de Ingenieria (2023)

Interactive e-Books in Mathematics Learning: A Systematic Mapping Study

Santiago Jácome-Guerrero, Marcelo Román, Norma Barreno, and Lorena Parra-Gavilánez

Abstract Learning mathematics is one of the main problems facing education. The current digital era makes the use of several alternative learning environments possible. One of these environments is achieved through the use of interactive e-books, which allow learning mathematics using various resources so that the audience can interact with more than text and images, with videos, hyperlinks, interactive questions, worksheets, assessments, increasing the motivation to learn at any time and place through fixed and portable electronic devices, thus improving their academic performance. The objective of this study is to determine the contribution of the use of interactive e-books in the learning achievements of mathematics. A systematic mapping study was carried out by searching the main scientific databases. The search retrieved 45 articles, and 19 articles were retained as primary studies. The papers were categorized according to criteria related to the contribution of interactive e-books in mathematics learning: learning theories, didactic strategies, digital educational resources, and incidence of personal factors in education. In conclusion, the use of interactive e-books in the learning of mathematics encourages the student's ability to reason, interact, and manage their knowledge.

Keywords Interactive e-Book · Learning · Mathematics · Systematic mapping study

S. Jácome-Guerrero (✉)
Departamento de Ciencias de la Computación, Universidad de las Fuerzas Armadas ESPE, Sangolquí 171103, Ecuador
e-mail: psjacome@espe.edu.ec

M. Román · N. Barreno
Departamento de Ciencias Exactas, Universidad de las Fuerzas Armadas ESPE, Sangolquí 171103, Ecuador

L. Parra-Gavilánez
Departamento de Ciencias, Humanas y de la Educación, Universidad Técnica de Ambato, Ambato, Ecuador

G. F. Olmedo Cifuentes et al. (eds.), *Emerging Research in Intelligent Systems*, Lecture Notes in Networks and Systems 902, https://doi.org/10.1007/978-3-031-52255-0_24

1 Introduction

The Organization for Economic Cooperation and Development (OECD) is an international organization that works to build better policies for better lives. PISA is the OECD's Programme for International Student Assessment, and it measures 15-year-olds' ability to use their reading, mathematics, and science knowledge as well as what they can do with what they know [1]. The report presented in [2] that the PISA test results point out the quality and equity of learning outcomes achieved in some countries worldwide. This report allows educators to learn the policies and practices implemented in countries that managed to exceed the learning indicators. In turn, in [3], the low levels of student learning, especially in mathematics subjects, are evidenced. When reviewing the databases of the PISA test results and referencing the works presented in [4], it is determined that the academic performance of Latin American students is, on average, lower than that of students in North America, Europe, and Asia.

Mathematics education is developed mainly based on learning, teaching, and evaluation processes. It is mentioned that mathematics learning lies mainly in problem-solving, and when problems are not well formulated or understood, it may be the origin of learning difficulties. Similarly, in [5], it is argued that teaching mathematical problem-solving in a real context efficiently develops logical-mathematical thinking skills. In [6], discovery learning is implemented, where students are encouraged to search for different ways and mechanisms to solve problems without previously establishing a script or solution mechanism. In this line, in [7], it is stated that different types of learning should be implemented for problem-solving.

People live in a globalized world where scientific and technological knowledge development has generated significant resources focused on strengthening the didactic-pedagogical aspects of the teaching–learning process [8]. Advances in technology have revolutionized traditional education. Advances in computer technology have led to digital devices becoming smaller, faster, cheaper, and more interconnected. These devices, in turn, make it possible to run software for educational purposes. One of the ways to enhance education using technology is through the so-called e-books, one of the most widely used digital educational environments in recent years. An e-book is an electronic representation of a book. The term is a contraction of the phrase electronic book. E-books handle limited resources, such as text and images. As noted in [9], the production and consumption of digital books have been increasing steadily. On the codex, the computational environment is now the latest invention to affect literature significantly. However, about ten years ago, due to technological advances, so-called interactive e-books became available, with more elements with which the audience can interact, such as videos, links, audio, and other interactive features. This supports the multiple advantages of interactive e-books in increasing students' motivation to learn and improving academic performance. In addition, interactive e-books are easy to access, better organized, and can load the book on any fixed or mobile device, making it easy to use anytime, anywhere [2].

In this sense, researchers generally point out that technology in education allows the development of various teaching methods, learning support strategies, and the generation of interactive digital educational environments [10]. Our research aims to use a Systematic Mapping Study (SMS) to identify the contribution of interactive e-books in mathematics learning. The search for information is guided by four research questions determining the pedagogical and technological aspects of interactive e-books.

2 Related Works

The researchers of this study found only one systematic review on using interactive e-books in mathematics learning [11]. In this work, the authors pointed out the different alternatives to improve students' scientific literacy by using various resources, such as learning models, without achieving the average standard of scientific literacy skills. Therefore, the article points out that the interactive e-book can be used as an alternative learning resource to enhance scientific literacy by identifying six types of content that help students increase their literacy skills through animated images, videos, hyperlinks, interactive questions, student worksheets, and assessments.

3 Research Method

This paper seeks to answer four research questions related to determining the contribution of interactive e-books on learning achievement in mathematics. (RQ1) What are the learning theories in which it is framed? (RQ2) What are the didactic strategies used? (RQ3) What are the digital educational resources used? and (RQ4) What is the impact of personal factors on learning? To answer these questions, an SMS was conducted to identify and classify these aspects in the published literature.

3.1 Search String Selection

The first step was to identify the relevant search strings and keywords. Several options were tested to do this, and the best one was chosen. Several initial articles were read, and the keywords and basic topic knowledge were obtained. After combining the opinions of two experts in mathematics education, one expert in software development, and one expert in pedagogy, the search string chosen was: ("interactive e-book" OR "interactive e-book" OR e-book) AND (learning OR teaching) AND (mathematics OR mathematical OR math).

3.2 Databases and Search Protocol

The search was performed sequentially in Scopus, WOS, ACM, and IEEE Xplore. The options provided by each database determined the search fields used. Since interactive e-books have been used in education for approximately ten years, the search range was from January 2015 to May 2023. The search was ordered considering the database that yielded the most results. The search fields were selected to ensure that all databases were similar.

- Inclusion criteria: The paper is written in English, AND the abstract OR title OR keywords mention an issue regarding interactive e-books' contribution to learning mathematics.
- Exclusion criteria: The paper does not present any issue related to interactive e-books' contribution to learning mathematics, OR books, OR book chapters.

3.3 Paper Selection

The searches were run using the search string defined. The number of papers returned by the first search was 45, called *retrieved papers*. Then, by inspecting each recovered paper's title, keywords, and abstract, 35 papers were filtered into the group of *candidate papers*. The whole group of candidate papers was screened for duplicates. When duplicates were found, only the first occurrence of the paper was counted and maintained. The others were deleted. The final group has 21 papers, which are called *non-duplicate candidate papers*. Each paper of the Non-Duplicate Candidate Papers group was read to determine if they described aspects of using interactive e-books in teaching mathematics at any level of education. The results were cross-checked by one expert in pedagogy, two experts in mathematics teaching, and one expert in software development, and any disagreement was discussed and resolved in our meetings. Finally, 19 papers were identified as *primary studies*.

Figure 1a shows the number of primary studies by year of publication. Publications on the topic started to increase in 2015, and since then, several articles have been published each year, confirming the interest in this field. Figure 1b shows the number of publications according to the classification criteria to answer the research questions. More detailed information on the selected primary studies and their contribution to the research questions is described in Table 1.

Learning theories.

The results show that the educational models are centered on learning approaches based on behaviorism, cognitivism, and constructivism [12–17]. In addition, didactic models such as instructional design are visualized [18, 19], and the adaptive scaffolding model [20]. In addition to the refinement of classical models, referred to as the ASSURE model by their authors [18] and Lee and Owens' model [21], see Table 2.

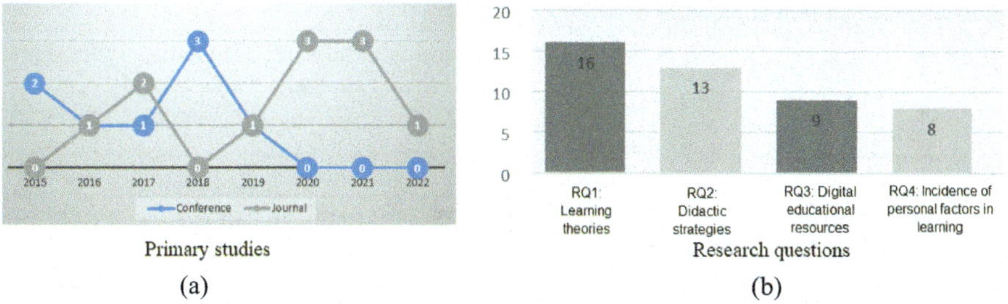

Fig. 1 **a** Primary studies. **b** Research questions

Once the didactic-pedagogical theories incorporated in the research have been identified, knowing the research design used in them is of interest. It would be expected them to be qualitative designs. Still, the study is framed in quasi-experimental designs because it resorts to control groups and experimentation research phases by analyzing pretest, test, and posttest [12, 20]. The quantitative method also uses didactic techniques to assess learning [22]. Validity of knowledge focuses on the development of knowledge tests and knowledge retention tests [18], progressive sequential analysis to explore the behavioral learning patterns of the experimental and knowledge control groups [28], and monitoring of class participation [17]. Researchers report difficulties in choosing appropriate learning technologies [17, 18, 28], teaching models, and strategies that stimulate students' attention, inspire active participation in learning, and implement different operational and representational strategies [20]. Another issue was the improvement of students' reasoning approaches [25], not to mention teachers' and students' lack of experience in using educational technology as a tool for learning [19]. Interactive e-books in mathematics learning are an open topic in academic research. Therefore, it is suggested in future work to record the learning processes in real-time to provide the researcher with more complete research data, as well as determine efficient and practical factors to assess the improvement of learning mediated by appropriate technology.

Didactic strategies.

The didactic process in the educational field is conceived as the procedure to guide learning. There must be elements such as design, planning, and execution of different academic activities that contribute to achieving learning outcomes. In contrast, the didactic technique is conceived as the resource that the teacher executes to carry out the purposes planned from the strategy. In turn, the objective of instructional design is to determine the most pleasant and digestible way to offer content and educational resources that help the student develop the ability to achieve specific tasks. See Table 3.

Several learning strategies were identified in the analyzed studies, such as situated learning, autonomous learning, and active learning [17, 20]. The use of the inverted classroom has also been identified [15] as gamification and flipped learning [23, 24,

Table 1 Selected primary studies

Id	Title	Type	Year	Learning theories	Didactic strategies	Digital educational resources	Incidence of personal factors in learning
1	The interactive e-book and video feedback in a multimedia learning environment: influence on performance, cognitive and motivational outcomes	Journal	2022	RQ1	RQ2	RQ3	RQ4
2	Effects of gamified interactive e-books on students' flipped learning performance, motivation, and meta-cognition tendency in a mathematics course	Journal	2021	RQ1	RQ2	–	RQ4
3	Blended learning based on ebook integrated Youtube in learning mathematics	Journal	2021	RQ1	RQ2	RQ3	RQ4
4	The Development of Hypermedia Based E-Book for Integral Calculus Subject	Journal	2021	RQ1	RQ2	RQ3	–
5	Examining the effects of Geogebra applets on mathematics learning using interactive mathematics textbook	Journal	2020	RQ1	–	RQ3	RQ4
6	A project-based learning into flipped classroom for ePUB3 electronic mathematics learning module (eMLM)-based on course design and implementation	Journal	2020	RQ1	–	RQ3	–

(continued)

Table 1 (continued)

Id	Title	Type	Year	Learning theories	Didactic strategies	Digital educational resources	Incidence of personal factors in learning
7	Learning fractions with and without educational technology: What matters for high-achieving and low-achieving students?	Journal	2020	RQ1	–	RQ3	–
8	Pre-school children's behavioral patterns and performances in learning numerical operations with a situation-based interactive e-book	Journal	2019	RQ1	–	–	–
9	Touchy Feely affordances of digital technology for embodied interactions can enhance 'epistemic access'	Conference	2019	RQ1	–	–	–
10	Using Brainwave Characteristics for Exploring the Effect of Integrating Graduated-Prompting Strategy into Interactive e-Books on Students' Learning Attention	Conference	2018	RQ1	–	–	–
11	Effects of problem-solving strategy based interactive e-book on measurement concepts	Conference	2018	RQ1	RQ2	–	RQ4
12	Interactive Personalized eBooks for Education	Conference	2018	–	RQ2	–	–

(continued)

Table 1 (continued)

Id	Title	Type	Year	Learning theories	Didactic strategies	Digital educational resources	Incidence of personal factors in learning
13	Effects of knowledge-Based Scaffolding Interactive e-Book on Multiplication Concept	Conference	2017	RQ1	RQ2	–	RQ4
14	Facilitating and bridging out-of-class and in -class learning: An interactive E-book-based flipped learning approach for math courses	Journal	2017	RQ1	RQ2	RQ3	–
15	A Flip Blended Learning Approach for ePUB3 eBook-based Course Design and Implementation	Journal	2017	RQ1	RQ2	RQ3	–
16	Towards analytics for educational interactive e-Books: The case of the reflective designer analytics platform (RDAP)	Journal	2016	RQ1	RQ2	–	RQ4
17	The Development of an eBook for Teaching and learning Mathematics	Conference	2016	–	RQ2	RQ3	–
18	An interactive e-book approach to supporting flipped learning in an elementary school math course	Conference	2015	RQ1	RQ2	–	RQ4
19	Towards Configurable Learning Analytics for Constructionist Mathematical e-Books	Conference	2015	–	RQ2	–	–

Table 2 Learning theories

Learning theories	Primary studies
Educational model	[12–22]
Research design	[12, 20, 22–25]
Validity of knowledge	[15–17, 22, 26–28]
Difficulties encountered	[17, 19, 22, 28]
Proposes improvements in research	[12, 13, 20, 22]

Table 3 Didactic strategies

Didactic strategies	Primary studies
Learning strategy	[12, 15, 17, 20, 23, 24, 26, 27, 29]
Teaching techniques	[13, 20–22, 27, 30]

26]. In addition, there is evidence of using the personalized educational model in the research as Blended Learning [26]. From the position of [12], it is important to consider strategies for understanding the problem and focusing on solving problems in real-life situations [20, 22]. In terms of didactic techniques, they mainly use problem-solving [20, 22, 27, 30]—rote, repetitive, algorithmic, and game-based cooperative learning [13, 20].

Finally, there is no clear evidence of academic activities such as exhibitions, debates, forums, or others. However, in [21], the design of didactic activities and materials should be related to previous knowledge and strategies for understanding the problem [24]. It is noted that flipped learning is a well-recognized mode that reverses the traditional classroom instructional arrangement by offering learning content outside the classroom and engaging students in more in-class activities. However, it remains challenging for students to understand the learning material independently, particularly when learning abstract concepts such as mathematics. In summary, when researchers present student learning results, they limit themselves to referring to developing their problem-solving skills. It is argued that students can create mental representations of mathematical concepts, objects, and forms, but there is no evidence of any indicator that would allow verification of this statement.

Digital educational resources.

It is indisputable that today, people live in a digital society. The authors state that the implementation of new technologies has entered the classroom at different levels of education, and their motivation is to innovate and improve educational processes, generating a change in all the elements involved in the educational process [17, 21, 24]. See Table 4.

A common denominator is evident when the researchers emphasize the design of their e-books, mentioning that they are highly interactive with multimedia features of images, sound and movies, contextualization, and personalization. They emphasize that, by using them, they break the unidirectional learning of paper textbooks. They

Table 4 Digital educational resources

Digital educational resources	Primary studies
Multimedia resources	[15, 17, 24]
System architecture	[18, 19, 21, 27, 29]
Specific software	[16, 18]

indicate that students can interact with the learning content to visualize complicated content [18].

The technological didactic resources used are mobile and multimedia technologies [15], specific mathematical software such as the Geogebra applets [16], SimMAGIC system architectures [19, 27, 29], as well as technologies such as HTML5, CSS3, and JavaScript. The convergence of the works reviewed fit in understanding that the digital revolution in the teaching–learning processes does not only mean providing computers and digital blackboards to educational centers. It is pointed out that their efforts are focused on the analysis, design, implementation, and evaluation of digital materials. It reflects on the characteristics of an interactive e-book. It poses the challenge of implementing an interactive web system composed of structured libraries of mathematical concepts. These questions motivate reflection, critical thinking, and space for solving exercises to encourage ingenuity and creativity. It should work on all modern devices through an Internet browser (cell phones, tablets, and computers with different operating systems). This multimedia structure must be well planned, structured, and effectively integrated into the mathematics teaching–learning process.

Incidence of personal factors in learning.

The incidence of individual factors on student learning is one of the critical aspects of this systematic review. In this context and considering the criteria of learning theories and didactic strategies analyzed above, we argue that the teaching–learning process should be examined from the theoretical approach of the didactics of mathematics, also called Mathematics Education, which the teacher is expected to be able to encourage and motivate the students' work, leading them to reflect on the acquisition of knowledge. To master the students' social and emotional aspects and be skillful in generating integrated learning environments of previous expertise with the new ones. Also, to be able to design models that adapt to the uncertain and changing learning conditions that occur in a classroom. Finally, he must know how to prepare his students to develop the ability to solve real-life problems using different strategies and technological resources.

With this background, it is conceived that a series of factors, such as the student's attitude towards the subject, the pedagogical practices of the teachers, and the school environment, influence student learning and their school performance. Therefore, in this systematic study, Table 5 describes what has been analyzed in the different research works, where the aptitude and attitudinal aspects that contribute to the learning process stand out [1, 31].

Finally, our interest is to determine how the different research works evidenced the learning of mathematical concepts to be used in problem-solving through interactive

Table 5 Incidence of personal factors in learning

Incidence of personal factors in learning	Primary studies
Learning	[12, 15, 16, 20, 22, 26, 27]
Aptitude	[15, 20, 26]
Attitudinal	[15, 23]

e-books. It is essential to understand that learning is generally influenced mainly by aptitude and attitude. The factor related to the aptitude seeks to bring into play the student's previous knowledge for learning the new concept. The attitude is related to their affective-social-emotional environment and the organization of the study.

Learning material allowed students to creatively approach mathematical problems, develop creative mathematical thinking, and encourage decision-making by answering questionnaires and interviews to foster self-efficacy [12, 16, 20, 22, 26, 27]. For example, it is stated that students suggest that assignments with interactive applets and new types of learning materials inspired them to learn more [16].

In [15], it is argued that the findings revealed students' motivation and cognitive load because the interactive e-books provided feedback, and the video significantly reduced students' intrinsic and external cognitive load. In turn, in [23], a student learning motivation questionnaire was applied, and its results showed better motivation in those students who used the interactive e-book than those who did not. It complemented his argument by indicating that the motivation influenced the metacognitive part of it.

4 Discussion and Validity Threats

The analysis of results presents the information gathered from the selected primary studies. This information was classified into four categories. Each category responds to the research questions raised in the systematic review.

Table 2 shows the analysis of the criteria related to the learning theories used in the interactive e-books. According to [32], the development of mathematics teaching–learning and learning theories should be approached from a sociocultural perspective, and its research methodology should have a qualitative design. For this reason, the information search determined that 57% of the works indicate the educational model implemented, only 31% indicate the research design, 37% show how to validate the student's knowledge, and 21% expose the difficulties encountered in their work that could influence learning. Finally, 21% propose improvements as future works.

Table 3 analyzes the learning strategies and didactic techniques used to promote the development of mathematical thinking and participation skills by students. The results show that 47% of the articles refer explicitly to group work, especially to implementing tasks promoting reasoning and problem-solving. These have been developed in theoretical frameworks related to collaborative learning, rote learning, didactic scenario simulation, and inverted learning. The specific way of applying the learning methods or strategies has been sought for the didactic techniques. It is determined that 32% of the works show that this indicator is used through problem-solving. However, there is no clear evidence of how the development of cognitive skills and the construction of mathematical knowledge is carried out, nor is the classification of mathematical exercises and problems implemented by the researchers identified, nor is it possible to distinguish the type of exercise proposed, whether it is of an algorithmic nature or whether it has statements of real situations.

Regarding the digital educational resources used in designing and implementing interactive e-books, Table 4 shows that 16% use web technologies, 26% implement system architectures, and only 11% use specific software such as Geogebra applets. In Table 5, the intention is to conceptualize the level of student learning as a result of their profiles concerning the attitudinal and aptitude aspects of the individual. These aspects are closely related to the teaching practice and the generation of various educational environments [33]. 37% of the authors indicate that student learning is related to their satisfaction and motivation without specifying the parameters that allow them to assert such results. 11% indicate that records of the attitudinal part of the student are kept during the whole research process.

Regarding the validity of our SMS, it may be threatened by including only journal and conference publications written in English. Although the terms used in the search string are commonly used by other authors, related terms may have been overlooked. In turn, the SMS authors may have made errors in judgment when analyzing the publications. Finally, the publications were evaluated and ranked according to the judgment and experience of the authors, and other investigators may have evaluated the publications differently.

5 Conclusions

This paper describes an SMS study conducted to answer four research questions. RQ1. What are the learning theories in which it is framed? The research resorts to theories such as behaviorism, cognitivism, and constructivism, generating a theoretical symbiosis that allows the generation of models and didactic strategies that facilitate learning. RQ2. What are the didactic strategies used? The works mainly mention the active and collaborative learning strategy to help analyze and reflect the students' knowledge. Aspects of the inverted classroom are used where the student's performance in the learning process is sought. Additionally, strategies that link theoretical mathematical knowledge with reality are proposed to highlight its usefulness and importance. RQ3. What are the digital educational resources used? It is pointed out that interactive e-books are designed with multimedia features of images, sound, and movies where students can interact with the learning contents. The technological didactic resources are mobile technologies, specific mathematical software such as Geogebra applets, and SimMAGIC system architectures. In addition, technologies such as HTML5, CSS3, and JavaScript. RQ4. What is the incidence of personal factors on learning? The researchers argue that students' motivation and cognitive load are essential to consolidate learning through interactive e-books. Their conclusions are framed in conceiving students' attitudes as the interest to participate during the process in which they acquire new knowledge, and how they act in using the electronic book shows their predisposition and interest to progress in learning. Meanwhile, the researchers state that the student's previous knowledge is conceptualized as their aptitude, a personal factor that affects learning and the development of cognitive skills that contribute to the processes of mathematical comprehension, facilitating the execution of tasks proposed in the e-books.

It has been found that interactive e-books constitute a didactic strategy that favors mathematics learning since their interactive contents stimulate student participation and break with the old scheme of having printed texts. In addition, their ability to interact with the students facilitates the visualization and understanding of mathematical concepts that could be complex and difficult to understand.

Finally, in the works analyzed, the absence of qualitative research was identified, allowing a better understanding of the mathematics learning process. At the same time, it was identified as a low quantitative because sufficient research criteria are not used in mathematics education. The absence of effective learning record indicators could be seen, which demonstrates learning achievements in terms of the time spent using interactive e-books.

References

1. OECD.: Programme for International Student Assessment, https://www.oecd.org/pisa/. Last accessed 18 Feb 2023
2. Schleicher, A.: PISA 2018: Insights and Interpretations, https://www.oecd.org/pisa/. Last accessed 19 Feb 2023
3. Navarro-Martinez, O., Peña-Acuña, B.: Technology usage and academic performance in the Pisa 2018 Report. J. New Approaches Educ. Res. **11**(1), 130–145 (2018)
4. Agasisti, T., Antequera, G., Delprato, M.: Technological resources, ICT use and schools efficiency in Latin America-Insights from OECD PISA 2018. Int. J. Educ. Dev. **99**, 102757 (2023)
5. Szabo, Z.K., et al.: Examples of problem-solving strategies in mathematics education supporting the sustainability of 21st-century skills. Sustainability **12**(23), 10113 (2020)
6. Putriani, D., Rahayu, C.: The effect of discovery learning model using sunflowers in circles on mathematics learning outcomes. Int. J. Trends Math. Educ. Res. **1**(1), 22–25 (2018)
7. Darmawan, E., Suparman, S.: Design of mathematics learning media based on discovery learning to improve problem solving ability. Indones. J. Learn. Adv. Educ. (IJOLAE) **1**(2), 20–28 (2019)
8. Murray, J.: Routes to STEM: Nurturing Science, Technology, Engineering and Mathematics in Early Years Education, Taylor & Francis, pp. 219–221. (2019)
9. Sargeant, B.: What is an ebook? What is a book app? and Why should we care? An analysis of contemporary digital picture books. Child. Lit. Educ. **46**(4), 454–466 (2015)
10. Prasetyaningsih, P., Vitasari, M., Wilujeng I.: Development of science-oriented interactive electronic module guided inqury wave and disaster theme. Jurnal Penelitian dan Pembelajaran IPA **8**(2) (2022)

11. Firdausy, B.A., Prasetyo Z.K.: Improving scientific literacy through an interactive e-book: a literature review. Institute of Physics Publishing. (2020)
12. Huang, Y., Wu P., Chiang H.C.: Effects of problem-solving strategy based interactive e-book on measurement concepts, pp. 453–462. Springer Verlag (2018)
13. Droutsas, S., Patsilinakos, P., Symvonis A.: Interactive personalized ebooks for education. in 2018 9th International Conference on Information, Intelligence, Systems and Applications (IISA). IEEE. (2018)
14. Yu, Y., Chen, H., Wu, P.: Using brainwave characteristics for exploring the effect of integrating graduated-prompting strategy into interactive e-Books on students' learning attention. Institute of Electrical and Electronics Engineers Inc. (2018)
15. Yorganci, S.: The interactive e-book and video feedback in a multimedia learning environment: Influence on performance, cognitive, and motivational outcomes. J. Comput. Assist. Learn. **38**(4), 1005–1017 (2022)
16. Radović, S., et al.: Examining the effects of Geogebra applets on mathematics learning using interactive mathematics textbook. Interact. Learn. Environ. **28**(1), 32–49 (2020)
17. Tsai, T., Lin, J., Lin, L.: A flip blended learning approach for ePUB3 eBook-based course design and implementation. Eurasia J. Math. Sci. Technol. Educ. **14**(1), 123–144 (2017)
18. Ramadhani, R., Fitri, Y.: A Project-based learning into flipped classroom for ePUB3 electronic mathematics learning module (eMLM)-based on course design and implementation. Univ. J. Educ. Res. **8**(7), 3119–3135 (2020)
19. Reinhold, F., et al.: Learning fractions with and without educational technology: What matters for high-achieving and low-achieving students? Learn. Instr., 65 (2020)
20. Huang, Y., et al.: Effects of knowledge-based scaffolding interactive e-Book on multiplication concept. Institute of Electrical and Electronics Engineers Inc (2017)
21. Awaludin, B. Wibawa, Winarsih, M.: The development of hypermedia based e-book for integral calculus subject. Int. J. Online Biomed. Eng. **16**(4), 109–127 (2020)
22. Karkalas, S., Mavrikis, M., Labs, O.: Towards analytics for educational interactive e-Books: the case of the reflective designer analytics platform (RDAP). Association for Computing Machinery (2016)
23. Zhao, J., et al.: Effects of gamified interactive e-books on students' flipped learning performance, motivation, and meta-cognition tendency in a mathematics course. Educ. Tech. Res. Dev. **69**(6), 3255–3280 (2021)
24. Hwang, G., Lai, C.: Facilitating and bridging out-of-class and in-class learning: an interactive E-book-based flipped learning approach for math courses. Educ. Technol. Soc. **20**(1), 184–197 (2017)
25. Karnam, D., et al.: Touchy feely affordances of digital technology for embodied interactions can enhance' epistemic access. In: 2019 IEEE Tenth International Conference on Technology for Education (T4E). IEEE (2019)
26. Lai, C., Hwang, G.: An interactive e-book approach to supporting flipped learning in an elementary school math course. Asia-Pac. Soc. Comput. Educ. (2015)
27. Fitriani, S. Fatimah, Herman, T.: Blended Learning Based on ebook Integrated Youtube in Learning Mathematics. IOP Publishing Ltd. (2021)
28. Li, J., et al.: Pre-school children's behavioral patterns and performances in learning numerical operations with a situation-based interactive e-book. Interact. Learn. Environ. **28**(2), 148–165 (2020)
29. Figueiredo, M., Bidarra, J., Bostad, R.: The development of an ebook for teaching and learning mathematics, In: Antona, M., Stephanidis, C. (eds.), pp. 49–56. Springer Verlag (2016)
30. Karkalas, S., et al.: Towards configurable learning analytics for constructionist mathematical e-books. CEUR-WS. (2015)
31. Mazana, Y., Suero Montero, C., Olifage, C.: Investigating students' attitude towards learning mathematics (2019)
32. Das, K.: Realistic mathematics & vygotsky's theories in mathematics education. Shanlax Int. J. Educ. **9**(1), 104–108 (2020)
33. Feldon, D., et al.: Cognitive load as motivational cost. Educ. Psychol. Rev. **31**, 319–337 (2019)

Author Index

G. F. Olmedo Cifuentes et al. (eds.), *Emerging Research in Intelligent Systems*, Lecture Notes in Networks and Systems 902, https://doi.org/10.1007/978-3-031-52255-0

SPRINGER NATURE

GPSR Compliance

The European Union's (EU) General Product Safety Regulation (GPSR) is a set of rules that requires consumer products to be safe and our obligations to ensure this.

If you have any concerns about our products, you can contact us on ProductSafety@springernature.com

In case Publisher is established outside the EU, the EU authorized representative is:

Springer Nature Customer Service Center GmbH
Europaplatz 3
69115 Heidelberg, Germany

The manufacturer's authorised representative in the EU is Springer Nature Customer Service Centre GmbH, Europaplatz 3, 69115 Heidelberg, Germany. If you have any concerns regarding our products, please contact ProductSafety@springernature.com

Printed and bound by CPI Group (UK) Ltd, Croydon, CR0 4YY
01/10/2026
02225860-0009